Six Legs Better

ANIMALS, HISTORY, CULTURE
Harriet Ritvo, Series Editor

Six Legs Better

A Cultural History of Myrmecology

CHARLOTTE SLEIGH

The Johns Hopkins University Press
Baltimore

© 2007 The Johns Hopkins University Press
All rights reserved. Published 2007
Printed in the United States of America on acid-free paper
2 4 6 8 9 7 5 3 1

The Johns Hopkins University Press
2715 North Charles Street
Baltimore, Maryland 21218-4363
www.press.jhu.edu

Library of Congress Cataloging-in-Publication Data

Sleigh, Charlotte.
Six legs better : a cultural history of myrmecology / Charlotte Sleigh.
    p.   cm. — (Animals, history, culture)
Includes bibliographical references and index.
ISBN 0-8018-8445-4 (hardcover : alk. paper)
 1. Ants—Research—History. I. Title.
    QL568.F7S57 2006
    595.79'6072—dc22   2006019846

A catalog record for this book is available from the British Library.

CONTENTS

Acknowledgments vii

Introduction 1

PART I PSYCHOLOGICAL ANTS 21

1 Evolutionary Myrmecology and the
Natural History of the Human Mind 23

2 A (Non-)Disciplinary Context for Evolutionary Myrmecology 38

PART II SOCIOLOGICAL ANTS 63

3 From Psychology to Sociology 65

4 The Brave New World of Myrmecology 82

5 The Generic Contexts of Natural History 96

6 Writing Elite Natural History 119

7 Ants in the Library: *An Interlude* 139

PART III COMMUNICATIONAL ANTS 163

8 The Macy Meanings of Meaning 167

9 From Pheromones to Sociobiology 190

Conclusion   219

Notes   231
Essay on Sources   287
Index   295

Illustrations follow page 166

ACKNOWLEDGMENTS

Many people have contributed in one way or another to the writing of this book, and I should like to thank them for their often very generous help. I owe my greatest gratitude to John Forrester, for getting the whole project started and for his ongoing support and advice. For their oral recollections of some of the historical characters involved, I am indebted to Daniel Cherix, Stefan Cover, Ethel Tobach, and Howard Topoff. I have received historical advice from, and benefited from the critical readings of, David Birmingham, Peter Bowler, Joe Cain, John Clark, Alex Dolby, Raymond Fancher, John Forrester, Sally Horrocks, Nick Jardine, Karen Jones, Ron Kline, Abigail Lustig, Greg Radick, Harriet Ritvo, Ulf Schmidt, Anne Scott, Crosbie Smith, Marion Thomas, and David Van Reybrouck. For support and encouragement in writing and in applying for grants, I am grateful to Janet Browne, Rod Edmond, John Forrester, Nick Jardine, and Harriet Ritvo. During trips to archives, I have enjoyed the hospitality of David and Elizabeth Birmingham, David Ellis, J. C. Salyer, Heidi Voskuhl, and Paige West. There are no words sufficient to thank Nick Thurston for his many kindnesses and intellectual support; thanks too to Pascal Sleigh for literally kicking me into completing the project.

Archivists and librarians are the unsung worker ants whose scurryings make books such as this one possible. I am thankful to those in the following academic nests: the Archives of the History of American Psychology at the University of Akron, Ohio; Balfour Library, Cambridge; Kings College Archives, Cambridge; Magdalene College Archives, Cambridge; University Library, Cambridge; Whipple Library, Cambridge; the Department of Manuscripts and University Archives, Cornell University Libraries; Pusey Library, Harvard; Templeman Library, University of Kent; the Département des Manuscrits de la Bibliothèque Cantonal-Universitaire, Lausanne; the American Philosophical Society, Philadelphia; and the Library of Congress, Washington, D.C.

For financial support that materially aided the research and writing of this book, I thank the British Academy, for sponsorship of the original dissertation research and the project "The Natural History of the Human Mind," which grew into chapters 1 and 2. The Leverhulme Trust generously granted a research fellowship during the academic year 2003–4, giving time for further research and writing.

Chapter 4 appeared in slightly different form as "Brave New Worlds: Trophallaxis and the Origin of Society in the Early Twentieth Century" in the *Journal for the History of the Behavioral Sciences* 38 (2002): 133–56. I am grateful to the journal for permission to reuse this material.

Six Legs Better

Introduction

"Begone, vile insect." With these words, Doctor Frankenstein greets his monstrous creation on the icy slopes of Montanvert.[1] The creature, with admirable insouciance, replies, "I expected this reaction." Frankenstein's outburst attempts to downplay the monster's hideous threat by reducing him to the stature of a mere bug or beetle—a strategy that evidently was not unexpected to its recipient.

The risible insignificance of insects was well established by the time Mary Shelley published her novel in 1818. In the seventeenth century, the scheming relatives of one Lady Eleanor Glanville had plotted to cheat her of her inheritance by claiming that she was insane. Their evidence? She collected insects—patently the activity of a cracked mind.[2] Yet such little creatures—not even defined as six-legged until comparatively late—were also a humbler of man himself. Was not man as far below God as the insects were below man? Nature showed that "there are beings within . . . the orb of the fixed Starrs . . . which do [more] incomparably excell man in the sense of dignity and infirmity than man doth excell the vilest insect."[3] Robert Hooke's glorious *Micrographia* (1665) illustrated a flea, a mite, and a gnat in such size and detail that each covered a whole huge page, confounding assumptions about their simplicity and insignificance, notwithstanding Hooke's pro forma protestations that they were of inferior value to "an Horse, an Elephant, or a Lyon."[4]

By the end of the nineteenth century, a great variety of people had taken up the study of insects, and for equally varied reasons. These people included travelers, landowners, farmers, government employees, colonialists, doctors, psychiatrists, psychologists, engineers, schoolteachers, missionaries, sociologists, zoologists, gentlemen, and ladies. Each chose a different kind of insect, the instantiation of a particular interest. One might, for instance, become a lepidopterist, collecting butterflies in their myriad forms of beauty and speculating about their mimetic resemblances. Or one might become a coleopterist, with the appropriately Sisyphean task of classifying beetles—the largest order of creatures on earth—

and perhaps studying their predation on human crops. One might become a dipterologist, specializing in flies and their role in disease transmission, or an apiarist, perfecting methods of beekeeping, or a myrmecologist, unraveling the social arrangements of the ants.

The word *entomologist*, then, invokes a somewhat problematic category. All these people were interested in insects, but they came from so many different walks of life and approached the subject from such diverse angles that it is impossible to group them according to any conventional disciplinary history, such as historians of science have generally produced in the past thirty years.[5] The first international meeting for the discipline, the International Congress of Entomology, was not organized until 1910, indicating a lack of unity in the subject before the twentieth century. Even after this event, entomology was not a well-disciplined, professional field. By way of background to myrmecology in the period covered by this study, however, two categories might usefully be introduced to describe insect students of the latter nineteenth century: economic entomologists and traveling entomologists.

Economic Entomologists

Economic entomology, or applied entomology, as it was known in Britain, dealt with the control of insect populations where they interfered with the life of humans, usually in agriculture. Economic entomology is the only kind of entomology that can be regarded as a truly professional or disciplined science during the late nineteenth century and early twentieth century, and at that only in North America.[6] Economic entomologists participated in scientific institutions, including professional bodies; sat on government-funded committees; and published in their own journals. They were represented at the International Congress of Entomology by the Section of Economic and Pathological Entomology.

Economic entomology quickly became entrenched in North America as a result of the rapid demographic changes sweeping the continent. The European colonization and the westward expansion of the immigrant population in the nineteenth and twentieth centuries left unprecedented environmental change in their wake. New species were brought in, ecosystems disrupted, and monocultures imposed across the land. Imported insects were introduced to the extant flora, while native insects flourished on the new crops brought by the colonists.[7] A rapid succession of fresh agricultural problems was thus thrust upon Americans that Europeans did not have to face in the same way. Of course, it was not inevitable that these difficulties should have been seen as entomological. One can easily imagine how they might have been constructed as agricultural, botanical, or even climatic. As it happened, insect specialists successfully persuaded the authorities and the public that insects were the key to the problem and that they were the ones to solve it. Recent critical histories of medicine tell stories in which disease is constructed as a particular pathogen, which then provides a target for professional intervention.[8] Although we are most accustomed to thinking of germs in this role, insects could be substituted to yield an account that describes the development of agricultural entomology in the United States equally well.

The most celebrated American entomologists from the mid-nineteenth century were not professionals in any usual sense of the word. There is nevertheless a discernible pattern to the lives of many entomologists born in the 1830s and 1840s. After training in medicine and serving as surgeons and physicians in the Civil War, they often took up positions as secretaries of scientific and philosophical societies or as museum curators.[9] Although a number of independent, scattered payments were made to entomologists in the 1850s and 1860s, the major transition to paid professionalism started in the 1860s with the foundation of various institutions. Land-grant agricultural colleges and their practically oriented curricula, which typically included entomology, were established using money released by the Morrill Acts of 1862 and 1890.[10] Meanwhile, science students who had done graduate work in Germany returned to the United States impressed with the importance of facilities for original research. Thanks in large part to their agitation, the Hatch Act of 1887 provided for the foundation of state agricultural experiment stations all over North America.[11] By 1894, forty-two states and territories employed entomologists, and more than 300 publications had appeared on agricultural entomology.[12] Meanwhile, Charles V. Riley (1843–1895) lobbied for a national United States Entomological Commission, which he achieved in 1877. Some confusion with a sister entity, the Department of Agriculture, then ensued until 1894, when L. O. Howard (1857–1950) took over the Federal Bureau of Entomology, as the commission eventually became known, and got things in order through the exercise of his immense personal energy and determination.[13]

For the generation of entomologists who came of age after the Civil War—men born in the 1860s—there was a formal educational path and a well-established career pattern in entomology. These men received their initial training at the agricultural colleges founded around the time of their birth and were then able to take higher degrees in their specialties, often at major universities. After completing their studies, they often had very mobile careers, acting as consultants to experiment stations, local agricultural organizations, or conservation agencies, or taking up one of the state appointments available in entomology. They alternated such posts with teaching stints at colleges or professorships at the universities, occasionally holding both university and agency posts simultaneously.[14]

The new generation of economic entomologists was not, of course, a passive creation of the federal government and the universities. Entomology was professionalized as a paid career in North America through a combination of scientific ambitions and large- and small-scale political machinations. In order to cement their disciplinary status, economic entomologists formed the Association of Economic Entomologists in 1889 (replaced by the American Association of Economic Entomologists in 1909). Entomology was further professionalized as a socially venerable calling and source of indispensable expertise through the nationalistic, profile-raising and popularizing tactics of the early entomologists.[15]

By comparison with the drive toward professional status for entomology in North America, there was very little development of economic entomology in Great Britain. In 1939, the British Empire's budget for entomology was only a quarter of that allotted by the United States, although the total imperial population was four times that of the United States. Encompassing many tropical countries, the Empire might have been expected to have prioritized its agricultural challenges and framed them as "entomological," making the contrast with America all the more striking. The nondisciplinary status of British insect studies is most powerfully conveyed by the fact that its leading light was a woman, and thus perforce an amateur. Eleanor Ormerod (1828–1901) came from a wealthy and well-connected family and enjoyed links with Hookers of Kew Gardens and the Royal Horticultural Society. From 1877 until her death, she produced pamphlets and annual reports on insect ravages in Britain. Yet despite international recognition, she never gained a professional post in Britain. J. F. M. Clark has argued that Ormerod's success rested on a denial of her femininity.[16] I would suggest rather that her success was largely due to the status of applied entomology as an overlooked field of endeavor in Great Britain; there were no professional or institutional structures to exclude her on account of her sex. Instead, she operated, literally, in a no-man's-land between upper-class agriculture and amateur, non-utilitarian entomology.

Whereas Americans were focused on agricultural issues in entomology, Europeans were primarily concerned with the loss of life and labor in the colonies to tropical disease. British applied entomology was therefore more medical than agricultural.[17] Ronald Ross's attack on mosquito-borne malaria is the most celebrated such piece of science, but there were many more.[18] The British Empire employed various men in an entomological capacity, but their work was not much acknowledged at home or at the centers of elite academia, nor did they particularly identify themselves as an autonomous scientific group. The situation changed gradually. In 1904 the Association of Economic Biologists was founded in Birmingham—away from the centers of academe—with the intention of linking and supporting British scientists working in the colonies. In 1909 an academic zoologist, Arthur Shipley, petitioned the government to set up an entomological organization that would transcend the far-flung and localized colonial appointment system and bring workers under one administrative roof. His appeal was successful, resulting in the foundation of the Colonial Entomological Research Committee (Tropical Africa) that same year. In 1913, the committee was renamed the Imperial Bureau of Entomology, and the name remained until 1930.

The somewhat slow coalescing of applied entomology in Great Britain, in contrast to the pronounced drive toward professional organization and recognition in the United States, can also be ascribed in part to a specific post-Darwinian research focus. In Great Britain, systematic insect scholars dominated entomology and had little time for or interest in applied entomology. Edward Poulton, for example, who loomed over the field in the early twentieth century, kept minds trained on the problem of butterfly mimicry. Applied entomology was only to be encouraged insomuch as it often produced "discoveries . . . of the highest interest for pure Entomology."[19] In 1909, the year of the Entomological Research Committee's establishment, the president of the Entomological Society of London, the preeminent national organization, gave a retrospective address in which he rated Shipley's achievements of far less importance than Darwin's anniversary. Darwin had shown the way for all key areas of entomology, which comprised only "systematics, morphology, physiology, embryology . . . [and] . . . bionomics."[20] The half-century of scientific research in Great Britain following the publication of *On the Origin of Species* was conducted in Darwin's long shadow. It took a threat from the government in 1917 to commandeer the Natural History Museum for the Entomological Society to begin emphasizing the importance of its collections for "practical entomology."[21]

A typical "career" in British applied entomology combined military or civil colonial service with either medical or zoological expertise. Harold Maxwell Lefroy (1877–1925) was one of the few truly elite, economically minded entomologists in early twentieth-century Britain. After a Marlborough and Cambridge education, he went to the West Indies as an entomologist attached to the Imperial Department of Agriculture. There he worked on the moth borer, which attacked the sugar cane crops. Between 1903 and 1912 he was Imperial Entomologist for India, after which he was called to the newly created chair of entomology at South

Kensington. He had a stronger sense of the need for applied biology than many of his peers, helping to found the Association of Economic Biologists in 1904. He died in his own laboratory at Imperial College, overcome by the gas he was developing as an insecticide, an unlikely martyr of science.[22]

Applied entomology was organized and practiced (though less fatally) in a similar way by other European countries in their colonies. In Germany, meanwhile, zoologists became involved with forestry and conservation. Here, entomologists worked on insect pests whose abnormal masses disrupted the forest hygiene.[23]

What kinds of insects did economic entomologists study? Medical entomologists, mainly European, were chiefly interested in varieties of fly, including mosquitoes. When it came to crop diseases, the particular province of North American entomologists, there were large pests like locusts to deal with, and small sucking destroyers such as scale insects and aphids. As these examples suggest, economic entomologists might have been justified in their frequent complaints that the scant scientific attention their insect subjects received owed to a lack of aesthetic appeal.

Where did ants and related insects fit into the study of economic entomology? Although termites were well-known to chew through all wood-based materials, including houses and books, and biting army ants could force residents to vacate a house altogether once they invaded, by and large, ants and termites were not the subject of economic or applied entomology. They spread no known disease, and were more household pests than agriculturally significant agents. One thing that did link ants to the economic insects was their numbers. Academic zoologists were keen to emphasize the huge numbers of insects known to man, and the even vaster numbers as yet unknown.[24] A single ants' nest or termite mound might contain millions of inhabitants and was a potent reminder of the numerical threat posed by the insect world. In Europe and America, a Malthusian discourse linked the language and even the treatment (segregation, gassing) of degenerate masses, both insect and human.[25] Though producing some short-term gain in insect control, this approach brought murderous results when applied to the human realm.

Traveling Naturalists

A second group of entomologists, mostly Europeans working in the nineteenth century, could be called traveling naturalists.[26] The exotic insects turned up on their travels intrigued, amused, and educated the Victorians. They were bigger, more colorful, and often far more poisonous than the kinds encountered at home, and they were so much more in evidence. Iridescent beetles, glittering butterflies, and swaying mantises flitted before one's face and fascinated underfoot. Clouds of flies buzzed and settled, mosquitoes whined and pounced, moths crowded and immolated themselves by night. Ants bit, termites consumed. Scorpions and spiders, irresistibly grouped with insects in the imagination, lurked in every dark corner.

Alfred Russel Wallace (1823–1913) was just one of the many collectors sent to foreign lands who returned with a case full of specimens and a head full of theories about them; his study of butterfly patterns and mimicry helped shape his theories about evolution.[27] Men like Wallace did not merely collect insects but observed them in life, too. H. W. Bates (1825–1892), who had shared an Amazonian journey with Wallace, found ants so interesting that he chose to put one on the frontispiece of his seminal work of traveling natural history, *The Naturalist on the River Amazons* (1863). Thomas Belt (1832–1878) was another naturalist traveler; although geology was his profession, natural history was his love. Belt had been a member of the Tyneside Naturalists' Club since his youth, and as an adult he was made a fellow of the London Geological Society and became a corresponding member of the Philadelphia Academy of Natural Sciences. His high reputation as a mining engineer meant that, among his extensive travels, he was called to supervise the Chontales Gold Mining Company in Nicaragua between 1868 and 1872. *A Naturalist in Nicaragua* (1874) was written about this trip. In addition to observations of Nicaraguan life, both human and animal, the book laid out many of the ideas in natural history that Belt had spent his odd moments pondering. Belt was most struck by the insects of Nicaragua. He made a vast collection of beetles during his stay, but in the book, he devoted most attention to the habits of ants. Many such entomological enthusiasts lived and wrote in the style of the traveling naturalist. In fact, it would be more accurate to call them naturalists who happened to be especially interested in insects. The social structures within which their knowledge was made, discussed, and ratified comprised the general educated book-buying public and various related learned societies; their papers were often read out at society meetings while they remained abroad.

As the nineteenth century progressed, a new set of opportunities arose for adventurous would-be naturalists. The acquisition and maintenance of colonial possessions required European men to live in Africa, Asia, and beyond. Many of these men—members of the armed services, administrators, governors, doctors—found a source of inspiration, or perhaps consolation, in the insects that distracted them from their duties. Some were charged specifically with entomological duties, especially after the founding of the Colonial Entomological Research Committee in 1909, but their interest in insects often took them on a different path from the utilitarian tasks they were supposed to perform.

A handful of travelers created a reputation for themselves as observers of exotic ants in the early Victorian period and continued to be cited into the twentieth century.[28] The tradition continued with Alfred Alcock (1859–1933), who started as a zoologist in Aberdeen. After a spell occupying medical posts in India, he became superintendent of the Indian Museum and professor of zoology at the medical college in Calcutta. Alcock's best-known work, *A Naturalist in Indian Seas*, however, was composed over two monsoons while his survey ship was laid up in Bombay Harbor and he worked on dredged marine material.[29] By 1909 he had achieved the rank of lieutenant-colonel and was on the Colonial Entomological Research Committee. Another example from the early twentieth century of the traveling naturalist become insect specialist is E. E. Green, who was president of the Entomological Society of London in 1923–1924. Green often found his entomologizing passion in conflict with his professional duties of governance. In the *Entomologist's Monthly Magazine* of 1901 he gave an unintentionally poignant account of collecting spectacular moths caught in the arc lights of a Boer concentration camp. Regarding his thirty-two years in Ceylon, he commented, "I sometimes paid undue attention to the fauna and flora of the country to the neglect of the more monotonous duties of coolie-driving." His eventual appointment as government entomologist enabled him, in his own words, "to combine business with pleasure."[30] Green's recollections echo the obituary description of a contemporary colonial entomologist, always "more interested in natural history than in soldiering."[31]

Even those colonial officials charged with entomological responsibilities were generally more interested in insects from a natural history perspective than from the utilitarian viewpoint they were supposed to espouse. A few more thumbnail biographies underscore their contributions. Major R. W. G. Hingston (1887–1966), who went to India with the Indian Medical Service, ended up writing a number of books on natural history, insects, and animal psychology. In recognition of his autodidactic expertise, he was appointed official medical officer and naturalist to the 1924 Everest expedition. Other colonial entomologists included the Belgian Emile Hegh (1877–1950), an engineer who wrote on termites and mosquitoes. Hegh's mosquito work had economic significance (and was published in a series on agricultural biology), but his book on termites addressed nonapplied matters, such as social life and nest construction.[32] Emile Roubaud (1892–1962) was sent from Paris as a medical entomologist to the French Congo and Senegal to study *Glossina* flies, such as the tsetse, and their transmission of trypanosomal diseases to man. He also worked on diseases spread by mosquitoes. Yet he insisted that the most important part of an entomologist's work was to observe insect behavior, not only because understanding this behavior might help combat the transmission of disease but also for its own sake, as his numerous publications on instinct, behavior, and the social insects attest. Roubaud developed a theory about the evolution of sociality by looking at the lives of wasps and considering the behavioral continuities between solitary and social forms. Paul Marchal (1862–1942) also performed double duty, as pest controller for the French and as traveling naturalist on his own account.[33]

The incidental nature of Alcock's, Green's, and Hingston's work in natural history illustrates well the sense in which these men were a continuation of the traveling naturalist tradition of the nineteenth century. Even those who were employed in an official entomological capacity did not always find that their jobs coincided with their interests in natural history. For example, the medical entomologist for the government of Palestine wrote on factors influencing seed-gathering in the ant, hardly a topic of medical relevance.[34] *Entomologist's Monthly Magazine* and its more scholarly cousin, *Transactions of the Entomological Society of London*, continued to carry reports relayed from far-flung amateur observers of insects well into the twentieth century. The interests of these men were represented at the 1910 Brussels Congress in two sections, the Section on Evolution and Mimicry and the Section on Nomenclature, Bibliography and Papers of General Interest.

The traveling entomologists, though they journeyed for a variety of reasons, looked for similar things in the insects they studied; they were a metaphor for the whole foreign experience of these men, who saw in insect behavior things that seemed pertinent to their adventures abroad. For Belt and those who did not have administrative or martial responsibility for the lands in which they traveled, insects presented an interesting diversion and an opportunity for moral reflection. Ants in particular, with their alternative societies, were a perfect Lilliputian object of study. They, along with their cousins, the bees, had been a staple of Christian meditation and didacticism for centuries, thanks to the writer of *Proverbs* and Aesop. The nature of Victorian formic reflection was often to shore up the naturalists' sense of civilized superiority. Belt chronicled the Nicaraguan ants' extraordinary skill in dealing with life's vicissitudes, their apparent foresight, and their achievement of something that seemed to him to approach Thomas More's Utopia. Furthermore, he compared the native humans unfavorably to the native ants. The British reader of *A Naturalist in Nicaragua*, looking through Belt's eyes, would have identified more with the prudence of the ants than with the laziness and profligacy of the native Nicaraguans and immigrant Hispanics.

It was possible to achieve such insect-based allegorical insight into human affairs without stirring so far abroad. Those who traveled in imagination only might perhaps be regarded as armchair-traveler naturalists. The English gentleman John Lubbock, Lord Avebury (1834–1913), was representative of that category.[35] Although he did not travel significantly, he drew moral lessons from the ants similar to those of his more itchy-footed contemporaries. Ants, for Lubbock, were, like all animals, capable of learning. (Besides taming a wasp to prove his point, he devoted the final two chapters of *On the Senses, Instincts and Intelligence of Animals* to an account of teaching a dog to read and various animals and birds to count.) Lubbock's optimistic doctrine of improvement echoed his paternalistic attitude toward "primitive societies" like those described in his book, *Prehistoric Times* (1865). His natural history connected the exotic human world in a reassuring fashion with the life of insects by proposing that all were part of the great process of progress and civilization.

For colonial administrators, governors, and officials, insects were rather more threatening. They produced a discomfiting awareness of the colonialists' fragility, for they seemed so much better suited to the landscape. In 1909 the chief entomologist of the Imperial Department of Agriculture for India expressed his doubts about understanding, and hence combating, insects, saying, "The senses, the instincts, the modes of expression of insects are so totally diverse from our own that there is scarcely any point of contact . . . a locust swarm may be the product of a blind impulse . . . just as a blind impulse ranges through a crowd of human beings . . . were [insects] possessed of higher forms of mentality . . . no-one can say what might be the course of the world's history . . . a combination of the red ants could probably drive human beings out of India . . . and human methods of warfare would require to be revolutionized to deal with it."[36]

For this entomologist, the insects embodied the threat posed by Indian humans and suggested it might be vain to assume the automatic ascendancy of the English. Nothing could be further from Belt's view. Economic and traveling entomologists alike, however, found in insects a powerful and imaginative means of representing their own professional hopes and cultural fears.

Ants Historically Viewed

Ants emerge from their pupae as fully formed adults. In this respect they are unlike many other insects, such as grasshoppers, which bypass the pupal stage and go straight from the egg through several molts, each successive imago resembling more completely the mature form. Biologists refer to the form of life cycle expressed by the ants as completely metamorphic, the grasshoppers' as incompletely so. Yet from a cultural perspective, ants too were incompletely metamorphic in the hundred-year span from 1874 to 1975. In the wake of the traveling entomologists and their moral readings of the colony, changing cultural contexts framed various reenvisionings of the ants. What aspects of human life did they permit scientists to model? Did they represent a social ideal to which humanity should strive, or an anathema that humanity should avoid at all costs?

During the hundred years covered here, ants metamorphosed through three main forms, appearing sequentially as psychological, sociological, and informational entities. In other words, they were used successively to model the human mind, society, and communication. For each period, one figure stands out from the scientific milieu. For the era of psychological modeling it is the Swiss psychiatrist Auguste Forel (1848–1931). For the sociological era it is the American academic and coiner of the term *myrmecology*, William Morton Wheeler (1865–1937), and for the information era it is the American sociobiologist Edward O. Wilson (1929–). These were the dominant interpreters in each era, the myrmecologists who established in each case the appropriate "scientific" way to see the ant. The achievements of the three men frame the time period covered by this study. At one end is the 1874 publication of Forel's *Les Fourmis de la Suisse*. This book combined the taxonomy of ants with the study of their behavior for the first time, a starting point in scientific naturalism that Forel, as a psychiatrist, hoped could be used to model the natural history of the human mind. The marker for the end of the study is Wilson's *Sociobiology* (1975), in which Wilson established ants as the exemplar for his "new synthesis" of all biology, now a cybernetic science. The structure of this book reflects these three periods of myrmecology, and for each one describes both disciplinary reformulations and broader cultural issues.

The scientists and naturalists discussed in this book studied ants for their own sake, and often did so with remarkable passion. They did not merely adopt ants instrumentally as vehicles for social and political agendas. Yet neither could they step outside the cultural frames within which they operated. In each case there was a two-way traffic between science and broader culture, with the culture shaping the questions posed by scientists and the scientific answers in turn directing cultural views, reinforcing or slowly altering conceptions of the natural and its significance for the human condition.

This bidirectional influence is evident in the psychiatric theories of August Forel. Dividing his time between professional responsibilities at an asylum and avocational pursuits in the Swiss countryside, Forel established to his own satisfaction that ants had psychic capacities and were valuable for the lessons they could teach humanity. Though not everyone accepted his human analogies, he cemented a new tradition for studying insect psychology in the context of evolution. He united the old tradition of collection and taxonomy with the new, natural historical approach of observing and understanding behavior; under his influence the two became mutually supportive approaches. Classification was pointless and dull when considered in isolation from animals' conduct, and conduct was meaningless without a grasp of phylogeny.

Forel published his magnum opus, *Le Monde Social des Fourmis*, with its explicit lessons of pacifism and internationalism, at the beginning of the 1920s. His evident horror at the carnage of the Great War, however, was the reaction of an old man, and Forel's consequent philosophical realignment (or perhaps retrenchment) did not have the impact that it would have had, had he been a researcher at the peak of his powers. Though barely a generation younger than Forel, William Morton Wheeler reconstructed Forel's European-influenced knowledge, with its focus on society, the body politic, and its evolution, in a thoroughly modernist, postwar context. For Wheeler, the interesting questions lay not so much in the psychological qualities of the individual ant as in the properties of formic society as a whole. He drew connections between the mass behavior of ants and of humans, and he did so within an active circle of sociological colleagues at Harvard University.

Wheeler died just before the Second World War, another watershed in the animal sciences. Comparative psychology, ethology, and ecology were all in fluid form in the years leading up to the war, with a number of different approaches from Europe and North America coexisting. Theodore C. Schneirla (1902–1968) was perhaps the only specialist in ant behavior to make an impact around these years, although Karl von Frisch was garnering considerable interest with his work on bees.

It was after the war that another figure arose willing to make claims for the ants that were of equal magnitude to Wheeler's and Forel's. More of a collaborator than either of his predecessors, Edward O. Wilson worked especially with mathematically inclined colleagues to produce informational accounts of evolution in general and ant behavior in particular, based on the pheromonal code he first started to unravel in the 1950s. At the same time, his peers at MIT, a mile away from Wilson's home at Harvard University, were working on the engineering of complex systems, and navy-funded zoologists were figuring out the communication of other animals with even more potential military significance, such as bats and whales.

The transition to the communicational construct of ants, the final metamorphosis in the ant century is of importance for the history of biology in general. Lily E. Kay has pointed the way to an understanding of molecular biology and genetics in the latter half of the twentieth century as information science, but perhaps we should not too quickly grant molecular biology paradigmatic status within the field of information biology. A history of theoretical biology in the cold war has yet to be written, but it will certainly reveal a wider horizon for the themes of information, misinformation, and life.

The history of myrmecology bucks the general trend of the history of biology in the period 1874–1975. During this time, biology increasingly became a laboratory-based science, and fieldwork correspondingly acquired a dubious status, associated with amateur ornithologists and, latterly, televisual natural history.[37] Yet Forel, Wheeler, and Wilson all had ambitions much larger than their focus on tiny, overlooked subjects might suggest. With varying degrees of success, each attempted to forge a new field from a particular construal of the ant. For Wilson and Wheeler, the disciplinary ambitions were partly reactive: each found himself practicing an unfashionable and potentially unworthy science. Wheeler's problem was a sometimes willful confusion of field science and "mere" natural history on the part of his laboratory-based colleagues, the emerging elite in biology. A side issue was Wheeler's desire to establish entomology as a pure science, separate from the applied insect work sponsored by the U.S. Department of Agriculture and various state authorities. Wilson's foes were the biochemists and molecular biologists of the postwar era, such as James D. Watson, who belittled his work and attempted to dominate biology at Harvard in terms of both funding and personnel. Both Wheeler and Wilson concocted new disciplinary names in response to their sense of being professionally cornered, and both of these names stuck: myrmecology and sociobiology.[38] Forel, by contrast, did not aim to establish a discipline in the way that Wheeler meant to put myrmecology on the map, and Wilson sociobiology. Forel's aim was to create utopia, admittedly a harder task.

Throughout the book, the story shifts around the world, from continental Europe to North America, following the ants wherever voices were raised loudest in discussion of them. The three main characters in the story were the best-known and most influential students of ants during the century of study, both within and outside their specialist realm, and it is their agendas—contested as they sometimes were—that form the backbone of this tale.[39]

Themes and Variations in Myrmecology

Myrmecology was not (and is not) a neatly constrained discipline like other areas of biology, such as molecular biology or genetics, that have emerged over the past fifty or one hundred years. And just as the essence of the ant changed over the century from 1874 to 1975, so too did the epistemological desiderata of myrmecology. The contexts, audiences, and opponents for the three figures we will be concerned with were all different, shaping reciprocally their scientific subjectivity and the object of their science. Many of the apparent continuities were superficial, having arisen to serve different ends. Although it might be tempting to say that ants' cultural evolution has been toward a final form, like the grasshoppers' physical ontogeny, this would be historiographically untenable. In fact, ants have meant so many different things to so many different investigators that at times it seems purely coincidental that they all studied the same organism. Nevertheless, there were thematic continuities in the work of Forel, Wheeler, and Wilson, continuities that arc across the study of natural history and yoke it to intellectual history, on the one hand, and cultural developments on the other.

There are many overlaps and continuities between the work of Forel, Wheeler, and Wilson, some of which initially appear as extraordinary coincidence if one eschews a history of ideas. Was it chance, for example, that Forel's interest in creating an international language should seemingly resurface in the cold war communication theory of Wilson's ants? Is there something intrinsically holist about ants that caused both Wheeler and Wilson to rail against reductionism? And why did both Forel and Wilson reach such apparently similar conclusions about the naturalized status of ethics, based on their observations of the six-legged creatures?

The methodologies developed by Gillian Beer and N. Katherine Hayles suggest how and why these paths of influenced may be traced. A strict sociology of science permits no discussion of ideas in and of themselves, and of course it is ridiculous to think of them as free-floating entities, drifting through history and looking for minds to colonize. But as metaphors, scientific descriptions do, to a certain extent, have a life of their own, as Beer's work on Darwin has shown. The scientist reaches for a metaphor in order to describe a process in nature (indeed, that metaphor may even condition how he or she sees it). That metaphor has sticky edges—cultural resonances that reach beyond its immediate application to the natural world and suggest all kinds of unintended connections, images, and analogies to readers (in the broadest sense) of the first scientist's work. In this way the metaphor goes on to shape new exploration, experimentation, and representation. Hayles describes a process by which metaphors, "like bureaucrats, [grow] less lively" and move "from transgressing boundaries to constituting them."[40] These models suggest the complex processes by which insects have been used to represent different aspects of humanity, and how those representations have suggested new connections to writers and scientists thereafter.

In fact, the relative historical proximity of the three main protagonists, alive in the same century, meant that many themes, questions, and anxieties *were* shared across decades and continents. Those shared matters were both entomological and general. Eugenics, for example, is an obvious discourse linking Forel and Wheeler, although there were differences in their exact interpretation of the issue—alcoholic degeneration of the germ plasm in the first instance, and immigration, race, class, and gender issues in the second.

A certain amount of stability of context also comes from the institutional background for the science: both Wheeler and Wilson worked in universities, while Forel's institutional connections echo throughout the story. Indeed, Wilson and Wheeler have both been remarkably well-read and philosophically and historically informed, aware of the scientific contexts for what they have done. This reflexivity on the part of its subjects gives the book a somewhat looping structure; thanks to the disciplinary-historical awareness of the actors in the story, each one retrieved various old readings of ant phenomena and made them "new," entailing a reiterative approach in this narrative. In particular, Wilson's astonishing ambition means that a teleological approach to his science is often warranted, in that he himself has palpably searched for the big theory that will establish his reputation for posterity. As will become clear, however, the university context is problematic as a unifying historical backdrop for the ants.

The most important entomological continuity was the process of collection and taxonomy. A network of letter-writers and specimen-swappers extending across time and continents linked all three protagonists and many, many more. A body of ant knowledge was constructed and maintained through expertise in taxonomy, reified in specimens and types traded, loaned, and given between specialists. From the foundations of this knowledge more might be built, though anyone who did not keep his hand in with the business of collecting and classification might be cast beyond the scientific pale. Edward O. Wilson in particular suffered as a scientist from a disinclination to engage in the activities of specimen collecting and exchange.

With some of the cultural background for continuities in mind, we can turn to the major thematic issues in myrmecology. Ants get everywhere, as every householder knows who has tried to eradicate them from the kitchen, and the same seems to be true of their infiltration of diverse and often surprising areas of cultural life. Here, issues of instinct, crowds, language, and analogy all turn out to have influenced and in turn have been influenced by the study of ants.

The theme of instinct forms an important constant throughout this account. In one way or another, the leading actors all responded to a deeply embedded, historical construct of the animal mind, and of the relation between instinct and intelligence. This construct, rooted in Thomist philosophy, proposes that where man has intelligence to guide his actions, animals are endowed with instinct. The social insects, with their complex lives and social structures, have long been held as the acme of instinct, whether that instinct was created or evolved. This is evident in Spanish depictions of ants during the Renaissance, through the economic portrayal of bees in the eighteenth century, and on to the natural theology of the nineteenth century and the scientific naturalism that came about toward the end of the century. There was general agreement among scientists, naturalists, and theologians that behaviorally, ants, bees, and wasps were among nature's most extraordinary creatures.[41] The need to explain their complex repertoire of behavior struck many men of science as paramount; in *The Origin of Species*, Darwin commented that their social coordination was "ranked by naturalists as the most wonderful of all known instincts" and thus one of the principal things for which a theory of evolution had to account. Chapter 2 of this book discusses the tensions that arose in attempting to reconcile the mental parallelism of instinct and intelligence with the hierarchical picture consequent on evolutionary theory. Myrmecology, with its psychological roots in Neolamarckism, played a crucial part in the post-Darwinian debates about instinct.

The traditional association of insects with the most impressive exercise of instinct in the animal world kept insects at least in the background, and often in the foreground, of any discussion about instinct. French psychologists, for example, focused largely on insects from the 1930s to the 1950s as their objects of study.[42] Odder contexts for instinctual discourse included literary theory and cybernetics. For the linguist and reader of myrmecology I. A. Richards (1893–1979), the role of instinct in interpretation was perhaps the main problem of literature. During the cold war, cyberneticians took the nonconscious nature of instinct and placed it in the robot, as creative writers such as Karel Čapek had been doing for several decades. One of their most important models for the robot was the ant. Thus, in the popular imagination, instinct once again became a kind of mindless intelligence, just as it was for Aquinas.

The particular identification of instinct with *social* insects gives the trope an additional characteristic: it is essentially constituted by the mass. A concern with the group or the crowd therefore forms a second major theme in this history. The book takes its title in relation to this thematic context, alluding to the Orwellian maxim from *Animal Farm*, "four legs good, two legs bad." This ideological pronouncement is famously reversed by the end of the novel, when the pigs have taken on the characteristics of their erstwhile overlords, the humans. The years from 1874 to 1975 saw a similar change in perspective regarding the ant mass. At the end of the nineteenth century, ants were regarded as laudable models for human life because of their hard work, social responsibility, and even the natural-theological reminder they provided about the relative importance of man in the natural world: the ants' apparently remarkable intelligence was a gentle prompt that human achievement was no cause for arrogance. Six legs were better than *both* four and two.

The trope of the mass, and the massed society, had its dark side, with its most ugly outworking on the field of nationalistic politics. Around the turn of the twentieth century, a degenerationist discourse had begun to reveal a dark and bestial underbelly in the emergent psyche of the crowd. As the twentieth century progressed, that lowest common denominator seemed to acquire some basis in reality in fascist and communist societies. Concerns about the mass manipulability of the crowd gave an uncomfortable edge to myrmecology between the wars, and ants became a considerably less desirable template for human life. They came to symbolize the unthinking mass, among which one's individuality would be meaningless. The ant was the arch-organism of modernism: as Orwell commented in *Nineteen Eighty-Four*, "the proles were like the ant." Six legs were better than two in a cruelly ironic sense: easier to rule, to organize, and to send to war, without any sense of the meaning of these actions in terms of individual lives.

After the Second World War, scientists become reconciled once again to the thoughts of the crowd. The concept of group intelligence that emerged now recommended ants as a model for complex task-solving machines. Instinct, reconstructed as drive, became a knowingly teleological description for purpose-driven cybernetic technologies. Ants once again took on desirable characteristics in a strictly instrumental sense.

Cyberneticians were concerned with the flow of information, that is to say, communication. The work of Forel, Wheeler, and Wilson, characterized by an overt interest in language, also brings to the fore this third theme of myrmecological history: some of its principal transitions are mediated not by scientists but by linguists. Understanding Wilson's communicational reading of ants entails going back to retrieve a surprising history that connects him with Forel; it turns out that, contrary to the expected trend of the twentieth century, scholars in the humanities have played an active role in shaping science, specifically cybernetics.

Wheeler's friend, the linguist and publisher C. K. Ogden (1889–1957), shared with him elitist anxieties about the inability of the masses to protect themselves from the seductive powers of language. While Wheeler strove to carve out an expert language of natural history, Ogden developed his own version of nonemotive, "Basic" English. Ogden's chief collaborator, I. A. Richards, then went on to contribute to the cybernetic conversations on language and communication, which featured ants as an important exemplar.

Language was, therefore, a mutable tool. In Ogden's hands, Basic English was an aid to mutual international understanding and an intellectual game par excellence. To Richards and more particularly to his Rockefeller backers, it was more utilitarian, a tool of industrial organization. Although Ogden's vision for the language perished, Richards' was successfully transmogrified in the context of the cybernetics circle of the postwar era. When Richards arrived in Boston in 1939, joining Wheeler's colleagues, he carried the Ogden approach into the heart of the new ergonomic sciences of management. The result was a powerful combination that shaped and was shaped by the postwar concept of the worker.

A final theme, underlying all others in this book, is the nature of analogy. Perhaps unsurprisingly, Herbert Spencer's and Ernst Haeckel's relations between body and society are never far from the surface of much myrmecological discourse; the formicary is seemingly an entity metaphysically in between the ordinary, unitary animal body and human society. Besides this specific set of analogies, there is also the question of the validity of analogy *tout court* in biology. Throughout the century examined here, myrmecologists had different degrees of confidence regarding the power of analogy, and what it could or could not provide for the scientist. Nevertheless, discussion of analogy, a key feature inherited from the study of ants as a branch of natural history, persisted.

Raised on Haeckel, Auguste Forel had no difficulty forging human-ant analogies. The name of his home and private asylum, La Fourmilière (the ant colony), was testament to his sometimes simple faith in the salutary lessons offered by the ants. Meanwhile, in the United States, nature-lover and essayist John Burroughs (1837–1921) was exploring and expressing a theory of analogy that captured the American relationship to nature around the turn of the twentieth century. Burroughs' readers, absorbed also in a Spencerian ideology of the frontier, presumed a natural grounding in the organic world: a natural comprehensibility of the landscape mediated through "natural" acts such as hunting and farming.

Although Burroughs' approach inflected the work of many of William Morton

Wheeler's contemporaries, most notably those whom I term the "domestic" entomologists, Wheeler himself claimed to be cautious in drawing overt analogies. His laboratory-based colleagues were following a positivist agenda that did not comport well with the traditional analogizing of natural history, and Wheeler had to be careful to research and write in a manner they would consider appropriately "expert." Yet even though it would not be fair to say that analogy was for Wheeler a scientific method, it takes very little digging to show how completely interwoven human and formic issues were in his work. Wheeler's uncompromising opinions on life during the Depression revealed themselves in his economic understanding of food exchange in the formicary. Finding ruthless analogy between ant and human appealed to Wheeler's acerbic wit and was one of his chief pleasures in myrmecology. *Foibles of Insects and Men* (1928) gathered together several such papers; Wheeler's caveats that they were all a joke served to highlight rather than efface the significance of his analogy.

Wheeler's ostensive eschewal of analogy was more whole-heartedly followed through by T. C. Schneirla. Schneirla was ruthless in identifying anthropomorphism in the work of fellow animal psychologists and had no truck with those who attempted to draw comparisons between human and apish behavior, let alone human and formic. This approach almost certainly reflected his disciplinary background in psychology rather than natural history. It was also a short-lived, even futile stance within the history of myrmecology. In their appropriation of von Frisch's work on bees, the cyberneticians, including Edward Wilson, rapidly reacquired an astonishing confidence that the study of ants could reveal truths about humans. Theirs was not a Burroughs-style faith in nature but an equally fervent belief in the concrete communication of abstract information in the zoological realm. Thus a programmatic confidence in the power of analogy—notwithstanding the changing nature of analogy—remained a notable feature of myrmecology throughout the century. It is perhaps this natural historical trait of analogy more than any other that marks out myrmecology among the scientific disciplines of the late nineteenth to late twentieth centuries.

And now the egg vibrates, the first imago emerges. . . .

PART I

PSYCHOLOGICAL ANTS

Insects have long provided grist for philosophizing about the human mind. Ants and bees in particular, with their complex ways of life, have provided a fertile comparison with human reason, by which we achieve many of the same things. Auguste Forel (1848–1931) pursued these questions using the neurological and psychiatric methods in which he was trained, paying attention to the key nineteenth-century concepts of instinct and intelligence. Thus the ant was encouraged through its first metamorphosis, becoming a psychological, evolutionary model of the human mind and suggesting, moreover, possible futures for the human race in an era framed by the Franco-Prussian War and the Great War.

Forel's work with ants successfully established a milestone in what was to become known as myrmecology, bringing together behavioral and taxonomic studies for the first time. It set the measure for future studies, placing behavior always within the naturalistic context of evolution and proscribing the value of classification without the study of live insects in nature. As Forel explained in his 1874 book, *Les fourmis de la Suisse*, "I have constantly studied the ants from the dual perspective of their classification and their behavior, which has never yet been done by anyone in a manner of any consequence. These two studies, when they are thus reunited, complement one another on a host of points. I insist upon this fact, for it is by this reunion that the present work distinguishes itself from the mass of its predecessors."[1]

Within about fifteen years of expressing this view, Forel had worked up a model, drawn from his neurological and psychiatric training, to explain how classification and the study of behavior were related. Moreover, he expressed it in such a way as to describe humans as well as ants. It was also at this time that he became fully confident in the theory and practice of hypnotism. Forel's psychoevolutionary theory of the engram, notwithstanding its similarities to other theories of instinct as degraded intelligence, was assembled from a set of cultural components unique to Forel in his particular cultural and scientific context. Forel's myrmecology (and psychiatry) cannot be separated from the issues that he explored and promoted so tirelessly: society, race, internationalism, eugenics, monism, pacifism, feminism, socialism, work, education, antialcoholism—the list goes on and on. One of Forel's friends quipped upon his graduation, *"Forel s'occupait de fourmis; maintenant il passe aux fous à remettre!"* ("Forel was busy with the ants; now he's going to cure the mad!"—a play on the sound-alikes *fourmis*, ants, and *fous remis*, recovered madmen).[2] Without knowing it, he was exactly right. Between these twin poles, sane ants and crazy people, Forel made sense of his entire world.

Giving context to Forel's work only emphasizes the difficulty of treating myrmecologists as a unified group, still less a discipline. Simply grouping those who worked on ants would artificially yoke a disparate variety of approaches. Of these, the simplest and best established in terms of correspondence and publishing networks was collection and taxonomy. There were also those who focused on ant behavior, but they included reactionary natural theologians and archmechanists—hardly a culturally plausible cluster. Connections across these groupings did exist in terms of correspondence and citation (whether positive or negative). But although individuals could forge links based on a shared concern with ants, their associations did not follow any particular or disciplinary pattern. Even so, Forel's vision of integrating taxonomy and the study of behavior through a progressive phylogeny of learning was successful to some extent, thanks to his personal persuasiveness and the fact that his theories pushed the right cultural buttons for many readers. Insect psychology became, for a time, a surprisingly substantial field of inquiry for science.

CHAPTER ONE

Evolutionary Myrmecology and the Natural History of the Human Mind

The bucolic valley basin view from the bedroom window was a mixed pleasure for Auguste Forel. At times it made him furious. For this fanatical abstainer's home overlooked mile upon mile of vineyards. The grapes against whose alcoholic product he spent his life inveighing grew practically up to the front door. Forel's home was called La Fourmilière—the Ant Colony—reflecting his conviction that ants provided the best model for a decent, sober life, in contrast to the endemic alcoholism of his countrymen in the Swiss canton of Vaud.

Forel's science, like his life, integrated his twin interests in the psychiatry of alcoholism and the life of ants.[1] He was born in the auspicious year of 1848 to a well-off and genteel family near Morges, a little west of Lausanne, and began his ant watching as a small child; later in life he commented freely that this had been his only escape from his mother's suffocating religious neuroses. Auguste's great-uncle Alexis Forel encouraged the boy in his studies despite Auguste's mother's doubts about the boy's safety in the garden.[2] Forel began medical studies at Zurich in 1866, and during his final year he grew more and more interested in psychiatry and the anatomy of perception. After failing the Vaudois medical examination in 1870, Forel turned to his ants for consolation and began to prepare a manuscript on the ants of Switzerland. He successfully passed the examination on a second attempt and entered the medical profession.

Forel's first professional position, in Bernhard von Gudden's Munich asylum, enabled him to develop his interest in brain anatomy, and here he helped to develop a new method of preparing sections of the brain. Meanwhile, Forel's manuscript, *Les fourmis de la Suisse* (*The Ants of Switzerland*), had been awarded the Schläfli prize of the Swiss Natural History Society and was published in the society's *Memoirs* in 1874. In 1879 Forel returned to Zurich, where he was appointed

Privatdozent at the university and director of the attached Burghölzli asylum. After some confusion, he was given a professorship of psychiatry at the university's medical school. In 1883 he married Emma Steinheil. Shortly after the nuptials his young wife persuaded him to give up alcohol, at first for a trial period, and then permanently. The result of this was Forel's great crusade against alcohol and the development of his psychiatric theories and techniques for treating alcoholism. In 1889, Forel oversaw the opening of the Asile d'Ellikon, just north of Zurich, which was to be a sanatorium for alcoholics. Forel supervised his new institution by means of monthly visits; he claimed a one-half to two-thirds success rate (defined as permanent abstinence) for his patients, with the remainder largely written off as "ethically defective." His views were beginning to harden, despite his early experience of Gudden's liberal asylum. During his professional career Auguste Forel published prolifically. He is best known for his 1889 book on hypnotism, which was translated into English in 1906, and for the eugenic *Die sexuelle Frage* (*The Sexual Question*), published in German in 1905 and in English three years later.

In 1898, the Forels returned with their children to Chigny, a small village just up the mountain from Auguste's birthplace in Vaud. Having handed over the directorship of the Burghölzli to Eugen Bleuler, Forel continued in private psychiatric practice, now using hypnotism as his major therapeutic technique. The Forels also opened their home to a small number of boarding patients. Some nine years later they moved again within Vaud, this time to a small village named Yvorne, a few kilometers to the south of Lake Geneva's eastern tip. Here they established La Fourmilière, with its vineyard views. Forel's semiretirement enabled him to concentrate once again on his first love. Over the course of 1921 and 1922 he published his five-volume magnum opus on ants, *Le monde social des fourmis*. This was translated and published in English in 1928, three years before his death.

The Ants of Switzerland

Carlo Emery, born in the same year as Forel, spent his professional life in Italy as a professor of zoology, a renowned expert on formic classification. Only after he struck up a friendship with Forel in adulthood, however, did the pair realize that as boys they had been watching ants separated by a distance of only nine kilometers. An even more remarkable coincidence concerned Forel's grandmother, who deplored Auguste's childish study of insects. She had read and been much affected by the definition of an entomologist as "a naughty boy who wastes his time catching insects, sticking them on pins, and then laughing to watch them wriggle in a box," and as a result, Forel was forbidden to collect any but dead insects.[3] Yet when Auguste's great-uncle Alexis Forel encouraged him in his interest, she suddenly "remembered" that she had a book on ants that had been presented to her as a young woman by a suitor, its author. This personage turned out to be none other than Pierre Huber, and the book an inscribed first edition of his celebrated *Recherches sur les mœurs des fourmis indigènes*.[4]

These coincidences underscore the existence of a remarkable tradition of studying social insects in Switzerland. Despite its relatively small size, Switzerland has contributed a disproportionate number of celebrated students of the discipline.[5] Its famous early names are Charles Bonnet (1720–1798), Henri de Saussure (1829–1905), and father and son, François and Pierre Huber (1750–1832 and 1777–1840). Later myrmecologists included Auguste Forel (1848–1931), Edouard Bugnion (1845–1939), Félix Santschi (1872–1940), Rudolph Brun (another psychiatrist; 1885–1969), and Heinrich Kütter (1896–1990). In part the strength of the tradition can be attributed to the direct influence of one Swiss on another, but there are also some remarkable coincidences in the geographic proximities of these ant enthusiasts.

Although it would be hasty to ascribe to Switzerland itself something that inspired a love of ants in its citizens, there is certainly no doubt that Forel interpreted the insects in the light of his nation's political condition.[6] Despite its independence in 1798 (with sovereignty granted to its government in 1803), Switzerland's identity throughout the nineteenth century was unclear. It was periodically threatened from without by France and Prussia, while within its borders there was constant wrangling over the respective powers of national and cantonal governments. A federal pact of 1815 supposedly granted cantons autonomy, but this was widely perceived to have been overridden almost immediately in practice by the establishment of a national army. During the early part of the nineteenth century, while ruled by the Bernese, Forel's canton of Vaud was among the most pro-Helvetian (demonstrated by its mobilization during the Napoleonic crisis of 1838), but even so there was no doubt that canton came before nation.

In 1848—the year of Forel's birth—and in response to revolutions across Europe, a new federal constitution was established that aspired to government by representation on a national level, though this too would soon be rejected as too *ancien régime*. Meanwhile Vaud itself had finally managed to throw off the influence of the neighboring canton, Berne. In Vaud as in other cantons, the 1850s and 1860s were marked by religious tensions. A proposition to throw out all Jesuits was ultimately defeated (provoking a massive public demonstration against Catholics), though protestant pastors also had their powers considerably restricted. The Eglise Nationale Evangélique had its status protected by law; other denominations were properly granted freedom of worship in 1862. In the same year the national government was again overhauled, this time by a combination of liberals and radicals, who rewrote a range of constitutional laws.

The Franco-Prussian War temporarily distracted Swiss attention from constitutional matters, but they were soon at it again. In 1874, the extant military arrangement was formalized; cantonal militias were abolished once and for all, and a national force was established. The Vaudois had voted against the new constitution (proposed in 1872), but it was accepted nationally by a narrow margin, with further compromises written in to appease the significant minority of antifederalists. Federalists soon became alarmed, as cantons were permitted to reject by referendum any law that displeased them, a right for whose exercise Vaud quickly gained a reputation, for it had moved away from its outward-looking stance of the 1830s.[7] Pro-federalism in Vaud gradually matured as its people saw the advantages offered by central government and grew convinced that not too much had been ceded by way of local control. The Vaudois did, however, reject by a crushing majority that old bugbear, federal military law. Their attitude, expressed in a 1903 history, was *"unification et non centralisation."*[8]

Following the overthrow of the Bernese aristocracy, haut-bourgeois landowning families such as the Forels became the most influential class in Vaudois society.[9] Forel could trace well-connected relations in the region back to the sixteenth century, particularly on the side of his mother, Pauline Morin. Intermarriage with other eminent lineages was common; his cousin also married a Morin and became professor at the University of Lausanne.[10] During the late nineteenth century, Lausanne grew as the metropolitan center for the new elite. Two obvious marks of this change were the expansion of the university (advancing from mere *académie* status) and the opening of the grandiose Palais de Rumine, incorporating a library and five museums and galleries, in 1906, just missing the centenary celebrations of Swiss independence.

After twenty-nine years away in Munich and, mostly, Zurich, Forel felt a suffocating sense of parochialism upon his return to Vaud. He thought that the locals in Yvorne, though equally as conservative as those in Morges, were at least more amiable and trustworthy and, especially, more self-reliant, having remained free from the infantilizing influence of the old and new aristocracy.[11] Forel could not help but agree with the self-deprecating Vaudois saying, *"bieau pays, pouettes zens"*—nice place, lousy people.[12] The sense of frustration was a larger affair than a feeling toward his hometown; to Forel, the history of Morges represented in miniature the history of Vaud, which in turn bore the same relation to the history of Switzerland as a whole.[13] These synecdochic relations were paralleled by the miniature history provided by the ants' nest—a history that stretched back through evolution, reliably demonstrating certain natural laws in action. "The resemblance between a society of ants and a society of men is no mere matter of appearances," he wrote. "Both depend on profound causes, hereditary or acquired."[14] Paradoxically, Forel escaped the claustrophobia of his native Vaud by looking to something even smaller, *la fourmilière*.

Thus a number of distinctively Swiss cantonal issues pervaded Forel's context, all of which were directly addressed by his adult writing, including those on ants. These issues were religious tension, the political relationship of the part to the whole, and, related to that, the irony that peaceful existence within Switzerland's national borders had been secured at the cost of the establishment of a unified army for international warfare.

The Canton and the *Fourmilière*

Whenever Forel watched ants, he saw potential lessons for humanity. One of Forel's earliest theoretical innovations, in *Les fourmis de la Suisse* (1874), helped to establish grounds for this: the definition of the *fourmilière* (colony) as a technical term covering all the inhabitants of the *nid* (nest).[15] The *fourmilière* contained a variety of inhabitants—not just the ants born in that nest but any who had joined, for example through capture in a slaving raid. The *fourmilière* also included non-ants, such as the aphids "milked" by ants, the many small species of beetle that often cohabit with ants, and species that actually parasitize ants.

Forel conducted many experiments on the miscibility of ant societies. As a teenager, he accidentally created a mixed community of *Formica sanguinea* and *Formica pratensis*. He placed *pratensis* cocoons near a *sanguinea* nest, fully expecting the latter to use the former for food. The following year, he was astonished to find adult *pratensis* had been raised and were now working together with their supposed enemies to repair the nest. Later observations proved this to be an unusual occurrence involving these particular species, but Forel was inspired to carry out a series of experiments in which he induced ants of different species to live together in one nest. He observed that such things also occurred naturally; alliances were more readily formed when the introduced species were eggs or nymphs, but they might also occur between adults of the same species, although from different colonies. Rarely, alliances might be achieved artificially between adult ants of different varieties and different nests. After a certain amount of "quarrelling" they sometimes settled down to work together, although Forel noted that this could not be done with ants of very divergent genera.[16] The experiments were written up in 1874, and more of the same were recorded nearly fifty years later, in *The Social World of the Ants*. In the latter work, Forel also cited similar experiments carried out by Adele M. Fielde, a former American missionary and latter-day convert to humanism. Her concern was to establish what factors caused ants to accept or reject one another's presence within the same nest. Fielde discovered that even conspecific ants originally from the same colony might be rejected by their peers if they were removed sufficiently long that their identificatory odor wore off.[17]

By contrast, according to Forel, men and women typically lived in much smaller, more protective units and did not learn to accept fellows of foreign origin. Humans had a "vain and brutal egoism"; though they might manage to extend partiality to their nearest relatives, man was all too often a "family-loving individualist."[18] This produced aggression in society as each individual sought to protect his own and his family's interests. Because the *fourmilière* was not a family-based unit but an interspecific community, the ants illustrated that life could be lived more successfully if such attitudes were abandoned. Every national state should be organized like a *fourmilière*, Forel thought, incorporating all members, not just the racially identical, on an equal basis.

Moreover, each nation-state should not consider itself a unit in any essentialist sense. Again, there was a precedent for this form of human organization among the ants. Forel identified two types of nest, the monodomous, where a group of ants lived in a single nest, and the polydomous, where the colony, or *fourmilière*, was spread over several nests, or *nids*.[19] This organization echoed the cantonal structure of Switzerland. The *fourmilière* of Switzerland included various languages and traditions and was spread over a number of cantons, which themselves approximated to *nids*. Forel supported a number of nationalist causes toward the end of his life (especially those of Lithuania and Macedonia), but he remained terribly aware of the potential danger of their independence, namely, the development of a nationalist supremacy or "racial" exclusivity in response to the oppression they had suffered. He exhorted his correspondents not to abuse their nationhood, should they achieve it, by becoming obsessed with "racial" identity, or by oppressing their own minorities, or by becoming militaristic and settling old scores. "I support you, *but* . . . " was the refrain of Forel's letters to nationalists. So anxious was he about the double-edged nature of nationalism that he often asked his correspondents not to publish his name.[20] The nation-state should never make the mistake of acting like a slightly enlarged human family but instead should behave like the *fourmilière*—Forel's own home—with its multispecific inhabitants living in harmony. Forel summarized these two conditions for the ideal society—cultural and "racial" variety and a federal constitution—in a 1927 letter: "In Switzerland we have two good things: total equality of languages, beliefs and races, and the right for popular initiative. Their fruits are excellent."[21]

If on the national level the ants' *fourmilière* indicated an organization akin to that of Switzerland or America, together these states would form an international superformicary, a polydomous nest spread across the globe. In other words, each nation should be like a canton under a global government, the "Etats Unis de la Terre."[22] Dreading the impending war of 1914, Forel was much exercised by the question of creating a harmonious world system. The key was to see things in terms of education, not race. At least that was how Forel saw it, although his exclusion of "obviously inferior" races such as "the Negro, the Mongol and the Malay" does not seem like a nonracial perspective to modern sensibilities. At the end of his life Forel was still pondering "which races can be of service in the future of mankind, and which are useless," and, given this, to work out "how [the useless lower races] can be gradually extinguished."[23]

Notwithstanding these opinions, Forel's claim that European and North American blood was a homogeneous mix of Celtic, Slav, Germanic, and Jewish stood in contrast to much contemporary discourse. In this period, for example, Wilfred Trotter made his "discovery" of innate difference between German and British, characterizing the latter as, among other things, apian, the former as lupine.[24] Again, Forel's canton provided the lesson for the contingent nature of difference in miniature: the problem with the Vaudois was that they thought in German and spoke in French. This intellectual handicap, palpably an effect of education, demonstrated why war between Francophone and German speaker was pointless. European differences were made by tradition, religion, social and political mores, and most especially by language.[25] The miscible ants were an even smaller illustration of the same. If even ants could be "taught" to overcome nest and species differences to work together, so could humans, provided they had the basic racial ingredients.[26]

Ant-Citizens and Their Educability

Forel, then, was interested in training the wild animal (the ant) and the uneducated individual (alcoholics, but also anyone else who did not subscribe to Forel's entire philosophy). Neurology provided the model for how this could be done. As he indicated in his references to "lesser races," there was an element of limitation in the starting materials for education. Some, like untreatable alcoholics, were simply "ethically defective." Forel's darkest eugenic moments are still kept under wraps in the various archives where his papers are deposited, but we do know that he carried out castrations and other sterilizations in order to "treat" some of his patients and, crucially, to treat by prophylaxis the health of the race.[27] In this sense, Forel's representation of instinct appears somewhat like the brutish social predestination described in the late nineteenth century by European social theorists. This instinct, an inherited behavior whose expression was more or less inevitable, informed criminal, legal, and medical planning. It crops up as a primitive and bestial force in the recommendations of Cesare Lombroso and in the naturalist novels of Emile Zola.[28] Yet for Forel, provided the right ingredients were in place, the exercise of such "instincts" was not always inevitable: "Predatory, egoistic and hypocritical though human nature may be in itself by inheritance, yet it can be tamed from childhood upwards *by social education*. My perception of this I owe in the first place to . . . the study of ants."[29] People, like ants, could be educated to live in peaceable socialism.

For Forel, the value of education did not lie merely in *what* was learned but also in the act of learning itself. Learning was work, a Spencerian struggle of evolutionary value, and the social consequences of this group effort were responsibility to the corporate body and communal achievement. Forel believed strongly that inheritance of money and estate was wrong in that it prevented work, and to his credit he practiced what he preached. On a number of occasions he refused money from his parents, and from an early age his family nickname was "Gueux" or "Gugu"—"the little hobo." His favorite epithet, *labor omnia vincit*, reflected the qualities of the ants in their ceaseless social labor, connected with their ceaseless evolutionary improvement.

Forel's understanding of educability was, at times, unconventional. Frustrated by a local artisan's refusal to commit to the Morges Good Templars' Lodge, Forel "took his arm and literally dragged him downstairs" from his own home and all the way to their meeting place, where the cabinetmaker took the pledge.[30] This approach, which might generously be described as pragmatic, was, however, related to Forel's training in neurology. Early in his human neurological research, Forel reached the conclusion that the nerves in the brain did not need to be connected by physical anastomoses, as was at that time presumed. "Why do we always look for anastomoses?" he asked himself around 1886. "Could not the mere intimate contact of the protoplasmic processes of the nerve cells effect the functional connection of nervous conduction just as well as absolute continuity?"[31] This functional rather than physical connection of nerves underwrote Forel's idea that the connections could be altered; in today's parlance, they were not hardwired. The term used by Forel was plasticity, meaning that new learned behavior could replace old forms as habit and eventually, over the course of evolution, as inherited instinct.

In a novel situation, the senses of an organism would receive certain stimuli, inscribing a permanent formation in its nervous system (usually the brain), called an *engram*. With repeated presentation of the stimulus, the engramic response could eventually be elicited even by a weakened form of that stimulus. This state of readiness was known as ecphoria and corresponded to the psychological condition of association. In the long term, it corresponded to the evolutionary endowment of physiological conditions that enabled the organism to react in a certain way to its environment. Engrams were "ecphorized," or harmonized in a complex with the experience of the senses, so that the appropriate response could be called forth in each combination of circumstances. The sum of the acquired and inherited engrams was called the *mneme*. Discord between the conflicting actions of new stimuli and preordained mnemetic excitation was immediately resolved in higher organisms by the introspective aid of attention—in other words, by a conscious decision whether to pay attention to one urge or another. In the long run, discord was resolved within the life of the organism by neuronal "regeneration"; this amounted to relearning habitual behavior and was the business Forel was engaged in as a psychiatrist. Over the course of evolution, discord between internal and external stimuli was resolved by "adaptation"—in other words, the inheritance of a new and more appropriate tendency.[32] Forel's theory involved minor choices that animals had made over many generations, which became fixed in the form of inherited psychology. A lot of knowledge was therefore accumulated in the complex instincts of the ants without overloading their small brains with the memory needed for constant decision making, or plasticity of response. Instinct was an acquired thing, fixed from the outcome of primitive intelligence over the course of evolutionary time.

Forel encouraged his brother-in-law Edouard Bugnion (1845–1939) in his study of the termites and published Bugnion's monograph as an appendix to his own *The Social World of the Ants Compared with that of Man*. Forel believed Bugnion's studies illustrated his own instinct theories perfectly, showing how, over time, a race acquired its fixed, adapted behaviors.[33] Just as it was for Forel, the knowledge of this process was provided for Bugnion by a comparative study of primitive extant forms of termite with their "more recent and highly-perfected" relatives. Hence, claimed Bugnion, "we have only to observe the habits and customs of each species in order to follow step by step the evolution of instincts and their ever-increasing complexity."[34] The ceaseless struggle between the ants and the termites provided the termites with the evolutionary impetus—according to "the laws of natural selection"—to develop numerous defensive strategies, of both a physiological and behavioral nature. Soldier ants had evolved with large heads to block entrances to the nest, and bellicose instincts to match. It made no sense to ask which of these developments came first, for neither made sense without the other. By looking at the diversity of behaviors, Bugnion proved that in advanced species, advantageous "reasoned actions" were merged into habits, thereafter becoming "automatic or instinctive."[35]

Forel's instinct, related to the instinct of drive, yielded pragmatic consequences. This was a sense of instinct as inherited behavior that was part of an evolutionary story. Thus an alternative way to reconstruct the evolutionary tree would be according to the relatedness of animal behavior. This should produce a diagram identical to the version produced by comparative anatomy. Where parts of the traditional tree were missing owing to underdetermination by anatomical evidence, they might be filled in by considering behavior of extant species, and vice versa.[36] The two methods would complement one another and provide a fuller image of evolution, and one, moreover, that connected animals with perennially interesting questions about human behavior. Forel sometimes put this claim extremely strongly; although he did not think there was a morphological homology between humans and the lower animals, he did claim an analogy proceeding from a "homology of functions" or from adaptation toward an "analogous end." Ants and humans responded to the same pressures of social organization; their societies had to evolve, by hook or by crook, to perform the same tasks necessary to sustain the life of the race. Because behavior and not anatomy provided the key to these analogies, the convergence was to be demonstrated by the naturalist, from living observation and experiment, and not in a laboratory by the morphologist.[37]

Anything that one discovered about ants could be applied to humans, and vice versa. Forel's neurological perspective, originally derived from human medicine, certainly extended to the ants. He researched and published extensively on the senses and nervous system of ant and corresponded with Wilhelm His, among others, on the subject. The senses and nervous systems of insects were a physical reflection of the evolved, adaptive aspects of the insect psyche.[38] What ants had perfected and incorporated into their inheritance over millions of years the recovered alcoholic could achieve in a few years. Hypnotism was an excellent way to inscribe new engrams, provided the patient had not been blighted by "blastophthory," the permanent damage of the germ line through his or her ancestors' abuse of alcohol. So long as this had not occurred, the patient might be reeducated. Living in a formicary-style asylum and learning the value of work for the good of society provided experiences that were ecphorized so that they would be continued even when the patient returned home. Plasticity was the key to formic and human behavior; it was the scientific conviction that gave Forel the confidence to proclaim, in relation to the nonaccidental nature of the resemblances between humans and ants, "Comparative anatomy [has] shown me the unbroken continuity of evolution between the animal and the human brain, as clearly as comparative psychology and physiology, and the ants [have] shown me the unbroken connection between the animal and the human psyche."[39] Thus Forel invariably proclaimed himself a monist: *l'âme*—the mind or the soul—was one in nature with the brain and the nervous system.

Monism

Monism was an innately political theory in its construction. Forel borrowed his mnemic psychological vocabulary from the nationalist German biologist Ricard Semon;[40] another of his correspondents, Ernst Haeckel, was even better known as an advocate of monism. For Haeckel and Semon, monism was about the relationship of the part to the whole, understood in the context of German unification. Marine organisms with their regenerative potential were akin to German states, requiring metaphysical unity in one organized whole.[41]

Forel had a certain amount of sympathy for the dream of German unification, but his experiences treating French soldiers during the Franco-Prussian War (not to mention the Swiss experience of being menaced by Prussia) put him off its consequences. Bismarck, he judged, was a brilliant unifier who had wanted to readapt feudalism wisely and moderately for the modern era. The problem arose because of the nature of the German people, who, like the Austrians, had never really been penetrated by the spirit of the French Revolution, despite the appearance of revolutionary fervor in 1848. Rather, they remained happy to obey. Having placed their faith in God and Bismarck, they had allowed their patriotism to grow unhealthily, yielding fruits such as *Simplizissimus* and the Nietzsche cult. (But, he allowed on the plus side, their good qualities included hard work, discipline, vegetarianism, and the protection of birds.) Bismarck's successors—the press, militarists, Prussians, feudalists, diplomats, and pan-Germanists—had spoiled his legacy by leading the German people badly astray.[42] The First World War caused Forel deep personal anguish, and it was undoubtedly for this reason that he decided against donating his specimen collection to Berlin. An open letter to Haeckel, written during the war, expressed Forel's concern that Germany wished to subsume smaller states, including Switzerland.[43]

Forel's was a different kind of monism, one that did not subsume the parts into the whole but rather allowed them to live independently in federation.[44] Haeckel's marine polyp was governed by the whole organism; if broken off, it regenerated according to the innate characteristics of the original. Forel's ants, however, would die if singly separated from the *fourmilière*—unless, of course, they were adopted by another, in which case they would change their life to suit the new colony. Thus the ants were simultaneously more dependent and more independent than Haeckel's pan-nationalist polyps. Besides, education was more important to Forel than a mistaken idea of nationality as inheritance or race; notions of national (or *fourmilière*) blood types were disproved by the educability of ants and humans to live in harmony. For Haeckel, monism was opposed to socialism, since it entailed a battle against "natural forces."[45] This was obviously unsatisfactory to Forel, and his model focused on natural force (educable instinct) to reconcile the two. His work contrasted with that of social entomologist Heinrich Ziegler, who worked with Weismann's unmodifiability of germ plasm to argue against socialism; Ziegler's primary unit of analysis among the ants and humans was the family, rather than society as a whole.[46]

Forel's boyhood neighbor Carlo Emery (1848–1925) did not see lessons for Italy's constitution in the anthill. Emery's take on the social nature of ants was more reductionist than Forel's (or Wheeler's, for that matter). Emery thought that ethology and physiology would eventually unite to form a chemical theory of everything, with formulas to explain every aspect of life from the bottom up. Besides, as his 1899 account *Sulla missione delle scienze nella vita* argued, it was not the aim of science to address the problems of existence.

Thus Emery, in his 1901 paper "Les insectes sociaux et la société humaine," was more circumspect than Forel in drawing conclusions about the psychology of insects. He was anxious to point out that the resemblances between certain human and ant behaviors, such as slave making, were largely superficial. The only true analogies were the degeneracy produced by parasitic dependency in either organism, and their group properties when in crowds. Both had "a social feeling of cohesion, of collective unity"; both were "subject to mutual suggestion and incitement by example." But the big difference in the latter case was that the mass of the formicarian crowd was infertile, the individuals held rigidly in their caste positions.

Emery could not escape the deeply ingrained nineteenth-century analogy between ants and "primitive" humans, which Forel used to make such disparaging comparisons about educability. For Emery, ant societies showed the maximal complexity that societies could reach in the absence of humanoid intelligence, and thus the savage human society provided the best parallel to ant life. The ant had a mere rudiment of intelligence, which highlighted the perfectibility of (European) man. This latter intelligence began with the more impressive "sense of the good and the beautiful." "Nothing like this," Emery wrote, "is known amongst the insects, nor amongst animals in general. Not only is individual initiative rarely evidenced amongst them, but the inventive spirit, which has put tools in the hand of man and perfected them . . . seems [also] to be entirely missing."[47] This inventive spirit was necessarily a product of individuality, and thus in general it would not be advisable to imitate the ants' cooperation, however utopian. To do so would be to lose the human spirit, "the insatiable desire for betterment," which, although the root of many troubles, was also the cause of human progress.

Monism was also a religious standpoint. Early in his life, Forel's conviction that mind and brain were the same shored up his antipathy toward religion, especially as exemplified by his ever-anxious mother ("Send me a grain of mustard," she begged him in one of the rare letters she got around to writing him while he was in Zurich[48]) and by the petty, unpleasant confessionary wars that surrounded him in Vaud. But Forel was never quite able to throw off his religious context. His favorite sister, Blanche, remained a believer, as did Auguste's friend and later Blanche's husband, Edouard Bugnion. In one letter to Auguste, Edouard gently refused his friend's socialism but pointed out that if anyone had lived out its principles, it was Jesus Christ. Blanche's letters to Auguste, meanwhile, reveal that Edouard suffered terrible black moods and found life a struggle; she begged Auguste not to press his secularist hypnotism on Edouard lest it destabilize his faith, his one anchor in life. Even Auguste's own wife, his beloved and respected Emma, was a believer, though not inclined toward proselytism. (One is reminded of her namesake, Mrs. Darwin, who also retained a quietly nonintellectual faith.) There are few anecdotes pertaining to Forel's adult life that indicate a sense of humor, but one that does concerns this situation. At Christmas in Yvorne, Emma would gather the children around the piano to hear Bible stories and sing; as an antidote to this religious expression, Auguste would place himself at the other end of the room and carol out wholly unsuitable student ditties from his Munich days.[49]

Nor could Forel avoid organized religion in his psychiatry or his myrmecology. The pastors and missionaries who sent him ant specimens were fellow-travelers and moral allies in various ethical and antialcoholic leagues. Pastors who drank were also among Forel's most despised foes of abstinence. Abstinence, in fact, became an evangelistic religion for Forel, complete with testimony, discipleship, meetings, a sense of international brotherhood, and a need to win converts. Perhaps it is not so surprising, then, that around 1920, Forel eventually subscribed to an organized religion, the recently founded Baha'i movement. He was still no believer in God, but something in him responded to an overtly religious framework for his social and political beliefs. "I am a monist in the following sense: I am certain that the functions of the brain and the human soul are but one inseparable whole," he wrote to Abdul Baha Abbas, founder of the faith.[50] "*Le socialisme sera moral ou il ne sera pas,*" ran one of Forel's favorite maxims: socialism will be moral, or it will not be at all.[51] Morals, like neurally encoded habits, were purely functional, not formal. As far as Forel was concerned, their worth was given by their value for biological society. In the case of human beings, the moral impetus was bound up with "hygienic" procreation. A sterile union was ethically neutral, whereas parentage should be qualified for by social worth and "intrinsic hereditary qualities."[52] Forel meant his eugenic, socialist morals as earnestly as the most fervent protestant Vaudois pastor.

Thoroughly Vaudois Ants

Like many other entomologists in their time, Forel has been dismissed as an eccentric. Although he was remarkably tireless in support of his chosen causes, none of these was eccentric in itself; his activities make sense when considered in the contexts of European politics and Vaudois civic life. These contexts are represented in all their wealth and variety among Forel's letters and papers: a letter from the International Federation of Eugenic Organisations is filed alongside a matinee announcement of the Lausanne section of the Ligue Antialcoolique promising "choirs, music . . . without forgetting tea [and] games" for all.[53] The social life of Lausanne's protestant bourgeoisie circa 1900 revolved around a multitude of similarly worthy societies, all putting on matinees, soirees, and other events. Some were local branches of national or international organizations, and some were purely regional. A small sample of the groups with which Forel was involved, whether peripherally or centrally, included the local Ordre des Bons Templiers; the Société Vaudois de la Paix; the Société pour la Developpement de Morges; L'Aurore, Société Artistique et Littéraires d'Abstinents de Lausanne; the Association Vaudoise pour le Souffrage Féminin; the Parti Ouvrier Socialiste Vaudois; and the Ligue pour l'Action Morale (this last being the innovation of

Forel himself). Some of these societies spawned subsocieties, such as the Lausanne Chorale des Bons Templiers (whose president, one A. Emery, may well have been a relative of Carlo Emery). International or foreign societies with which Forel was involved in one way or another included the Universal Esperanto Association, the Pestalozzi Association, and, of course, the antialcohol International Order of Good Templars. Clearly, causes and societies were a way of life for the upstanding metropolitan Lausannois. Many of their members and organizers were professors at the new University of Lausanne, while names associated with Forel's international societies were well-known figures from around Europe in the sciences, the humanities, and eugenics. Forel would have appeared eccentric to his friends, neighbors, and family had he *not* interpreted ants within this context.

CHAPTER TWO

A (Non-)Disciplinary Context for Evolutionary Myrmecology

Auguste Forel's reassuringly Swiss ants seemed anything but familiar to other myrmecologists. In 1918 the French insect psychologist Eugène Bouvier attempted to do justice to the bizarreness of the insect realm that so fascinated him, his colleagues, and his nonscientific friends and family:

> Insects are creatures which seem to defy the imagination with the strangeness of their form and their extraordinary habits . . . What can we think of the predatory wasps which paralyze with dagger thrusts? . . . What . . . can we think of the larvae that hatch from these eggs and scientifically devour their host, leaving its most vital organs untouched until the last? Everything about these animals surprises us even when, in the present stage of their evolution, they seem to come near us and to engage in activities which might be considered human, such as we observe in the social species . . . The old anthropomorphic school is, indeed, dead: we no longer attempt to explain insects by man; we rather try to grasp the mechanism that allows these animals to evolve mentally and to acquire activities which seem human.[1]

Social insects in particular exhibited behaviors so complex, so astounding, that one was compelled to ask, *how? why?* It was an American psychologist who credited C. O. Whitman with the first statement (in 1898) that "instincts and organs are to be studied from the common viewpoint of phyletic descent." Whitman's student W. M. Wheeler, embarking on myrmecology two years later, also stated, "it can hardly be doubted that there is a phylogeny of instincts."[2] But it is likely that Wheeler himself would actually have traced this psycho-evolutionary credo to Forel's statement of 1874 that ants must be studied "from the dual perspective of . . . classification and . . . behavior."

To understand Forel's study of ant psychology and the insect psychologists of his time, there are a number of disciplinary contexts (or rather undisciplined contexts) in which he must be placed. The first is as an ant-lover in a circuit of specimen exchange and taxonomy, the second as an investigator of animal behavior. Forel, Wheeler, and Emery formed a triumvirate in ant classification in the early twentieth century; of these, Forel and Wheeler were both interested in behavior, Emery much less so.[3] Forel's experiments on insect behavior commenced in the late 1870s. The earliest experiments concerned vision, but by 1887 he had reached some general evolutionary theories on the psychology of insects. These early experiments were published in rather obscure journals that were poorly disseminated and short-lived, but Forel's parallel account of the human psyche, adapted from Semon, was published in the widely read *Der Hypnotismus* of 1889. Around 1900, the connection between Forel's theories of the human and ant mind were made more public. His theories on myrmecology came to America by way of Wheeler's translation of a lecture given at the Fifth International Congress of Zoology in Berlin in 1901.[4] Forel's experiments of the period 1878–1906 were gathered some five years later in an original English-language publication, *The Senses of Insects* (1908). By then, however, there was already a healthy tradition of investigating "the psychic life of insects" in Europe, especially France.

This branch of animal mind and behavior studies was interpreted variously as zoology or psychology; ethological or physiological in method; laboratory- or field-based; synchronic or diachronic in scope; concerned with proximate mechanism or long-term evolution. These approaches did not map neatly into distinct categories (say, natural history versus natural science) but formed a spectrum of sciences, shaped by national, political, professional, and religious concerns. "Instinct"—usually mentioned in the same breath as its corollary, intelligence—was, however, one area of commonality, a key shared discourse generally acknowledged to be paradigmatic of the insects.

Ant enthusiasts working around the turn of the twentieth century treated instinct as an *evolutionary* story intended, ultimately, to reveal something about the larger process of psychological evolution that included humans. What was instinct?[5] One entomologist writing in 1916 meant by the term "innate complex acts . . . and nothing more precise."[6] Darwin had been scarcely more exact: "everyone knows what is meant by instinct," he airily assured readers of the *Origin*. Was it degraded intelligence or something altogether different in kind from intelligence? Like Darwin, George Romanes postulated two methods by which instinct evolved: by natural selection and also by the gradual stereotyping (or degradation) of originally intelligent acts.[7]

For the myrmecologist, instinct was a functional concept. The function of behavior was evolutionarily prior (or parallel) to its anatomical substrate or form.

Though similar to Romanes' version of instinct as stereotyped acts intelligent in origin, this theory did not carry the negative connotations of degradation. It was above all a behavioral typology that revealed something about evolution and, for the hymenopterists, progress toward eusociality. The metaphysical arguments of laboratory and especially human psychologists in the 1920s about the quiddity of "instinct" were of no concern to them.

To frame their concept of instinct (which I argue was central to myrmecological discourse), "myrmecologists" had to define themselves in several different ways and in contradistinction to various different groups. As ant lovers, they were defined by their networks of collection and exchange. As observers of animal behavior, French (or Francophone) entomologists were part of the disciplinary story of animal psychology. Within this, myrmecologists—observers of animals in the wild—had to distinguish themselves from those who pursued a mechanistic, laboratory-based approach. But among their natural history allies were also a prominent minority of antievolutionists from whom they had to create distance.

Disciplining the Ant Collectors

Insofar as the students of ant behavior were organized around 1900, it was as a rather Victorian network of letter writing and specimen swapping. The main nodes of this network were Forel and Wheeler, and to a slightly lesser extent Emery. Forel had acquired general scientific credibility through his psychiatric expertise, Emery and Wheeler through academic zoological positions. Specific myrmecological expertise, however, was built up through the process of collection and taxonomizing: by becoming a center to which far-flung collectors and "amateurs," museums and other "experts" would send their specimens. The greater part of Wheeler's and Forel's correspondence concerned classification, and a close reading of it reveals a complex etiquette regarding the exchange of dead ants.[8] John Clark, for example, corresponded with both Wheeler and Forel from Australia, but after a perceived slight by Wheeler, he transferred his allegiance—in material terms, his sending of specimens—to Forel. Clark's letters express numerous generous offers to send Forel specimens and types, and demonstrate allegiance by attacking classifications proposed by Wheeler. "I wish Wheeler would work as you did. If so our [Australian] ants would not be so complicated... I intend to follow your lead in this matter."[9] Knowing Forel's eyes to be weak (he had begun treatment for glaucoma in 1916), Clark anxiously checked whom would be an acceptable alternative expert to send his specimens.

By contrast, the Londoner W. C. Crawley failed to act correctly in the Maussian exchange. He was probably aware that Clark had sent Forel some Australian specimens, since Clark had sent some to him but had been dissatisfied by his response. Crawley wrote to Forel, baldly requesting him to post some of his exemplars. "I receive Australian specimens from time to time and have difficulty identifying them, which I can do with the aid of your descriptions," he explained nonchalantly.[10] Crawley's crimes lay in his weak praise of Forel's work, his protective obscurantism about his Australian source, his attempt to obtain specimens for the purpose of getting priority in naming, and his failure, so far as records indicate, to reciprocate by ever sending Forel any specimens of his own. There is no record of Forel responding to the request. Apparently in response to personal remarks initiated in a letter from Forel, Clark wrote, "Crawley is impossible, he is too fond of sport to do serious work." This was a serious charge in the eyes of Forel, whose motto was "work conquers all."[11] Horace Donisthorpe extended the tradition of Englishmen who failed to understand or refused to accept the etiquette of exchange. In 1918, Donisthorpe wrote requesting an enormous number of Forel's papers, which he had seen listed in a bibliography but of which he had been unaware.[12] Three years later he sent Forel a paper of his own, but unlike most correspondents he demanded that Forel return it, albeit "at your convenience."[13] Moreover, the purpose of sending the paper was to correct Forel, who had "deeply hurt his feelings" (not to mention Crawley's) in failing to mention either of them in the first volume of *Le monde social des fourmis*. Donisthorpe committed a faux pas in neither presenting himself as a docile supplier of specimens (or, in this case, suggestions) nor framing his professed equality through the accepted norms of reciprocal exchange. That the possibility existed of trustworthy specimen exchange between taxonomic equals or potential rivals is amply illustrated by Forel's correspondence with Carlo Emery. A veritable flood of ant gifts and loans passed between the two, together with more personal exchanges. Emery, for example, designed Forel's "Ex Libris" labels. Their cordial relationship lasted despite Emery's frank criticism of Forel's political perspective.[14] The crucial thing was to act properly within the etiquette of epistolary exchange.

Wheeler began by integrating satisfactorily into Forel's network. He sent Forel almost his very first collected specimens, which Forel identified and used as the basis of a paper.[15] Shortly after Wheeler's visit to Switzerland he sent Forel the skin of a musk ox, under the pretext that it was a present from "Mrs. Wheeler" to "Mrs. Forel" (perhaps thinking the ladies in question were unable respectively to write and read).[16] As Wheeler built up and maintained a similar network of his own, he ceased to participate on terms satisfactory to Forel and sometime around

1922 offended him, causing Clark to write, presumably in reference to that slight as recounted to him by Forel, "I . . . do not care too much for Prof Wheeler [either] and when he treats you, the greatest authority on ants rudely, how would he treat a beginer like me?"[17] Wheeler's fondness for whisky meant that he could never write the magic words employed by many of Forel's correspondents: "by the way, it may interest you to know that I too am an abstainer/member of the Independent Order of Good Templars / temperance society in my own country." As far as Forel was concerned, this was the instant password that earned one full and trustworthy admittance into the ant exchange.

Sometimes nonmyrmecologists would attempt to tap into the exchange system for their own purposes, threatening to destabilize the naturalists' intellectual and material economy. Auguste Forel, for example, wrote to his cousin Alexis regarding some ants promised by a Swiss pastor in Chile. The specimens turned out to be fewer in number than Auguste had been given to understand, and Auguste warned his cousin that though this man was now offering to visit with "hundreds" of ants for Alexis (had Auguste perhaps offended the pastor?), he might be after something. "Our colonists are hardly compliant," he complained, punning unintentionally on the ants themselves.[18]

The final effect of this exchange system was to render the ant collection of immense value, for it reified a whole system of trust, or acknowledgment of expertise: a literal accumulation of knowledge and possession of nature in its standard forms. Natural history collections are a good example of the "Matthew effect": The more specimens an individual or institution has, the more likely it is that a collector will have to send a new or possibly new species to them for checking against extant types, giving them the power to rule on its novelty. Forel's own collection comprised 3,500 ant species of the world, all described by him; this was around half of those known at the time. Two-thirds of the specimens had been obtained and donated by expeditions other than Forel's, or had been supplied by museums wanting identification in exchange for allowing him to keep the specimen. Hearing of Forel's worsening eyesight and general poor health in 1921, the curators at Museum of Natural History in Geneva wrote to express, rather briefly, their sympathy and to inquire what, by the way, would be the future of his ant collection? They would be very happy to have it, they generously assured him.[19] After a further of exchange of letters Forel agreed, choosing to overlook their disturbingly vulturine attitude. Before the Great War he had planned to give the collection away to Berlin; after the war he decided to sell it. His decision to favor Geneva over his native Lausanne was mostly determined by the fact that the curator at Lausanne was an alcoholic.[20] Just before his death, however, he made a family collection with one specimen from each species, and this collection was eventually given to the Lausanne Palais de Rumine. Likewise, an early collection (c. 1874) that embodied in material form his book *Fourmis de la Suisse* was also given in its entirety to Lausanne. His gifts to his protégé Heinrich Kütter formed the basis of the Kütter collection at Lausanne.[21] This collection is now regarded as the symbolic repository of Forel's tradition: as in the network of correspondence, a gift carries more weight than a sale.

It was through this system of specimen exchange and correspondence, then, that ant enthusiasts were "disciplined" around the turn of the twentieth century. But Forel's interests stretched beyond mere taxonomy and into the shared psychology of humans and Formicidae. In trying to persuade his fellow collectors of the validity of his project, he strayed into the disputed terrain of animal mind and behavior.

The Undisciplined Sciences of Animal Mind and Behavior

Around the time that Forel was working—and, arguably, right through the twentieth century—there were multiple disciplinarities in the study of animal behavior.[22] A number of traditions existed, often grouped artificially by contemporaries and subsequently by historians on a single procrustean bed of methodology.[23] Comparative psychology, animal psychology, experimental psychology, ethology, ecology, natural history, and behaviorism, and latterly behavioral ecology, sociobiology, and evolutionary psychology, have all been competing names for the science of what animals do. Additionally, there are named animal specialisms relating to those organisms considered by their fans to give a special insight into behavior: primatology and myrmecology, to name two. (No such naming exists for those animals constructed as the average, experimental "every-animal"; there is no rattology or caninology.[24]) In the earlier part of the twentieth century it was not even clear whether those who studied animal behavior should be called scientists or naturalists.

During Forel's lifetime there were two major trends in studying live animals, both of which started around the mid-nineteenth century. Laboratory scientists, particularly in Germany and France, were working on a completely mechanistic approach to animal behavior. Inspired by the Bernardian approach to physiology, they rejected terms such as "psyche" and restricted themselves to a positivist investigation of animal behavior. By virtue of being in laboratories, these researchers were relatively well disciplined.

In Germany, radical materialists gave purely automatic accounts of insect behavior. Albrecht Bethe (1872–1954), for example, posited that ants were reflex machines that experienced no mental life as they went about their tasks.[25] He claimed to show that ants returning from foraging expeditions obeyed a reflex in following an odor trail home, and used no powers of memory. Thus, for him, ants had no chance of experiencing success and learning its methods.

Jacques Loeb (1859–1924) developed a mechanistic theory in detail. Like Bethe, Loeb was born in Germany, where he trained as a physiologist. In 1892 he went to Chicago at the invitation of C. O. Whitman. There Loeb achieved considerable recognition and success, even though his reductionist approach clashed with that of his boss.[26] Loeb created completely mechanical models to account for the behavior of simple organisms. He called their movements "tropisms," which he considered to be reactions to directional factors such as light, chemicals (including food), air currents, gravity, or heat. Periodic variations in sensitivity might also occur, which would explain why winged ants, for example, swarmed upward into the air (that is, expressed negative geotropism) only on one day in the year. If one of these tropisms did not predominate, behavior was considered to be the complex result of an equilibrium of different tropic reactions and sensitivities.

The second trend was an explicitly evolutionary, natural historical approach. Following the general acceptance of transformism, the desire to explain human psychology in the light of animal evolution grew quickly. Darwin's *Descent of Man* (1871), his *Expression of the Emotions in Man and Animals* (1872), and the work of his disciple George Romanes (1848–1894) are obvious examples of the widely held ambition to tell an evolutionary story about the human psyche. John Lubbock (later Lord Avebury) focused on the hymenoptera in order to answer such questions, finding in higher and lower ants echoes of human development through hunting, pastoral, and agrarian stages.[27] Such psychology was intrinsically comparative across species and had to rely on the interpretation of behavior for its data. Because it was framed by a natural historical approach, it was less disciplined than the psychophysiological approach to animal behavior.

Adherents of these various approaches often clashed because they were looking at the same animals. For well-disciplined sciences, this kind of overlap was not a problem. From 1905, physicists lived quite happily with the notion that light could be both wave and particle, amenable as such to two different research methodologies. But it is harder to divide a beetle into "physiological entity" and "evolutionary entity" than it is to divide light into waves and particles when a beetle seems to us a patently unitary and, crucially, a tangible item.[28] The persistent influence of this "natural" classification of animals made it hard to partition them metaphysically for study by different methods and their respective experts.[29] The competing students of animals were also possessed of a shared vocabulary, and often found their arguments being twisted because they could not make their terminology obedient to their respective programs of research. (*Instinct* was perhaps the worst of the quisling words, as I explain later.) Proponents of each approach further differed among themselves as to whether they should attempt to corral all students of animals into their own respective paradigm or whether, as the physicists did with light, they could parcel out the animal kingdom between them. Some engaged in combative debate across the laboratory / natural history divide, while others simply got on with their own way of doing things.

Disciplinary questions about animal behavior were worked through rather explicitly in France, and a recounting of this story provides vital context for Forel's myrmecology.[30] In France, Forel's contemporary, the biologist Alfred Giard (1846–1908), developed work on behavior into a kind of zoological psychology. In 1887 he protested against laboratory-based science and advocated studies that focused more on organisms interacting in and with their natural habitat. His most basic interest was evolution: providing ultimate explanations for alteration in organisms with respect to the environment.

Giard promoted his approach at the Institut Générale Psychologique, which had grown out of the Fourth International Congress of Psychology in 1900. The institut was not a research institute in its own right but a forum for people to exchange ideas. One of its four sections was devoted to zoological psychology, and here Giard met and talked with zoologists, psychologists, doctors, and all sorts of interested parties, among them Yves Delage (1854–1920), Georges Bohn (1868–1948), Edmond Perrier (1844–1904), Eugène Bouvier (1856–1944), Henri Piéron (1881–1964), Pierre Janet (1859–1947), and Pierre Hachet-Souplet (1867–1947). Hachet-Souplet had been the backer of Perrier's failed laboratory of zoological psychology at the Parisian National Museum of Natural History; when it did not come to fruition, he instead created the Institut de Psychologie Animale, based on his idiosyncratic "taming method."

Giard's program was initially developed thanks to the participation of Henri Piéron and Georges Bohn. Shortly after his death, however, Giard's erstwhile supporters divorced themselves from his legacy. Both Bohn and Piéron moved toward a more mechanistic approach, criticizing the anthropomorphism and religious teleology of Darwin, Romanes, Lubbock, and the amateur French entomologist Jean-Henri Fabre. They eschewed Giard's ultimate questions about evolution, preferring instead to focus on the proximal causes of behavior. Both became involved with American (but not German) science, Bohn forming an alliance with Loeb and Robert Yerkes, Piéron with Herbert Jennings.

Bohn and Piéron, moreover, fell out with one another. Bohn criticized Piéron for not going far enough with positivism, objecting to his appeals to "intelligence" and "will" in animals. Bohn chose Watson and Yerkes' *Journal of Animal Behavior* as his preferred vehicle of publication, reflecting what in hindsight could be termed his behavioristic approach. Paradoxically, this approach demonstrated a greater commitment to Giard's ethological methodology than Piéron's: Bohn regarded the laboratory as nature in miniature or, conversely, nature as a giant laboratory. Piéron, meanwhile, went a little more below the surface in his explanations of behavior, employing a diversity of methods that included the more traditional laboratory science of physiology.

Bohn was professionally sidelined as Piéron's star rose, a success that was symbolized by his succession to Binet's old post at the Sorbonne in 1912 and, in 1920, his founding of the Institut Psychologique de Paris. As Bohn failed, so his "Galileo of biology,"[31] Jacques Loeb, also fell from favor, and Piéron's more psychological approach triumphed. French "psychology," not zoology, emerged as the discipline that swallowed up animal behavior studies. (In the United States, by contrast, the sciences of animal behavior were, largely thanks to Whitman, counted as zoology rather than psychology.)

These developments in the discipline of animal psychology were contemporaneous with Forel's researches, but to what extent can they be considered context for his myrmecology? It was seemingly his natural milieu, as his myrmecological writings tended to be in French. Indeed, Forel had work published in Giard's *Bulletin de l'Institut Générale Psychologique,* alongside Perrier, Piéron, Bohn, and Bouvier. But Forel's choice to publish on ants in French may have been more of a biographical matter than a deliberate targeting of scientific peers. Forel's boyhood language was French, and his love of ants originated and was firmly rooted in this period of his life. He was self-taught in myrmecology, and therefore perhaps continued to think about ants in French, whereas his psychiatric training was conducted in German, which remained the primary language in which he published on mental health.

Forel sustained remarkably little correspondence with the French set. He was little concerned with comparing ant behavior to that of other insects, still less other animals, and hence deriving a general theory of animal psychology. Forel's general psychological theory was very much drawn from the human realm, even from the more philosophical end of the spectrum (such as Ribot had successfully sought to divorce from psychology in France). Forel loved the ants in and of themselves. Wheeler too, who had an even better claim to be a myrmecologist, only referenced the French animal psychologists when they published specifically on ants.

The relationship between Forel and the French animal psychologists was, then, somewhat one-way: though he was much cited by them, he did not trouble himself too much with their work. In fact, Forel's greater affinities lay with the three Frenchmen born within four years of himself, Edmond Perrier (1844), Alfred Giard (1846), and Alfred Espinas (1844), all of whom reached maturity around the time of the establishment of the Third Republic (or, in Forel's case, Swiss federalization). All three were natural historians, field scientists (or not a researcher at all, in Espinas' case) who told an evolutionary story about psychology. But—and this is the point that the younger generation of workers reinterpreted to their own positivist advantage—their analogies were actually social or organizational, not psychological.[32]

Even so, it would be wrong to count Forel as a proto-sociobiologist. His background in psychiatry, together with his continued reference to the human mind as a material and curable individual entity, meant he was primarily a psychologist, not a sociologist. To this extent we can group Forel together with his audience, the French animal psychologists. A common origin for their work is most of all evident in the predominant role played by insects, especially the Hymenoptera, from Giard's generation to the *Année Psychologique*.[33]

Preeminent among the psychological entomologists was Eugène L. Bouvier (1856–1944), Perrier's student and later chair of entomology at the Muséum d'Histoire Naturelle. Bouvier began publishing on behavior and instinct among the insects in 1900 and concentrated on the subject after his retirement. His most famous book, *La vie psychique des insectes* (1918), dealt, like the rest of his work, primarily with ants. Its title paid homage to Binet's *La vie psychique des micro-organismes* (1889), thus aligning Bouvier with the psychological tradition represented by Binet (1857–1911) and his laboratory of psychophysiology at the Sorbonne. Some of *La vie psychique* was inspired and directly contributed to by Bouvier's student Georges Bohn, and concerned proximal explanations for simple behaviors. But Bouvier was also explicitly indebted to Forel, and the narrative of the book led toward conclusions that covered more complex examples of instinct and bigger, evolutionary explanations.[34] Such higher instincts, less amenable to error than the moth's simple, fatal phototropic attraction for the flame, were the best examples of adaptation in nature. For Bouvier, just as for Forel, these instincts were fixed by heredity on the basis of primitive ancestors' simple intelligent choices. This process was, according to Bouvier, the principal factor in the evolution of the articulates. And like Forel, Bouvier did not consider memory or instinct metaphysically identical to, or necessarily located in, the nerves.

Another of Bouvier's students was Henri Piéron, with whom Bouvier remained more strongly allied than did Bohn. Even this self-proclaimed positivist often talked in the psychological terms shared by Bouvier and Forel. Besides being taught directly by Bouvier, Bohn had come through an education in French animal psychology whose foundational figures—Giard, Espinas, Perrier—shared a great affinity with Forel. Moreover, Bohn had read Forel's experiments in the *Bulletin*. Thus he wrote that "Psychology . . . is the science of behavior [*comportement*] . . . [based upon] researches carried out in a parallel manner on lower animals and man, showing the common laws of memory."[35] Piéron sketched out a continuity between reflex and instinct and urged that such acts be recognized among humans as well as among their usual exemplars, the insects. The million-dollar question concerned the origin of instinct and reflex, and here Piéron hedged his bets. Having considered the indisputable phenomenon of "mnemonic acquisition" in life, he added, "Biologists in general refuse today to admit the inheritance of acquired characteristics, of adaptive modifications arising in the course of an individual's life. But there are some results which point to the hereditary transmission of certain individual modes of behavior."[36]

Piéron himself went on to sponsor Etienne Rabaud (1868–1956), who, though older than Piéron, had not enjoyed the same precocious success. Rabaud focused on questions of insect orientation: how insects found their way about. Such questions had historically been answered through experimentation on bees and ants. Rabaud deployed the observations of Charles Ferton (1856–1921), an amateur contemporary of Bouvier's in the field of insect psychology who published prolifically on hymenopteran instinct between 1891 and his death.[37] Rabaud was immersed in the literature of ant orientation, comprising work dating back to Lubbock and on through the Swiss psychiatrists Brun and Forel, and Rabaud's contemporaries Santschi (Forel's myrmecological protégé), Victor Cornetz, the mechanist Albrecht Bethe, and the American Charles H. Turner.

Rabaud's ants found their way about by assembling a holistic picture comprised of multiple cues (chiefly olfactory and visual; Rabaud rejected Forel's implication of ultraviolet sensation, apparently on the grounds of its susceptibility to supernatural interpretation). That these cues were not linked in a sequential form of memory was demonstrated by the following test: if an ant were chased some meters to the north of its nest, then to the east, it would take a direct southwesterly route back.[38] Though Rabaud insisted that no consciousness need be attributed to animals to explain their powers of orientation, he could not help but compare their abilities with those of primitive humans. For example, an "Indian" boy of twelve had guided Bates, hopelessly lost, through the forest; another jungle-dweller had astonishing powers of reconstructing a route in reverse from tiny cues, yet was "incapable of counting beyond two or three, or of constructing a logical argument." By comparison, civilized man possessed deeply inferior skills of orientation.[39] Evidently the neurological substrate for route-finding was for Rabaud, as it was for Forel and Bouvier, of little significance.

One should not overstress the direct or unique influence of Forel on the French insect psychologists, although his channeling through Bouvier was certainly significant. Their shared interests in insect psychology were also largely due to long-standing and pan-European traditions of analogizing animals and humans, and the particular place of insects within those traditions. A well-established metaphor relating to insects—"instinct"—was taken up by all the insect psychologists and reinterpreted to explain and relate animal and human mind and behavior.

The Insectan Discourses of Instinct

There were two animals with particular significance for the enterprise of recovering the evolution of the human psyche: apes and ants. These two animals reflected traditional zoological typologies dividing the animal kingdom in two. Such typologies had their roots in Thomist philosophy and ascribed to animals instinct in place of the rationality given by God to humans.[40] During the Enlightenment, bees had come to symbolize the peak of that instinct, whose exercise, according to many, produced an apparently Smithian economy.[41] Apes, which during this time also provided an analogical reflection for humans (vide Swift's Yahoos), dramatically changed in representation when they were identified as direct ancestors.[42] So, for the late Victorians, apes represented one step down from humans in the development of rationality or the slightly weaker "intelligence,"[43] while ant and bee societies continued to represent the acme of an alternative evolutionary branch whose members were distinguished by their use of instinct. The ape mind was homologous to man's, the ant's analogous.

The degree to which this two-kingdom metaphor permeated scientific and social thought cannot be overstated. The familiar tropes of man and ant at the heads of their respective phyla, exercising respectively intelligence and instinct, were utilized by writers of natural theology and hard-line mechanism alike. The metaphor was used by entomologists, of course, but also by psychologists, educationalists, and social theorists. It was even used by those Victorians who *did* ascribe some form of intelligence to insects, such as the explorer-naturalist Thomas Belt and the archaeologist and entomologist John Lubbock. Belt wrote, "The Hymenoptera standing at the head of the Articulata, and the Mammalia at the head of the Vertebrata, it is curious to mark how in zoological history the appearance and development of these two orders (culminating in the one in the Ants, and the other in the Primates) run parallel."[44] In Belt's memoirs, the ants functioned to highlight the incompetence of the native Nicaraguans; in their self-improvement, economy, and planning, the ants appeared curiously English to the Victorian reader. When Belt called the ants' behavior "intelligent," the term simply underlined the otherness of the ants and their alternative supremacy.

After the Victorian doctrine of perfectibility lost currency, the parallelism of the two phyletic groups remained, now underscored by the distinctiveness of the two qualities, instinct and intelligence. Belt's two "crowning points" were unchanged some forty years later. In 1913, a populist book on insect life averred, "Regarded as a machine, an insect is more perfectly designed, more perfectly equipped, than any other invertebrate type. Just as man is chief of the animals with backbones . . . so insects are the leading race of the invertebrate class."[45] The author of this book gave little or no credence to tales of flexible insect behavior or to insect intelligence. Instead, insects' inflexible behavioral patterns made them more like machines. In 1918, Bouvier's academic book on insect psychology claimed similarly:

> Man occupies the highest point in the vertebrate scale, for he breaks the chain of instincts and thus assures a complete expansion of his intellect. The insects, especially the Hymenoptera, hold the same dominating position in the scale of articulates, where they are the crowning point of instinctive life. These two groups represent the actual extremes of the two paths followed by psychic evolution in the Animal Kingdom,—the articulates toward instinct, the vertebrates toward intelligence.[46]

Ten years after this, the naturalist R. W. G. Hingston opined that

> The psychological tree has two great branches, the branch that represents the growth of intelligence and the branch that represents the growth of instinct. Man stands at the summit of his own branch and thus dominates all creation. But the insect crowns the other branch. In it instinct has reached the highest development. In fact many acts performed by instinct are as wonderful as the acts of reason.[47]

And in 1924, an educationalist wrote that "the best examples of pure instinct are seen in lower animals, for example in the spiders."[48] This complemented his thesis that it was the very vagueness of man's instincts that enabled his education; man was intelligent, while insects were instinctual. In 1930, Julian Huxley emphasized the instinctive foundation that underpinned even the most astonishing apparent displays of intelligence among ants. This fact made their complicated actions all the more remarkable. "The enormous gulf between the intelligence of ants and our own," he wrote, "is most readily realized by reflecting that no ant receives any education . . . their capacity for profiting from . . . instructions could not very well transcend that of an earthworm." Instead, they succeeded through the exploitation of instinct. The supreme effectiveness of this tactic meant that "the social insects in general, and ants in particular" were "with man, the fine flowers of the tree of life."[49]

In 1931, W. C. Allee produced a diphyletic tree, making the point of the two kingdoms very powerfully in visual terms. The tree illustrated a paper titled "Cooperation Among Animals" and thus underscored Allee's theme that insects and mammals had the most highly developed social groups and that one could therefore trace the origin of sociality back to their common causes.[50] Allee was less concerned with the psychological aspect of his research than with the sociological. Nevertheless, for many readers the diagram would have been readily comprehensible, for it echoed the pervasive two-kingdom trope whereby ants, with their complex behavior (or mind), topped the invertebrate branch of the animal kingdom, while man, the summa of intelligence, did the same among the vertebrates. As Bergson put it only the following year:

> Instinct and intelligence are two forms of consciousness which must have interpenetrated each other in their rudimentary state and become dissociated as they grew. This development occurred on the two main lines of evolution of animal life, with the Arthropods and the Vertebrates. At the end of the former we have the instinct of insects, more especially the Hymenopterae; at the end of the second, human intelligence.[51]

Thus, instinct was the key thing to focus on for the study of insect behavior and its evolution, which in turn was closely related to a study of insects as the most helpful exemplar of instinct. The Forellian myrmecologists, engaged in their evolutionary quest and naturally choosing the Hymenoptera as a meaningful piece of the larger jigsaw puzzle, needed to exclude any discourse of instinct that did not entail evolutionary concepts.

Instinct and the Nonevolutionists

Pace Darwin et al., some myrmecologists, particularly those in Britain, had little interest in ants' behavior and its evolutionary significance. Instead, they effectively continued to pursue a structuralist taxonomy. Forel's correspondent John Clark complained in 1922 that W. C. Crawley had provided an unsatisfactorily dry identification of the Australian specimens he had sent him: "I am more interested in the life histories of ants than in describing new species [as Crawley is]."[52] Yet so powerful was the connection of instinct to insect behavior that even those who were not interested in evolution used the concept to shape their studies. Antievolutionist natural theologians retained the divine aspect of instinct intended by Aquinas. For them, instinct was a polar opposite to intelligence, and to find anything but instinct in animals (or, perhaps, anything but intelligence in humans) would be tantamount to blasphemy. The most prominent natural theological writer on instinct was the independent Provençal writer Jean-Henri Fabre (1823–1915). Fabre was an immensely popular writer for a general audience, especially after his promotion by Maurice Maeterlinck, and entomologists could not simply ignore his opinions and observations. Though not conventionally religious, his antievolutionary views were shared by those of his less well-known contemporary, the Jesuit priest Erich Wasmann.

Fabre's most celebrated account of instinct, a description of the solitary wasp *Philanthus,* is one of those natural history tales that has sunk deep into the public consciousness. The female, marvelously, would sting her prey at a precise point so as to paralyze and not kill it. She would then carefully place the victim in the small nest in which she had already laid an egg. When the larva hatched out, it would be able to eat the paralyzed prey, which had stayed fresh until required. The instincts of Fabre's insects were innate and unchangeable, and they were placed there by a higher power. Usually this quasi-divine provision worked to the good of the insect, allowing incredible "planning" for circumstances about which it could have no foreknowledge.

"Instinct," Fabre considered, "is omniscient in the unchanging paths that have been laid down for it: away from these paths it knows nothing."[53] To prove his point, Fabre performed experiments in which he would disrupt the female wasp in the process of nest provisioning. He discovered that if the order of her routine was disturbed, she would be unable to grasp the need to undo or redo a particular action. For example, the solitary mason wasp *Pelopoeus* would first create a rough cylindrical pot from mud, then kill a small spider and place it at the bottom. The egg would be laid on top of this, and then the cell would be filled with another eight or so spiders. If Fabre removed the first spider and the egg while the wasp was hunting the second spider, she would not notice the change upon returning but instead would deposit the second spider in the empty cell and go straight off to find another. If Fabre removed subsequent spiders, she would continue to do this, fetching up to twenty or more without noticing that the cell, which anyway lacked an egg, never filled.[54] Another example was provided by the

Sphex wasp. When Fabre removed the contents of the *Sphex* wasp's egg-cell just after she had completed closure of the cell's mouth, the returning wasp busily mended the rupture, ignoring her egg and prey that had been placed outside.[55] These and other *expériences* demonstrated to Fabre the utterly fixed nature of instinct. Animals and man had psychological resources and inspirations that were innate and not acquired, inbuilt and not a matter of experience and conscious repetition. The collection of essays in Fabre's *The Wonders of Instinct* gathered together many of his observations that repeatedly made this point. If anything, animals were less like people than they were machines, where life was the "firebox that warms the animal and fuels all action."[56]

Fabre, moreover, rejected the idea that insects could display any kind of intelligence, whether considered in the strong sense of rationality or the weak sense of plasticity. Fabre dismissed those who tried to "find the origin of reason in the dregs of the animal kingdom."[57] Ants, so often the subject of speculation about insect intelligence, were barely touched on by Fabre in his writings; he declared they were unjustly admired creatures.[58]

To demonstrate how instincts were arbitrarily pre-ascribed to insects and had not evolved in tandem with their physiology, Fabre compared the praying mantis with the equally grotesque *Empusa* larva.[59] Despite the similarity of their appearance and physiology, the *Empusa* did not share the bloodthirsty habits of the mantis. To Fabre, this demonstrated that their common physical form did not produce an identity of needs; in other words, the same evolutionary forces could not explain the emergence of identical organisms with different behavioral propensities. His observations showed him "that propensities and aptitudes do not depend exclusively upon anatomy" but that "high above the physical laws that govern matter rise other laws that govern instincts." To the very end of his life, Fabre denied the transmutation of species, and the priority that he gave to instinct was interwoven with this belief; the preordination of behavior showed, according to him, the impossibility of evolution.

Erich Wasmann treated insectan instinct in much the same way as did Fabre. In keeping with his religious persuasion, Wasmann maintained an explicitly theistic interpretation of instinct into the twentieth century.[60] Unlike Fabre, Wasmann was unafraid to study the ants in order to do so. His *Instinct und Intelligenz im Thierreich* (1897) and *Comparative Studies in the Psychology of Ants and of Higher Animals* (1905) made the point repeatedly. The entire argument of the latter book may be summarized as follows: Ants are demonstrably creatures of pure instinct. Yet, lowly as they are, they approach nearer to humans in their life and behavior than any other animal, which just goes to show that apes certainly do not have intelligence. Wasmann upbraided Belt for describing "individual assistance" among the ants, claiming that it was merely the promptings of instinct that caused them to gather up stragglers during migration. Likewise, Wasmann criticized Darwin's disciple George Romanes for crediting the *Eciton* ants with "higher sympathy for their companions" when it was in fact "merely a manifestation of the instinct of sociableness."[61]

Fabre and Wasmann were solitary workers, outside any kind of scientific institution and only partially integrated into the epistolary networks of entomology. Yet such was the prominence of Fabre that refuting his work was a necessary (perhaps also desirable and convenient) point of departure for the instinctual evolutionists. Ferton, like Fabre an amateur, debunked Fabre's celebrated account, alleging that some species of solitary wasp preserved their prey by stinging to paralyze and not to kill. Ferton was cautious when it came to defining instinct, but he certainly differentiated instinct and intelligence. He characterized the Hymenoptera as primarily instinctual, but without committing himself to Fabre's views on the fixed nature of their instincts. Etienne Rabaud drew extensively on Ferton's work, as did Emile Roubaud, another scientific entomologist who owed a great deal to Fabre and Wasmann. Despite their lack of respectability, their way of seeing insects continued to shape expert observation.

Curiously enough, the nonacquired version of instinct espoused by the naturalists Fabre and Wasmann apparently had a great deal in common with the mechanistic automatisms of Jacques Loeb et al., which, confusingly, were also often referred to as instincts.[62] Loeb's theory of tropisms shared inflexibility with their definition of instinct, and expressed it in a form as strong as Fabre's. Loeb posited physiochemical explanations for tropistic "instincts," thus evading a certain amount of the teleology that many considered to be the besetting sin of Fabre and Wasmann. But the origin of tropisms was a matter of debate, and Loeb did not look into them any more than Fabre or Wasmann questioned the demiurge. Loeb was interested in tropisms as purely physiological phenomena, just as Fabre treated his observations merely as a source of wonder. Like Fabre's instincts, some of Loeb's tropisms appeared to be useless, or could be produced only in the laboratory. Some were even injurious, such as the moth's persistent positive phototropism, which led it to perish in the flame of a candle. This bears comparison with the mason wasp *Pelopoeus*, which Fabre tricked into building useless nests simply by taking advantage of its stereotyped instincts. For Fabre, instinct's immutability was usually to the advantage of the organism, but for better or for worse, Loeb considered that tropisms were not adaptable, and in this respect his belief has a striking theoretical similarity to Fabre's theory of the fixity of instinct.[63]

To summarize, there existed a complex set of alliances in the arguments over insect behavior and instinct: the natural theologians of entomology produced models superficially similar to the mechanists' models but for completely different reasons. They shared with Forel, Giard, Bouvier, and Bohn a natural historical approach to their subject matter and despised the laboratory studies of the mechanists. Fabre and Wasmann, meanwhile, sided with the ant psychologists to attack laboratory mechanists on the grounds that they lacked animal expertise. Wasmann, for example, criticized Bethe's mechanistic model on naturalistic grounds as well as religious ones.[64] For one thing, Wasmann pointed out, Bethe had based his work on only three species, all of which happened to follow their outward paths very strictly on the return. This behavior was not true for all species. Forel argued a similar case, even though—unlike Wasmann—he shared Bethe's disdain for the dogmas of religion.[65] And Piéron's ally Herbert Jennings rebutted Jacques Loeb's nonevolutionary, tropistic theory of instinct. Loeb's fundamental error, he declared, was that he had none of the naturalist's grasp of the living organism: "I [Jennings] couldn't help but feel that [Loeb] suffers a little from his lack of acquaintance with animals—their structure, etc.— . . . His theory of tropisms depends on the symmetry of animals and when I incidentally mentioned that the Infusoria were as a rule unsymmetrical [sic] it seemed to strike him very suddenly that there was a difficulty somewhere."[66]

The studies of Fabre and Wasmann, and the enormous popularity of Fabre in particular, thus caused problems for the ant specialists. Their shared interest in the natural historical approach was not a happy similarity for Forel et al. The natural theologians yoked their organismic know-how to an antievolutionary program, while evolution and the construction of a behavioral phylogeny were central to the study of the ant psychologists. In this respect the latter were only too glad to lay into Fabre and Wasmann, just as they attacked the mechanists.[67] The crucial point for Forel and the myrmecologists was that neither the mechanists nor the natural theologians had any place for evolution in their schemes. The mechanists did not discuss the origin of insects' fixed behaviors, and the anthropomorphists, allowing insects the same faculties as man, had no space for the gradual evolution of these higher faculties over time, since more primitive organisms had them, just as humans did.

Instinct and the Evolutionists

Henri Bergson identified instinct almost mystically with the creative fluidity of nature itself; metaphysically speaking, the creative impulse in humans was the same as the *élan vital* driving the universe. Bergson's fashionable ideas inspired biologists and the nonscientific public alike around the turn of the twentieth century. This kind of instinct did not, however, cash out in any very specific pragmatic terms. It did not, for example, suggest an experimental methodology. Its value lay rather in inspiration, and as such it played an important part in the development of biology and other subjects (notably Freud's psychoanalysis). Besides underwriting evolution as a whole, Bergson was read by entomologists as connecting with a French engineering metaphor for biology—one that kept an economic focus on the energetics of evolution. These entomologists represented instinct as drive, a quasi-Lamarckian "oomph" underlying the process of evolution. Rendered in German by the word *Trieb,* this type of instinct was used in a variety of senses by natural scientists and philosophers to account for both phylogenetic and ontogenetic change and the interrelation of the two.

Thomas Huxley famously declared, "There is very little of the genuine naturalist in me. I have never collected anything, and species work was a burden to me; what I cared for was the architectural and engineering part of the business."[68] A significant cadre of researchers could have said the same, typically colonial workers such as Emile Roubaud or agronomists like Paul Marchal (1862–1942), and engineers (including Forel's friend and posthumous father-in-law, Edouard Steinheil, Thomas Belt, the Frenchman Charles Janet, and the Belgian Emile Hegh). These men were intrigued by the behavior of insects but did not have the time or interest to pursue taxonomic studies. Emile Roubaud, for example, wrote a beautiful natural history paper on solitary wasps around his hut in the Congo, considered worthy of translation and reprint by the Smithsonian. Yet he relied on a friend not only for bibliographical background but even to identify the species of wasp in his study.[69] Unsurprisingly, given Roubaud's work on malaria and trypanosomes, one obituarist placed him in a Pasteurian tradition of biology, although Roubaud had trained under Bouvier. Roubaud's behavioral work was wedged uncomfortably at the end of the obituary: "[There was] another part—and by no means a lesser one—of Roubaud's study; he insisted upon this point that the taxonomic and morphological study of insects were merely the premises of the entomologist, and that his principal task must be to observe their behavior attentively, which determines their transmission of various diseases to man."[70] Even this gave a misleading impression of the interest Roubaud took in insect behavior for its own sake, for which he was and is known in nonapplied entomological circles.

The engineers were fascinated by insects' nests in particular, which often seemed like the best-constructed buildings in their colonial outposts.[71] They brought an engineer's perspective to these strange edifices of dried mud and other substances, asking questions about the energetics of their design, construction, and maintenance. Charles Janet (1849–1932), an exact French contemporary of Forel, was one such. He spent his working life in Beauvais, just north of Paris, as an industrial engineer; by philosophical inclination and a love of nature, however, he found himself drawn to the social insects.[72] His earliest studies concerned the improvement of artificial formicaries for rearing and studying ants, one model of which was a great attraction at the 1900 Exposition. Janet's engineering perspective on construction extended in all directions: to artificial and natural nests and to the philanthropic construction of cheap housing for humans. His *Observations sur les guêpes* (1903) was mostly about wasps' nests, with a particular interest in how their typical hexagon shape was achieved. His answer was that it was an automatic solution to a problem of engineering geometry. "A cell becomes hexagonal simply because it is surrounded by six cells and because in consequence six partitions are necessary to separate the cavities thus juxtaposed. If that is the case, then a cell must pass from a circular to a pentagonal form in the case where, in special circumstances, it is surrounded not by six but only five cells."[73] (Janet went on to prove that this was in fact the case.) No agency was entailed in the language of his description; he similarly approached the problem of the strength of nest suspension. Perhaps the clearest example of his engineering approach was his work on the conservation of heat in wasps' nests.[74] It was no surprise that Janet's entomological obituarist wrote, "one senses [in all Janet's methods] the engineer's spirit of clarity and precision."[75]

Janet took his engineering approach beyond the obvious questions of nest architecture and applied it to questions about the evolution of the instincts producing these feats, questions about the energetics of evolution itself. Typically, this revealed itself in an account of optimization without agency: "The duplication of the superior wings is an acquisition of the family Vespidae. It *must have been* produced as a consequence of the [wasps'] need to circulate, in a most active manner, in very rough spaces between the [nest's] outer casing and the cells with their jagged partitions. The duplication constitutes an advantageous disposition to whose maintenance natural selection *must have contributed,* before fixing it in a definitive fashion [emphasis added]."[76]

Janet made his most important discovery, so far as evolutionary myrmecologists were concerned, around 1906.[77] It was already well-known that after a queen ant (in this case, from the genus *Lasius*) completed her aerial mating, she returned to earth and shed her wings prior to digging a hole, laying her first batch of eggs, and raising the larvae. Janet's histological skills enabled him to demonstrate that during the process, the queen digested the muscles that had formerly powered her flight. He inferred that it was this energy, and only this energy, that allowed her to raise that first batch in isolation, without the need to forage for prey, as Roubaud's semisocial wasps had to do. Wing muscles were the physiological prerequisite to fully socialized life, defined as the ability of the mother to "stay at home" and feed her first batch of larvae from her own bodily supplies before they went out, fully grown, to support her and their nestmates as the colony expanded to incorporate further generations. Janet's "discovery" of fuel conservation, inspired by his engineering approach, explained the evolution of eusociality. Wheeler summarized matters some three years later as follows: "to bring up a family of even very small children without eating anything and entirely on substances abstracted from one's own tissues is no trivial undertaking. Of the many thousands of ant queens annually impelled to enter on this ultra-strenuous life, very few survive to become mothers of colonies. The vast majority . . . start out with an insufficient supply of food tissue . . . I know of no better example of the survival of the fittest through natural selection."[78]

Wasps were an important key to the study of the evolution of eusociality (full social life). Although no solitary species of ant existed, wasps exhibited all gradations of sociality from solitary to eusocial. They thus constituted a behavioral "missing link," and entomologists' attention turned toward maternal behavior in solitary and semisocial Hymenoptera, having rejected the idea that insect societies had their evolutionary origins in consociations of adult insects.[79] The idea was that in a species half-evolved toward sociality, the mother would remain near the nest and continue provisioning her young to a certain extent as they grew. The fully social ant raised her first brood right to adulthood. Fabre's observations on larval provisioning, though not widely publicized until after Maeterlinck's promotion, were well-known among insect enthusiasts and played an important role in the corpus as reliable raw data.

These ideas were worked out in concrete terms by Emile Roubaud. Billeted in the French Congo, Roubaud was struck by the different types of wasps' nest in his neighborhood, and observed the respective provisioning habits of their builders.[80] What he saw could not help but remind him of Fabre's observations of *Philanthus*, that accurate stinger, and other solitary wasps. Roubaud categorized the basic types of provisioning as four in number:

1. Rapid and massive provisioning before egg hatches
2. Slower and massive provisioning, continuing for some time after hatching
3. Direct, overseen raising of progeny by living paralyzed prey, delivered little by little

4. Direct, overseen raising of progeny by malaxated prey (i.e., softened and rendered as a pellet), delivered little by little

Unlike Fabre, Roubaud looked at these behaviors—respectively displayed by different species of the same genus—in an evolutionary light. He saw them as representing successive evolutionary stages in the development toward full insect sociality. The least perfect dumped food with the egg and left the rest up to chance. The female *Synagris cornuta,* the most "developed" of the three, continued to bring prey to her larvae as they grew. She was thus able to check for parasites, which, besides damaging paralyzed prey before the hatched larva required it, might also be transmitted from prey to larva. In this way she made sure that her energies were not wasted on provisioning a larva that would die.

Like Fabre, Roubaud watched and saw that the mother wasp went through a very predictable string of actions in creating a nest, laying an egg within it, provisioning the nest with prey, and then sealing it up. But it was Fabre's firm belief that instinctual behavior such as provisioning was utterly fixed and could not be altered to suit circumstances. Roubaud decided to question this.[81] If the egg were removed, would the wasp still provision the nest and seal it up? If the nest were damaged, would she go back to the first step and repair it? Roubaud found that the wasps *were* in fact able to alter their patterns of behavior if they were interrupted, and concluded that insects were in general frequently able to dissociate strings of habits when conditions demanded. Moreover, he suggested that insects could manifest facultative, flexible behavior and switch from one type to another. Under the right conditions, actions that had previously been associated could be dissociated and performed in a different way. Mother wasps were able to tailor their efforts to the needs of the growing larvae, rather than going to a lot of effort that might be wasted. For example, they might choose large prey in preference to small during times of abundance, since this took less effort overall. Each instinctual act was changed for reasons of economy of effort, and eventually became fixed as the norm for the species. This resulted in phylogenetic progress from solitary to social behavior, since in the fourth and final feeding behavior, the young were raised in a cohort. A few years later, Roubaud summarized: "In the totality of the tribe of the Eumeninae, one can pick out the traces of a continuous evolution of the raising instinct, which has perfected itself in certain types up to the stage now observed among social wasps, in accordance with certain essential influences."[82]

Even if they had no particular background in engineering, a considerable number of entomologists based their evolutionary studies in insect psychology along similarly adaptational lines. Individual insects changed their behavior in order to act most efficiently under any reasonably permanent new circumstances, and these changes became fixed in future generations. Alfred Giard saw natural selection as a complementary, secondary force to adaptive transmutation; Eugène Bouvier was somewhat mischievously but not inaccurately described by Wheeler as "a sane and catholic Neolamarckian."[83] In Switzerland, Forel's brother-in-law Edouard Bugnion helped establish his adaptive views on instinct, and the Swiss-born psychiatrist Rudolph Brun gave an authoritative reformulation of Forel's theories as applied to insect orientation. North Americans who were convinced that function took priority over structure included Philip Rau and Charles H. Turner; the U.S. entomologist Alphaeus S. Packard devoted his final major work to Lamarck.[84] Meanwhile, in Britain, the ethologist William Thorpe was beginning his career as a Lamarckian entomologist.

Historically, then, the instinct question was completely intertwined with the study of insects. Roubaud's work was part of a proliferation of late nineteenth- and early twentieth-century writing on ants and wasps, particularly in the francophone literature, dealing with nest building, provisioning, and nutritional exchange, all from an engineering perspective on the energetics involved in these activities. Because these activities, located in the vespid "precursors" of the ants' sociality, were considered exemplary features of evolution, researchers created a physical economy of energy underpinning evolutionary potential.

In Forellian insect studies, the instinctual discourse defined evolutionary myrmecology. Instinct was framed more specifically by Forel's contemporaries—especially the French—as the energetics of evolution itself. Here was a theory explaining instinct that could only be based in a natural historical knowledge of behavior in the wild, not in a reductivist psychology dependent on the laboratory. Moreover, it promised to yield support for, and perhaps corrections to, the ants' phylogeny, a taxonomic question of perennial interest to collectors.

Myrmecology before "Myrmecology"

Not everyone who studied ants was a "myrmecologist." The term, suggested by Wheeler around 1906, captured a specific program of research that had by then coalesced, distinguished on the one hand from mere ant taxonomy and on the other from laboratory sciences of animal behavior. "Myrmecology" was disciplined from around 1874 to the early 1920s by Forel, thanks to his taxonomic expertise. (A complementary circle was disciplined by Wheeler from around 1910 to the mid-1930s.) Forellian myrmecologists were by choice students of ants rather than any other animal. "Myrmecology" was constructed as an evolutionary study of behavior, and its adherents used the insects to address questions about animal behavior (or otherwise comparative psychology) from within a naturalist, evolutionary perspective. Studying ants entailed, historically, an engagement with the discourses of instinct, although it allowed the possibility of constructing such an engagement afresh as a mixture of learning and inheritance. Forel constructed a new sense of instinct and intelligence based on his particular social and political reading of the ants, combined with his understanding of neurology. The ultimate Victorian aim of comparative psychology—to reveal something about the human condition—was, besides underlying the representation of ants, often not far from the surface of the myrmecologists' discourse, particularly in the case of Auguste Forel.

Forel was largely responsible for developing the acquisition version of insectan instinct that rivaled, in different spheres, both Fabre's and Wasmann's version of nonacquired innateness and Loeb's and Bethe's version of the same. Forel's raw materials for building a different understanding of instinct came from a number of sources. His ant data, which supported the plasticity of insectan behavior, came from written sources and his own research. He differed from most ant observers in having at his disposal the theoretical discourse and anatomical skills from neurology and a recognizable authority to pronounce on the human mind. He differed from human psychologists who missed out on his animal focus; the instincts that they identified, outside of a thoroughgoing evolutionary discourse, would proliferate and ultimately break down in the 1920s.[85] The positivist standards of most human psychologists had no space for the proof of such nebulous concepts as the evolutionists' concept of instinct-as-drive.

The lack of modern disciplinarity on the part of the evolutionary psychological ant enthusiasts around 1900 (the "myrmecologists"), however, left them vulnerable to attack. Animal psychologists and behaviorists were always keen to try and pigeonhole the opposition, lumping together methodology, theory, and "extrascientific" worldview in their criticisms. In particular, some of the younger generation—Bohn, Piéron, Rabaud—willfully caricatured the work of the behavioral evolutionists (Giard et al.) as being in the tradition of Romanes and Lubbock, and portrayed their analogies as crudely psychological in nature. Hymenopterists using instinct-as-drive as the basis for an evolutionary story were often accused of being "vitalists." This is a classic instance of winners' terminology biasing the perception of the debate. Almost certainly it was the undisciplined nature of the insect evolutionists that laid them open to the charge of vitalism. Compared to the disciplined, laboratory-based physiologists and experimental psychologists, they had no strong platform from which to respond.

An alternative way of phrasing the terms of the debate would be naturalist evolutionists versus laboratory nonevolutionists. Laboratory experimentalists looked absurd to ant and other animal enthusiasts, since they had no idea of the contextual significance of the behaviors they studied, whether synchronic or diachronic. As one pair of entomologists, Phil Rau and Nellie Rau, remarked, "The theories of Loeb and Bethe make not a beginning of an explanation of the activities in Waspdom. Had either of the gentlemen spent some time in the field with these creatures, his mechanical theory, if formulated at all, would presumably have been so modified as not to be all-embracing in its scope."[86] With these words they captured pithily the predicament of myrmecologists. To understand its perspective, one had to enter into the minute world of Waspdom or Antdom—something that was hard to persuade self-respecting scientists to do. In this sense, the history of myrmecology is the history of a disciplinary loser. Nevertheless, it is a fascinating and subversive loser, instructive to the historian. Something about looking at ants produced fertile science, even if it did not achieve a disciplinary identity. The relations of "myrmecology" to other disciplines illuminate unexpected dimensions of the latter.

Forel's influence lasted longer than that of Espinas and Giard because the little circle of ant collectors was somewhat sheltered from the professional wrangles of the other animal psychologists. Forel had a foothold in Germany, unlike the French, and in human psychology too. When Wheeler, as close in age to Rabaud and Bohn as Forel was to Giard, took up Forel's ideas, he enabled Forel to jump the generations in a way that Giard could not. Moreover, Wheeler transplanted evolutionary behavioral myrmecology to North America, an entirely new climate where the approach would grow very differently.

PART II

SOCIOLOGICAL ANTS

Irascible, brilliant, philosophical, depressive, anti-Semitic, elitist, obscene: William Morton Wheeler (1865–1937) was a memorable man to all who encountered him. He could not have been more different from the serious, idealistic, and patriarchal Forel—not least because of his fondness for the bootlegger's goods. A talented linguist (thanks in part to his bilingual schooling in Milwaukee), he was also extremely well-read in history, philosophy, literature, and many other fields. Alfred Whitehead remarked that he was the only man he had ever known "who would have been both worthy and able to sustain a conversation with Aristotle," a comparison that would have pleased Wheeler immensely.[1]

A number of themes marked Wheeler's career. One that has already emerged in connection with Forel concerns the question of disciplinarity. Wheeler rose to prominence thanks to the high profile that economic entomology had acquired in North America, but he wanted to work in nonapplied entomology. A corollary of these professional issues was Wheeler's relationship with laboratory science and the nascent disciplines of genetics and behaviorism. Wheeler's disciplinary milieu vis-à-vis the ants differed from Forel's in that French animal behavior studies were subordinate, at least after the Great War, to psychology, whereas in North America C. O. Whitman claimed them for zoology, perhaps even "biology."[2] Throughout his career Wheeler gave considerable thought to the question of what his activities with ants should be called. At various times he advanced "ethology," "natural history" or "ecology" as suitable names.[3] What he sought was something that respected the methodological freedom of the naturalist, unaffected by the fads of academe and the laboratory. He was also quite clear that he wanted a title indicating his aim to investigate the animal mind, and not just a neurological version of morphology. A few years later, around 1906, Wheeler coined the term *myrmecology*, focusing attention on the animal exemplars of the processes that most fascinated him in nature. Wheeler's insistence on the study of whole, live organisms required almost constant justification, as did his studies in the field. Such disciplinary worries also spilled over into more public domains, for Wheeler's work and writing were in frequent danger of being written off as "mere natural history." In chapters 5 and 6, the generic contexts of writing about live animals and the landscape are explored, together with Wheeler's attempts to define and practice an "elite" natural history.

A second major theme in Wheeler's life was the increasing amount of attention that he paid to sociological issues among the ants. Steeped in the European literature of evolution and group-thought, he translated this work—literally and culturally—for an interwar American context. This trend reached its apogee with Wheeler's enthusiastic embrace of the elitist philosophy of Vilfredo Pareto. Pareto's cynical account of individual psychology and its role in the stable circulation of money had, during the Depression years, more than a chance mirroring in the circulation of food among Wheeler's ants and in the electorate of his acquaintance, Herbert Hoover. The economic elements of Wheeler's view of the colony are also reflected in Aldous Huxley's *Brave New World,* itself based on the myrmecological reportage of Julian Huxley.

A third theme was Wheeler's philosophical concern with the nature of science and the savant. The word savant captures better his image of the expert than scientist; besides the fact that he preferred natural historians to scientists, Wheeler also had respect for sociologists, philosophers, and journalists who conformed to his general outlook. Constructing savants in direct antithesis to the Formicidae, Wheeler liked them to be radically conservative: iconoclastic and masculine in their tough-mindedness.

CHAPTER THREE

From Psychology to Sociology

What made Milwaukee famous was anathema to Forel, but very much home to Wheeler. The Wisconsin town of Wheeler's birth was populated and lubricated by German émigrés—beer-drinking liberals rather like the Munich inhabitants Forel had found so uncongenial.[1] As a youth, Wheeler became involved with Henry Ward's exhibition gallery in the town; he then followed Ward to Rochester to identify and list the items of natural history in the collection there. From 1885 to 1887, Wheeler taught physiology at the Milwaukee High School, whose principal, George W. Peckham (1845–1914), encouraged unusually advanced biology courses. Peckham's particular favorites were arachnids and wasps, and he encouraged Wheeler's interest in these groups. At around the same time, C. O. Whitman, then at the short-lived Allis Lake Laboratory in Milwaukee, inspired Wheeler to study embryology. After a stint teaching, Wheeler became director of the public museum back in Milwaukee, but after reflecting on his new-found pursuits of entomology, morphology, and embryology, he decided he wanted a formal university training.

In 1890 Wheeler accepted the offer of a fellowship at Clark University, in Worcester, Massachusetts, from the recently appointed Whitman, and commenced doctoral research on insect embryology. Soon thereafter Clark began experiencing financial and organizational difficulties, and Whitman was poached by the newly established University of Chicago. Wheeler deserted shortly afterward, following Whitman to Chicago in 1892; his summers were spent at Whitman's Woods Hole Marine Biological Laboratory in Woods Hole, Massachusetts.

After seeing his doctoral thesis published in the *Journal of Morphology*, Wheeler spent the years 1893–1894 studying in Europe. During his six months with Theodor Boveri in Würzburg, he also got to know the entomologist Karl Escherich, with whom he kept in contact for the rest of his life. The second long period of Wheeler's tour was spent at Anton Dohrn's Zoological Station in Naples.[2]

The subject of embryology, in which Wheeler received his initial training, was probably the most cutting-edge of the sciences around the turn of the century. The German laboratories and the *Stazione* were, moreover, its most prestigious centers of study. Thus, Wheeler was provided with impeccable scientific credentials. Embryologists were working through debates metonymic to those in biology at large, entomology included. Should one interpret natural phenomena phylogenetically or merely ontogenetically? Should one observe and describe, or should one intervene experimentally?[3] Such questions remained relevant to Wheeler throughout his life, though his materials for answering them changed considerably.

Toward the end of his time at Chicago, Wheeler met and married Dora Bay Emerson, who very quickly gave birth to the first of their two children. Distantly related to the celebrated transcendentalist and nature writer from Concord, Massachusetts, Dora Emerson came from a wealthy and educated family. Her first degree was in chemistry, and she received a second degree from Columbia University's Teachers College in 1898, just before her marriage. The college was at that time promoting new and exciting ideas about nature study in teaching, and it is probable that William's subsequent forced engagement with nonprofessional forms of natural history—its gendered division of labor between amateur and expert—was framed by Dora's opinions and experience in this regard.

Wheeler's departure from Chicago, along with that of two precocious students in entomology, C. T. Brues and A. L. Melander, is shrouded in mystery. In theory, Wheeler had been invited to the University of Texas to organize the biology department, a step up for the young biologist. It has also been suggested that feuds with faculty members and their friends elsewhere, including at Whitman's summer camps at Woods Hole, had made Chicago too hot for him. In particular, Wheeler had taken to calling Jacques Loeb "that God damned sheeny."[4] The new Mrs. Wheeler was also alleged to have had "a very unpleasant time with the other faculty wives at Austin," a possible hint that the couple had arrived with a black mark against them.[5]

After taking up his post at Austin in 1899, Wheeler suddenly developed an obsession with ants. In large part, he was making a virtue of necessity. Since there was practically no physical provision for laboratory research at Austin, he was forced to take up studies in the field. According to his own account, Wheeler had something of an epiphany about the little six-legged creatures, and for the next three years he wrote ferociously, averaging ten myrmecological papers per year. The feat was enabled in large part by his ruthless sloughing off of teaching responsibilities to assistants and graduate students. Wheeler also wrote to entomological experts in the United States and abroad and got them to send him as much literature as possible. In this manner he built up a comprehensive library. Auguste Forel and the Italian zoologist Carlo Emery were particularly generous with their help and advice. The well-known Cornell entomologist John Henry Comstock visited Wheeler in 1903, an event that seems to have aided the growth of Wheeler's reputation. Having settled on ants as a topic of interest, Wheeler finally had an answer to the question Whitman's students were said to have used in introducing themselves to one another: "What is your beast?"[6]

Quitting his post at the University of Texas later that year, Wheeler moved to New York and became curator of invertebrates at the Museum of Natural History. For a while he was filled with zeal for "the advancement of science and education in New York City and the country at large," a pronouncement that fitted in with the aims of Columbia's Maurice Bigelow, his wife, and the nascent nature study movement.[7] In 1906, at the founding meeting of the Entomological Society of America, Wheeler was chosen to give the inaugural address. At the next year's meeting, he was elected president. To cap his rapid rise, he was simultaneously offered the professorship of economic entomology at Harvard University's newly founded graduate school for sciences, formerly the Bussey Institute. Wheeler spent the summer of 1907 back in Europe with Forel (and Bugnion), collecting ants and talking with Forel at length. He returned and accepted the Harvard post in March 1908.

The Bussey Institute had been founded thirty-seven years earlier as an undergraduate school of husbandry and farming. In 1871 it was established as a department of Harvard University. The year that Wheeler joined, it was reorganized as a graduate school of applied science. Besides economic entomology, the institute included animal heredity, experimental plant morphology, and comparative anatomy. The department was reorganized again in 1915 as a graduate school of applied biology, becoming an independent faculty of the university. Wheeler was appointed its dean, a position he retained until 1929. According to persistent faculty gossip, the appointment at Bussey Institute, located in Jamaica Plain, a region of Boston, was a method to keep Wheeler out of Cambridge, where his constant feuding had become tiresome to his colleagues. The Bussey botanist Oakes Ames opined that "while Wheeler was a brilliant man it was a pity that he was not a gentleman."[8] Reorganizations notwithstanding, Wheeler remained at Harvard for the rest of his life.

According to his biographers, Mary and Howard Evans, Wheeler took his economic responsibilities seriously during the period 1908 to 1915 (or even 1926). Admittedly, he did once give a talk to the Boston Society of Natural History titled

"The Influence of Insects on Human Welfare," but the only paper of vaguely economic relevance that he published between 1908 and 1929 was "Ants and Bees as Carriers of Pathogenic Microörganisms"—hardly a pressing problem in applied entomology.[9] He also published once in the *Journal of Economic Entomology*—in the first issue, and almost certainly therefore for appearance's sake—a brief piece titled "A European Ant (*Formica levinodis*) Introduced into Massachusetts." This was not exactly a very important topic for applied science either. Even Wheeler's attack of typhoid—a disease for which lice had recently been identified as carriers—did not seem to persuade him that economic entomology was an urgent pursuit. Wheeler's commitment to his named post in no way compared with Howard's tireless advocacy on behalf of applied entomology.

Shortly after his appointment, Wheeler cemented his reputation with the publication of his vast monograph, *Ants* (1910), which was to remain the standard text on the Formicidae for around 50 years. The reviewer in *Science* judged: "Here we have morphology, anatomy, embryology, psychology, physiology, sociology, paleontology, zoogeography, taxonomy and even philosophy dealt with in an illuminating manner! The ant is presented to us as the hub of the universe, and if there is any biological subject which may not be suggested by the study of myrmecology, it is probably of small consequence."[10] Economics is notably absent from this list, which accurately presents Wheeler's life work in a nutshell. His two papers on applied entomology during his time as professor of economic entomology compare to around two hundred papers on nonapplied entomological matters.

In fact, Wheeler's writing during his first two decades as economic entomologist at Harvard demonstrates a great deal of ambivalence toward economic entomology. At the beginning of *Ants,* Wheeler lists a number of reasons why ants are interesting, citing their dominance in nature, their analogies with human societies, and the analogy between the ant colony and the multicellular organism. However, he finishes with the two matters that are probably the most important to him professionally:

> Two further matters call for consideration in connection with the dominant role of ants, namely, their importance in the economy of nature and their value as objects of biological study. The consideration of their economic importance resolves itself into an appreciation of their beneficial, noxious or indifferent qualities as competitors with man in his struggles to control the forces of nature. As objects of biological study their importance evidently depends on the extent to which a study of their activities may assist us in analyzing and solving the ever-present problems of life and mind.[11]

The economic reasons given here are vague, and Wheeler does not return to them elsewhere in the book. Similarly, in his obituary of Fabre some seven years later, Wheeler gives selective quotations that favorably highlight the hermit's disinterest in matters economic.[12] Wheeler's 1917 paper "On Instincts" has a half-hearted introduction in which he attempts to demonstrate the utility of his highly theoretical argument by stating that, in order to control insects, one must first understand their behavior, in other words, their instincts.[13] It is impossible to understand this statement as anything other than a sop to the institution that was paying his wages. In 1919, Wheeler went even further in a speech before sympathizers at the Symposium of the American Society of Naturalists, complaining that "many insects . . . have been misrepresented by the taxonomists or maltreated by the economic entomologists."[14]

Wheeler's lecture notes on economic entomology reveal a similarly grudging attitude toward their content. One typed manuscript, "The Influence of Insects on Human Welfare," starts in a standard enough vein with the statement that such concerns were the reason for the study of entomology. But by the lecture's end, Wheeler has digressed to one of his pet philosophical topics: how all living organisms could be regarded as parasitic upon one another. Unsurprisingly, ants then provide the principal illustration for his smuggled-in thesis. Another lecture, "Medical Entomology," rapidly turns into a discussion of the subdisciplines of biology and their relationships to one another. Wheeler slips smoothly from practical entomology into a discussion of the superiority of ethology, and its incorporation of all approaches to economic entomology, including his favorites, psychology and "biocœnoses."[15]

One should therefore correct extant biographies of Wheeler by saying that he never valued economic biology as a scholarly endeavor. Rather, for a long time—until the late 1920s—he was reluctantly forced to negotiate with economic interests in order to pursue his personal interest in theoretical entomology. Fortunately, Wheeler's terms of employment did not make this too difficult. Though the Bussey Institute historically was a teaching center for scientific farming, its mission was not as regulated as that of Cornell. The difference was that Cornell's extension work came about as a result of the Morrill Act, whereas Harvard's was enabled through a private grant. Wheeler was therefore less answerable for the public benefit of his work at Harvard than were the Comstocks at Cornell. In 1926 he persuaded the authorities to formally drop the "economic" part of his professorial title, becoming at last professor of entomology. His negotiating hand may have been strengthened by health problems, for he spent February and March of that year in a mental hospital near New York City with a "slight mental breakdown," as he called it.[16] At any rate, Wheeler was then able to pursue his true interests for the rest of his life, and moreover to shed much administrative responsibility. He and his students, along with the other Bussey staff, moved back into the heart of Harvard.

The other disciplinary negotiation that Wheeler had to perform was with the laboratory sciences of biology. His famous lecture, "The Ant-Colony as an Organism," given at the Marine Biological Laboratory at Woods Hole in 1910, has retained a certain cult status among thinkers on emergence and superorganismic organization.[17] More contextual aspects of the paper have been overlooked. By making such a comparison, Wheeler was aligning himself with the developmental, holist traditions he had encountered at the Naples Zoological Station and the quasi-Haeckellian monism he found in his mentor, Auguste Forel. Wheeler was also, however, staking a claim for the social importance of myrmecology, as well as attempting to place experimentation with the ant colony on a par with the increasingly popular laboratory-based approaches to more conventional organisms. The locus of publication had significance in this respect; the *Journal of Morphology* had been established by his former mentor, C. O. Whitman, during his Milwaukee days. It represented the best of the highly regarded Chicago tradition in morphology and embryology: bench-based work on specimens.

Calling the ant colony an organism suggested that analogous methods could cause it to yield results on a par with the prime science of the day. These methods had to be field-based; Wheeler was simply not interested in dead ants, or in the behavior of one or two ants in artificially constrained conditions. Only ants interacting in their full and natural social complexity were worthy of investigation. Meanwhile the taxonomic side of myrmecology also needed justification as means to a serious end; otherwise it was in danger of being written off as mere stamp collecting. Wheeler's "Organism" paper ranged itself against all the new breed of reductionists, be they geneticists, behaviorists, or physiologists.[18]

To a certain extent, Wheeler gathered a sympathetic scientific fraternity around himself, although it was never centrally disciplined. Throughout his career, he kept up a voluminous correspondence with entomologists from all around the world. They sent him specimens, observations, and asked him many questions. Wheeler also received and answered letters from amateur entomologists and agriculturalists. As time went by, he was more inclined to advise economic inquirers to seek information elsewhere, but he was unfailingly prompt in answering questions from naturalists and acknowledging the specimens they had sent. Wheeler had an extraordinary number of offprints made of all his papers, so that he could send his science out around the globe.[19] His most frequent correspondents were

Auguste Forel and Carlo Emery. Between them, these three men formed the pillars of international standards in methods of collection and the taxonomy of formic specimens.

Wheeler was more successful at building up his department and inspiring his graduate students to work on insects according to his own noneconomic framework. C. T. Brues and A. L. Melander went on to produce together a standard text on insect classification; C. L. Metcalf became a well-known economic entomologist. A number specialized in bees and wasps: T. D. Mitchell, O. E. Plath, George Salt, and Alfred C. Kinsey (later famed in the field of sexology). Myrmecological successors included George C. Wheeler (no relation), Frank M. Carpenter, William S. Creighton and Neal A. Weber. Other notable heirs were J. G. Myers, a parasitologist; William Mann and Marston Bates, both of whom wound up as popular nature writers; and Philip J. Darlington, curator at the Harvard's Museum of Comparative Zoology and later professor of zoology at Harvard.

Wheeler was a philosophical and reflective scientist who regularly contemplated the methods and aims of his work.[20] His affinity for holism took various forms, notably Bergsonism in the 1910s and early twenties, and after that, emergence theory. Ultimately, in the late twenties and thirties, Pareto fulfilled his desire to explain the behavior of high-level groups and societies. Although Wheeler was ruthless in ridiculing "queer creatures ... from the metaphysical barnyard" such as "entelechies" and "psychoids," he remained sympathetic to thinkers who attempted to unite intellectual disciplines, both within and without science. Wheeler took considerable pains when he entertained Hans Driesch in 1923, though he was by then highly critical of the extremes of his work. As a fellow-traveler in the journey to build up rather than break down the understanding of phenomena, Driesch was still worthy of respect.

Instinct was at the heart of the myrmecological science Wheeler had inherited from Forel. As such, instinct had to account for the social behavior of ants; equally, it was considered to lie at the root of human behavior. However, the instinctual account of human behavior became less and less palatable in the period after the First World War. Wheeler developed a new sociological account of ants that, in this context, more satisfactorily described both group and human behavior, also in this way complementing European developments in human sociology.

Psychological Approaches to Ants

As Forel and the French psychologists studied ants qua psychological entities, the nature of the "social instinct" came up for discussion, and with it the question of human sociality as an instinctual trait, or the existence of the so-called gregarious instinct. Forel's sanguinity in making psychological comparisons between the two-legged and the six-legged was not shared by all his peers and successors, but the insect-as-exemplar-of-instinct model was central to several psychological schemata.

The British human psychologist William McDougall was arguably the most important instinct psychologist attempting to answer these questions in the era encompassing the First World War. He judged that psychology was the foundation stone for the scientific study of society, and that this had to be a "comparative and evolutionary psychology" based on animal instinct.[21] The animal instincts that underpinned the actions of every human individual were most clearly expressed for McDougall by the insects.[22] In fact, he was a correspondent of Wheeler's, and the two shared in their letters the conviction that instinct psychology, modeled on entomology, was the most basic building block in the construction of knowledge about human nature. McDougall eventually moved to join Wheeler at Harvard in 1920, where the two dined frequently together.

W. H. R. Rivers was another British human psychologist with evolutionary leanings. His version of psychoanalysis, developed during the years of the First World War, described how the unconscious consisted of instincts that had been suppressed over the course of evolution. Under abnormal conditions, these thought processes would intrude once more upon the conscious mind; the natural fear instinct that rose to the surface on the battlefield created war neurosis.[23] Rivers used numerous references to insect psychology in order to make his case; for him too the Insecta were illustrative of the deeper recesses of the human psyche.

Wilfred Trotter's *Instincts of the Herd in Peace and War* (written between 1908 and 1914, but published in 1916) was an effort to apply the understanding of inherited human instincts in order to achieve a well-managed society. His belief in social instincts was based on a racial-animal model. According to Trotter, the Germans were naturally like wolves, while the English were like dogs. Besides comparing the innate savageness of the wolf to the domestication of the dog, Trotter's simile attributed a natural pack mentality to the Germans, for dogs, although sometimes induced to hunt in packs, were happily solitary creatures.

Trotter's analogy also hinted at a deep unease with the thought of complete sociality among humans. In the twentieth century Anglo-American world, humans were generally thought to be naturally individualistic.[24] They did not possess the true social instincts of the ants that subjugated the self for the sake of the whole. The self-sacrifice of soldier ants and especially termites created great resistance to the undue ascription of instinct to humans in the wake of the Great War. It was unthinkable that human self-sacrifice should be as explicable by instinct as the self-sacrifice of insects.[25]

Historians of psychology consider that the project of enumerating the various instincts eventually caused instinct psychology to collapse under the weight of its Ptolemaic complexity.[26] Forel and Wheeler, for instance, famously attacked Wasmann for his "instinct of amity" among the ants.[27] It seems quite possible that the unpalatable nature of the soldier-insect comparison may have been one important reason for the rejection of instinct psychology as exemplified by ants and their cousins; historians of instinct psychology have not as a rule paid attention to its roots in the insect analogy. Moreover, the writing off of instinct psychology after about 1920 also overlooks the fact that not all talk of instincts was by then conducted within a purely psychological paradigm.

McDougall's thoughts on the individuality of instincts were changing. In 1908, he had an atomistic notion of social psychology; the purpose of *Social Psychology* was to show that social behavior had each separate person's canon of instincts as its basis. After the war, McDougall developed a more sociological version of social psychology, accepting the axiomatic validity of collective psychology. The result of this shift was *The Group Mind,* published in 1920, just as he moved to Harvard.[28] McDougall's vision of society was converging with Wheeler's. Although Wheeler observed a group instinct at the phenomenological level, this was not inherent in ants *sui generis* but rather was a convenient way of describing emergent behavior within the social medium.

Sociological Approaches to Ants

W. M. Wheeler himself was quite explicit about the sociological tradition in which he placed his work.

> During the nineteenth century biology and sociology developed in rather intimate symbiosis. Though Comte founded sociology on biology, it is well known that certain important conceptions, such as the struggle for existence, the survival of the fittest and the physiological division of labor, were derived from sociological sources and later extended to the entire world of organisms in the Darwinian theory of evolution. If we may judge from the works of Spencer, Espinas, de Lilienfeld, De Greef, Worms, Waxweiler and others, this theory, after its first clear enunciation, seems to have been more heartily welcomed and embraced by the sociologists than by the biologists.[29]

Indeed, Forel's earliest interpretation and dissemination was at the hand of sociologist Alfred Espinas (1844–1921), author of the very influential *Des sociétés animales* (1878).[30] Espinas made substantial use of Forel's ants to illustrate his thesis that communal life was "a normal, constant and universal fact,"[31] and explained in *Sociétés animales* the formation of both human and animal societies. To do this, he used functional associations—an explanation very different from the ideology of natural progress that marked even Haeckel's earliest treatments of phylogenetic evolution.[32] Although Espinas claimed the phylogeny of divergent social forms was more like the branches of a tree than a hierarchy, he organized his book in terms of more or less ascending organizations. The first of these were accidental associations between different species and included examples of parasitism and mutualism. Next, he worked upward on same-species consociations, starting with simple-celled organisms (like *Radiolaria*) that formed colonies for nutritional purposes. These were mostly marine creatures, such as sponges, that did not exhibit any form of cellular differentiation but aggregated to form simple sacs or tubes in which digestion could occur, to their mutual benefit. Next came the communities that formed for reasons of reproduction. These were of three kinds: conjugal societies that came together simply for mating, maternal domestic societies (such as the ants) and paternal domestic societies (birds and mammals).[33] Finally, Espinas described the most truly social form of existence, the "relational" life of the tribe, which was exhibited by human beings.

The differences between intraspecific and interspecific societies were described in the final section of *Sociétés animales*, which spelled out the general laws of sociality. The two kinds of consociation worked by similarity and by a delegation of functions, respectively.[34] Emile Durkheim, who wrote in the same social-psychological tradition as Espinas, based his two kinds of societies in *The Division of Labor in Society* (1893) on almost identical concepts. "Mechanical solidarity" he defined as solidarity based on the similarity of a society's members, with the superior "organic solidarity" the result of labor divided among the differently able members of a society. Mechanical solidarity was the property of the most primitive communistic human societies. Durkheim in fact cited Espinas' *Sociétés animales* as explaining the causes of mechanical solidarity, meaning, presumably, aggregation for the purposes of nutrition and reproduction. Both authors analogized societies based on "organic solidarity" in similar ways, too. Just as Espinas compared the anthill to the mammalian brain, so Durkheim analogized the society with divided labor to the body with its organs, specialized for their different tasks. The comparisons were functional and quasi-teleological. In other words,

Durkheim and Espinas thought about component parts and the functions they played in maintaining the phenomenon of the stable society.

If Durkheim saw society as a body, Espinas saw it as a mind. The true society, resulting from functional associations, was defined by Espinas as "a living consciousness, or an organism of ideas." He considered the ant colony to be "truly, a single thought in action (albeit diffuse)."[35] He claimed that all societies existed as thinking organisms because their ideas and impulsions (the two components of consciousness) were communicable, and could be accumulated. Ideas could be imitated, and emotions could spread by sympathy, almost like an echo; the greater the number of individuals, the greater the force of these impulsive repercussions. By pooling these aspects of consciousness, an animal society could possess many more acts and specializations adapted to the exigencies of life than could an isolated individual of the same species. Eventually, the effect of the group was exercised as a function on the members whose ancestors had originally formed it. In a true group, the individual was the work of society rather than the author of society; "what is more," Espinas continued, "the individual's action is limited to a short time, while the collective action weighs upon the individual with all the weight of acquired instincts and changes in structure obtained during the past of the race."[36] Although subsequent authors took issue with Espinas for his supposedly mystical "group-mind" theories, it can be seen that his vocabulary drew on functional theories to explain the formation and persistence of societies.[37]

The behavior of the human crowd also began to be studied in France toward the end of the nineteenth century, with LeBon's book on the subject published in 1895. This positivist approach to sociology, bypassing questions of individual consciousness or instinct, suited the methods of the day. Very few texts on social psychology were translated into English before the 1920s, but the experience of the First World War created a need for new explanations of human behavior on the battlefield. Meanwhile, unfolding European politics demanded similar accounting for the conduct of the crowds that seemed to be seduced en masse by the rhetoricians of communism and fascism, seriously challenging British-beehive liberal utilitarianism in the process.[38] By this time Wheeler, who had been at work on ants for more than ten years, was ready to bring social ant psychology, or ant sociology, to an Anglophone audience.

Social life was consistently at the heart of Wheeler's biology from the time that he took up the study of ants. In 1902, he cast doubt on the metaphysical integrity of the single organism as it is commonsensically conceived, a hunch that was developed further in his 1910 lecture, "The Ant-Colony as an Organism." Here, his metaphysical skepticism about the nature of the organism was used to propose the "superorganism," an essentially social entity: "One of the fundamental tendencies of life is sociogenic. Every organism manifests a strong predilection for seeking out other organisms and either assimilating them or co-operating with them to form a more comprehensive and efficient individual."[39]

This postulate survived Wheeler's disillusionment with Bergsonism (an important stimulus to his science at that time) and, if anything, grew stronger later in his life. Five years after he professed to have left Bergson behind, Wheeler was still claiming that the whole of the organic realm constituted "one vast, living symplasm" whose fragmented parts formed one metaphysical whole of "Common Life."[40] An important corollary of this ontological blurring of organic entities was a non-Darwinian take on evolution, since competition among them was meaningless. For Wheeler, Darwinian evolution was essentially antisocial, a matter of competition between individuals that meant cooperation, whenever it was observed, necessitated a special explanation. His own study of complex interrelationships in nature persuaded him that things were, in fact, the other way around—that cooperation was the norm, and selfish individualism a surprising and noteworthy phenomenon that had to be accounted for on the rare occasions that it was observed. The Victorian notions of struggle were at most half the story of evolution, according to Wheeler, and cooperation was the more significant factor.[41] Wheeler made continued efforts to disseminate his sociological perspective on animals. Following the example of Harold J. Coolidge, a primatologist at the Museum of Contemporary Zoology who in 1931 ran a course with his wife on the evolution of animal sociology, Wheeler set up an undergraduate course at Harvard the year afterward on human and animal sociology, in conjunction with Pitirim Sorokin. Its provocatively unified approach to everything "from fish and insect groups to modern human societies" grabbed a good deal of attention; the *Boston Post* described it as "several degrees more radical than the internationally famous controversy upon which the Scopes trial hinged."[42] Rather than treating ants as psychologically interesting subjects that happened to live in societies, their social nature was rather for Wheeler their scientific essence.

But what about the insects themselves? How was this rather abstract history of sociological myrmecology borne out by the actual science and its subject, the ants? The mutual feeding of ants—an essentially social phenomenon—was the key to Wheeler and his work. Unpalatable as it might seem, the regurgitated spittle of the Formicidae was the food of love.

Making Ants Social: The Food of Love

The Parisian engineer Charles Janet noticed in 1895 that "One of the first actions [performed by the imago wasp after hatching from its cocoon] . . . consists in gently tapping her mandibles on the head of one of the first large larvae which she encounters, and in drinking the droplet of liquid which the latter disgorges."[43] Moreover, Janet, found he could induce the phenomenon himself: "If one lightly touches the heads of large larvae with a paintbrush or the tip of a pencil, one can see them spreading their mandibles, and throwing them backwards, as if to leave the necessary space between their mouth and alveolar partition to lodge an alimentary globule, and, at the same time, to disgorge a droplet of limpid liquid . . . a similar globule is disgorged whenever a worker has just nibbled at the head of a larva."[44] The surprise was that the young were sometimes feeding the adults, and not always the other way around, as one would naturally expect.

Janet's observations were read with interest by the medical entomologist Emile Roubaud, who had seen stages of social evolution in the various types of provisioning behavior exhibited by the potter wasps. Janet, it will be recalled, had had also posited that for fully social insects, it was the mother's digestion of her wings that permitted, in energetic terms, the raising of the first brood. Wing digestion was necessary to this, but was it sufficient? This rather odd behavior—the giving of sweet droplets from the larvae to the workers—suggested to Roubaud and Wheeler that it was not. Presumably the mother obtained additional energy from this process. First Roubaud and then Wheeler went on to extend Janet's energetic model of research into the general economy of food-exchange in the nest.

When Roubaud looked at social wasps, he concluded that the mutual feeding observed by Janet was the thing that fulfilled the possibility of keeping the fully social nest together: "One may see the [worker] females pass back and forth three or four times in front of a lot of larvae to which they have given nutriment, in order to imbibe the secretion. The insistence with which they perform this operation is such that there is a flagrant disproportion between the quantity of nourishment distributed among the larvae by the females and that of the salivary liquid which they receive in return. There is therefore a real exploitation of the larvae by the nurses."[45] In other words, he was suggesting a fifth behavioral level for his behavioral schema (pages 58–59). In nests where the first brood was raised together, there were some species in which the larvae offered salivary secretions to their adult sisters, and because they were so tasty the sisters remained in the nest, looking after the next generation of hatchlings—eusociality. Roubaud named the process of mutual feeding oecotrophobiosis.

Wheeler's book *Ants,* written some ten years before Roubaud's conclusions, treated feeding within the nest as central to ant life—a debt to the energetic approach of the French Lamarckians—but approached the phenomenon from a different angle, though still a social one. The principal significance of feeding for him at this time was that it was commonly supposed to produce polymorphism, either through varying the quality or the quantity of larval food.[46] Polymorphism— the existence of multiple ant castes—was the major focus of *Ants,* and feeding, explaining its proximal causation, therefore took a crucial part in the monograph.

The exact method of feeding larvae varied, Wheeler found, but it did seem somehow connected to the larger process of evolution toward sociality. Most ants fed their larvae on regurgitated foods with the exception of the Ponerinae, Dorylinae, and a few Myrmicinae, subfamilies whose larvae were fed on solid scraps. Of these, the first two were regarded as primitive types because of their incomplete manifestation of distinct castes. Thus, something seemed to connect feeding by regurgitation with higher forms of ant sociality. In 1910, however, Wheeler was inclined to hedge the question about evolutionary causes.

When Wheeler read Roubaud's 1916 article, it suddenly struck him that he had some strange observations dating back to 1901 that now seemed highly pertinent to the phenomena of sociality and feeding. At that time he had been studying the primitive subfamily of ponerine ants and noticed unusual behavior in the species *Pachycondyla montezumae.* The workers fed the larvae by placing insect fragments on their ventral surface; when they did so, they were sometimes inundated by a "copious, colorless liquid" exuded by the larvae. This liquid was a digestive agent, but when secreted, it was "eagerly lapped up" by the nurse in question. The larval behavior was to be expected; larvae fed in such a primitive manner needed to digest their rough scraps in some way. What was odd was that there was apparently no need for the adult nurses to partake of the secretions, and yet they did so with gusto. The subfamily Ponerinae, looking more like the earliest fossil ants, were regarded as less social: "the ancestral stirp of the higher subfamilies, and . . . the oldest existing expression of social life among the Formicidae."[47] They were considered to have evolved from solitary wasps, such as Janet and Roubaud had studied.[48] What Wheeler had in the ponerines, then, was a glimpse of the evolutionary path toward sociality, marked out by feeding behaviors in extant primitive Hymenoptera, that involved not just worker-to-larva exchange but, crucially, larva-to-nurse exchange.

Most ants belonging to the more highly evolved subfamily Myrmicinae fed their larvae regurgitated liquid rather than solid fragments. Yet Wheeler remarked that the newly discovered myrmicine species *Paedalgus termitolestes* also exhibited highly developed salivary glands in the larval form. This was more unexpected than when it was observed among the primitive ponerines. Because myrmicines were fed predigested liquid and not solid food, there was seemingly no need for glands producing digestive secretions in the larvae. Wheeler concluded that the larvae actually acted as a food store for the workers of the nest. Closer examination revealed a variety of strange organs in certain larval stages among a variety of myrmicine species, organs that Wheeler christened "exudatoria." All of these secreted substances were licked up by the nurses; their apparent care for the larvae was in their own interest. Social relations depended on the stable feeding possibilities that the larvae provided for adults rather than on any kind of adult duty toward the young or the group.

Wheeler had now recreated the evolution of sociality from the progressive maternal feeding behaviors of subsocial wasps,[49] through to the reciprocal feeding of the ponerines—the link between solitary wasps and modern ants—and found that this latter phenomenon remained a crucial feature of the most social or "eusocial" life. These last insect mothers raised the first brood themselves, sisters who then stayed on to care for subsequent hatchlings, bribed to stay on by that irresistible larval exudate, and becoming the worker caste. Wheeler had moved beyond Roubaud's purely "selfish" account of maternal behavior as the creator of society. From eusocial maternal behavior a novel entity emerged: society.[50] Like Espinas before him, Wheeler now had a model for the origin and continued function of society that was not just the sum of its constituent members. Even if the proximate cause of caste behavior was partially or totally genetic, the origin of the behavior of individual ants *within the context of the colony* could not be explained in terms of individual inheritance. Mutual feeding relations were the true and necessary cause of social forms of life, and to stamp his authority on the phenomenon, he renamed it *trophallaxis*.

The centrality of trophallaxis to myrmecology may be seen in the structure of the series of lectures given by Wheeler at the University of Paris in 1925.[51] After a general introduction, he defined sociality as the possession of a worker caste, which was the morphological expression of the division of labor within the nest. There followed two lectures on polymorphism, then one on trophallaxis that described the definitive tasks and exchanges carried out by social insects. "I believe," Wheeler claimed, "that [trophallaxis] constitutes the most essential characteristic of the social medium."[52]

Wheeler decided as early as 1918 that trophallaxis was not necessarily or essentially a larva-nurse exchange of food.[53] He anticipated a potential objection to his paper in the assiduous care expended by ants on pupae; these did not produce an exudate, and thus the centrality of trophallaxis to nursing behaviors might be questioned. Wheeler suggested in response that trophallaxis might also be considered to involve nonsubstantial exchanges. One could justifiably describe the "attractive odor" of pupae as a kind of volatile exudate; thus trophallaxis even explained the care of pupae. Wheeler even tried to make a connection to the functional equivalent of cuticular exudates among humans, suggesting that pubic hairs were responsible for diffusing sexually attractive secretions.[54]

As soon as Wheeler had described the exudatoria of ant larvae, he confirmed their similarity to those of certain symphiles (non-ant members of the nest to which ants were strongly attracted). This similarity was not based on anatomy. Some of the glands of these symphiles were similar to those of ant larvae; others, however, exuded liquid from the so-called fat-body, dispersed underneath the whole of the chitinous cuticle and communicated to the surface by means of small pores. Wheeler glossed over these gross anatomical distinctions; what interested him was function: "in these [myrmecophilous] larvae the voluminous fat-body *functions* as a huge exudatorium . . . This at once suggests that in many ant larvae the general fat-body may have the same *function* . . . [T]here is just as much reason for supposing that the fat-body may *function* as an exudate organ in the ant-larva as in the larvae of the Lomechusine myrmecophiles."[55] Wheeler's point was that there were three sources of liquid agreeable to worker ants—salivary glands, exudatoria, and the general fat-body—which were functionally equivalent forms of trophallaxis.

In later years, notably in his Parisian lectures, Wheeler continued to emphasize the broadness of the trophallaxis concept, stressing its applicability far beyond the business of food exchanges among ants. In support of this point, Wheeler argued that taste and smell were intimately related senses (and anthropomorphic ones at that), and that therefore one might as well keep the word "trophallaxis" to refer to both "senses," which were both chemoreception of one sort or another.[56] Furthermore, Wheeler countered earlier criticism that the exchange must be necessarily or immediately reciprocal. If the exchange was odor-based, it simply did not make sense to refer to the economy of individual exchanges as Roubaud had done with his food-based interactions.

In this context, Wheeler chose to discuss the researches of Fielde (page 28), which were now quite old, on the odor of the brood. He wanted to emphasize the fact that trophallaxis was a form of communication, performing a role in the protection of the nest besides one of reciprocal nutrition. The functioning of the ant colony superorganism necessitated a communication between its constituent parts, a communication that mapped precisely onto the chemoreceptive phenomenon of trophallaxis. This phenomenon was a microcosm of the relational loops among the broader community, including symphiles, and other animals and plants. Trophallaxis was, for Wheeler, any kind of functional exchange among a community that holistically construed, went beyond the nest. Trophallaxis was to become the paradigmatic phenomenon of myrmecology qua sociology for the next twenty years, and arguably for the remainder of the century.

CHAPTER FOUR

The Brave New World of Myrmecology

Trophallaxis and Depression Economics

While Wheeler was watching the ants feed each other, the human ants of Forel's *Etats Unis de la Terre* were doing the same. Amid the hardships of the Great War, the young Herbert Hoover was busy providing nutrition for the hungry millions of Europe. As chairman for the Commission for Relief in Belgium he fed ten million people for four years, accomplishing the task, moreover, with astonishing efficiency; when accounts were reckoned at the end of the operation, administrative costs were found to have constituted less than one percent of the total funds. Riding the wave of domestic respect earned for this feat, Hoover went on to achieve political success, culminating in his election to the presidency in 1929, before his disastrous mishandling of the Depression caused his rapid fall from grace. The success of both Hoover and Wheeler hinged on the circulation of food, and there is more to this connection than mere chance. Both were caught in a dilemma between the need for social management and a fervent belief in individualism. Both attempted to resolve the dilemma through an understanding of socioeconomics; specifically, both modeled recession in the same manner. Their rooting in a common political and cultural environment is embodied by Dora Emerson Wheeler's active participation in Hoover's presidential campaign, and a consequent acquaintanceship between the two couples.[1] Whether or not Herbert really read Morton's books, as his cordial notes claim, theirs was a shared worldview: that trophallaxis, with its twin features of economic distribution and worker behavior, literally made society. In Wheeler's case at least, the construction of ant workers as irrational economic architects of society created a problem. As an economic metaphor, the description naturally applied equally to human life. Yet Wheeler did not want to be an unwitting, antlike member of society. As he made ants social through his economic model, so he strove equally to exempt himself from social expectations, to make the scientist intrinsically antisocial.

Everything in Hoover's and Wheeler's economy came down to the worker. In Wheeler's words, the worker caste was the "necessary creator of the social medium";[2] without the worker class there was no society, and the raison d'être of the worker class was food and its exchange. Workers, Wheeler argued, were essentially defined by food; they were hunger forms of their kind. Although they might have some hereditary predisposition to develop their caste's physiology and behavior, a lack of feeding played a necessary role in their ontogeny. Even when fully developed, they continued to display their enormous appetite in their constant search for food. The size of the colony was a function of the trophophoric field, the area of food available to the ever-hungry workers. In other words, ant "society" was intimately related to the outer world: "The social medium obviously comprises not only the regular activities of the workers in the nest and mainly the collection of food and its distribution among themselves, the queens, males and larvae but also the relations to food-yielding insects or plants in the trophophoric field and to the various guests . . . present within the nest itself."[3] Yet colony size was also determined by those very workers in the inner world of the nest, through their limited feeding of future workers. The more larvae they partially starved, the more workers they would produce. The more workers to forage, the larger the tropophonic field that would yield food. The more food, the more eggs laid and larvae hatched. And so on.

Trophallaxis was thus a self-regulating system of stimulus and response that completed the circuit between the inner world of the nest inhabitants and the outer world. (In fact, when the superorganism and its environment were considered as the product of trophallactic interactions, the metaphysical distinction between them almost melted away.) Right from 1918, Wheeler considered trophallaxis to be an elastic social phenomenon covering interspecific, parasitic, and even animal-plant relationships. Indeed, Wheeler rejected Roubaud's suggested name for the phenomenon partly on the grounds that it implied intraspecific relations only, not the full interspecific range.[4] Wheeler's functional interpretation of mutual feeding allowed him to propose a nested hierarchy of trophallactic interactions, arranged according to their position of importance with relation to the nest. The most important form of trophallaxis, and the primary form in terms of evolution, occurred between queen and larvae and between workers and larvae. Adult-adult exchanges, which had been Forel's key behavior, were only on Wheeler's second level. Below this came exchanges with symphiles, then between ants of different species (which would occur when a different species had been brought as slaves into the nest), and finally with other insects and plants outside the nest. At the final two levels, ants would typically gain some food from the insect or plant in question, and in return afford it a certain amount of protection. Thus, "[t]rophallaxis, originally developed as a mutual trophic relation between the mother insect and her larval brood, has expanded with the colony like an ever expanding vortex till ... the ants have drawn their living environment ... into a trophic relationship."[5] Putting together all the levels, feeding was the key to society's evolution and maintenance.

Dora Wheeler already knew this fact to be true. Her interest in the distribution of food had commenced during the First World War, when she chaired the Department of Food Sanitation and Distribution for the Women's Municipal League of Boston. During the late 1920s, when Wheeler was actively developing his trophallaxis theory, she and her husband became acquainted with the Hoovers. Dora enthusiastically became involved in Herbert's presidential campaign; his own work in food distribution was proof of his suitability for the post.

Though starvation was Hoover's prompt for emergency aid, his workers, like Wheeler's, needed to remain hungry to keep the economy working. For this reason, Hoover has been described as the last of the great laissez-faire thinkers in America, before the moderate social engineering perspective of Franklin Roosevelt and his successors.[6] He wanted Europe to adopt liberal capitalism after the First World War, and recommended withdrawing aid unless its member nations would sign on to the United States' economic program. (As it happened, most of the food bought on the Europeans' behalf during the war, and which they would now have to buy for themselves, came from the United States.) Hoover's reason to oppose the terms of the Versailles Treaty was that Germany, if economically punished, would not participate in liberal trade exchange but rather would turn to communism or fascism. Herein lay an irony, in that state intervention was necessary to ensure that laissez-faire prevailed: again, a tension between the individual and the needs of society. Hoover was no great writer, but he did attempt to square the circle of individualism and society in his one notable book, *American Individualism* (1922):

> Our individualism differs from all others because it embraces these great ideals: that while we build our society upon the attainment of the individual, we shall safeguard to every individual an equality of opportunity to take that position in the community to which his intelligence, character, ability, and ambition entitle him; that we keep the social solution free from frozen strata of classes; that we shall stimulate effort of each individual to achievement; that through an enlarging sense of responsibility and understanding we shall assist him to this attainment; while he in turn must stand up to the emery wheel of competition.[7]

Hoover's arguments for relief were always moral, not economic, so that his laissez-faire principles were not breached. Although much of his career (indeed, the most successful part) was spent distributing food to prevent famine, he believed that relief was a job for voluntary agencies and a blip in the normal function of the economic machine. Communities should retain control of their own destiny through the laws of supply and demand. Thus, during the Depression, money for farmers often was made available in the form of a loan, not a gift.

Hoover's methods of dealing with the Depression and its problems bear an uncanny resemblance to Wheeler's account of feeding patterns in the nest under favorable and adverse conditions. Indeed, Wheeler asked his readers to consider the exchange of goods by coastal and inland people as a model for trophallaxis. The exchange did not even have to entail an immediate reciprocation; the important thing was that it had the ultimate function of maintaining society in equilibrium.[8] Wheeler's measure of the ant nation's wealth was in the amount of food collected, which ultimately was exchanged into the currency of the number of ants bred. Ants' hunger for wealth precipitated their seeking food, which they could then trade with one another in the process of trophallaxis. The ants themselves, though a "hung[ry] and greed[y] . . . proletariat," had no conception of the greater process of production on a national level but merely a "keen interest in raising larvae and in securing the food necessary for carrying on the business."[9]

Hoover blamed external forces for the American Depression; the chief villain was that termite society, Soviet Russia, for allegedly dumping its goods on the world market. Such a situation never occurred for ants in nature, but if too much food *were* present in the nest (for example, if symphiles required no favors in return for their aid), the workers would be temporarily satiated and would lose the impetus to maintain the trophophoric field. The cycle of accumulation would be fatally weakened. At any rate, the American trophic circle now overlapped with those of other nests and required "extraordinary thrift and economy" to survive.[10] Hoover's decision to lend funds to farmers was a good myrmecological tactic. In so doing, he maintained them in a state of hunger after the immediate hardship was over, to the good of the system.

Just as an economy could go into recession, so Wheeler's cycle could reverse, and this did occur in nature under conditions of hardship. The trophallactic loop closed, and the trophophoric field retreated inward to the nest itself; adults were slaughtered, and the survivors consumed their own young so that the superorganism might survive. Such a solution could not be contemplated in human society: though cannibalism was unlikely, the disorder and strikes that would result from wage cuts were not so far from the decapitation of the North American harvester ants ominously described by Wheeler.[11] Hoover's strategy was to insist that wage levels be maintained, hoping that by protecting the purchasing power of the workers, a gradual decline in the economy could be managed.

In putting forward an economic model for trophallaxis, Wheeler asserted that the only true difference between human and ant society was that humans, an evolutionary novelty compared with ants, had not had time to incorporate their functional division of labor into their heritable morphology.[12] But the implied possible evolution of physiology was unnecessary for Wheeler to perfect the analogy. His modeling of the hunger and greed of the human workers, and their functional role, was drawn from Italian sociologist and economist Vilfredo Pareto.

Wheeler began reading Pareto in 1925 or 1926, and responded to him with enormous enthusiasm and proselytic fervor.[13] Briefly, Pareto's sociology revolved around the assumption that most people do not live by rational thought but by nonlogical "residues." Their individual irrationality—their propensity to be swayed by emotional appeals—meant that their behavior could be predicted or even controlled en masse.[14] At the beginning of the Depression, Hoover acted in precisely this manner, giving empty reassurances that things would soon return to normal. By doing so, he hoped to maintain morale and prevent a catastrophic loss of confidence in the economy that would breed its own destruction—a most Paretian way of hoping to manipulate the emotional psychology of U.S. citizens. As it was for ants, so Hoover's ultimate policies left it up to the individual human to tough it out, in Paretian confidence that the overall equilibrium would be maintained.

Wheeler heartily concurred with Pareto's pessimistic assessment of humanity. He opened his series of Parisian lectures that discussed the trophophoric field with a brief meditation on the "strange analogies" that obtained between the societies of the social insects and human beings, indicating explicitly that he had a Paretian critique in mind.[15] Wheeler was able to give Pareto's work a distinctively biological reading, retaining an individual hereditary basis for the mass proclivities of society; evolution explained the formation of society, which in turn explained the "individual" instincts of its component members. The residues of the common man thus condemned him to a life that was functionally similar to the ant's. Wheeler seems to have been convinced by Pareto that the lack of insight on the part of the common man fated society to move down the degenerate path of fixed instincts.[16] Nor was he the only one to reach this conclusion, as the work of the Huxley brothers reveals.

The Huxleys and the Economic Metaphor

Aldous Huxley remarked that in *Brave New World*, he presented a "picture of society in which the attempt to re-create human beings in the likeness of termites has been pushed almost to the limits of the possible."[17] In many ways this comes as no surprise. With its rigid caste system, its merciless economy, and its dispensability of the individual, the New World is obviously very similar to life in the anthill or termite mound. But why was Huxley so inspired to write about humans as social insects?[18]

The immediate answer to this question undoubtedly concerns Julian Huxley's book on ants, published in 1930, and the information that the brothers shared on social life in the insect world. Julian may well have introduced Aldous to Wheeler's satirical paper, "The Termitodoxa." In this odd essay, written only two years after he had named the phenomenon of trophallaxis, Wheeler exploited the morally unorthodox element of his theory about the origin of sociality. The paper was originally given as a lecture to the American Society of Naturalists and was purported to be written by the leader of the termites (one King Wee-Wee), who told the history of his species. As a race they had been degenerating until some of their biologists made recommendations about the running of society, implementing eugenic guidelines that corresponded to Wheeler's representation of actual termite behavior. King Wee-Wee emphasized that trophallaxis was the very mechanism that enabled the emergence of transgenerational societies. Unlike the idealistic Forel, Wheeler, a kind of Mencken of science, positively reveled in the suggestion that societies evolved for selfish reasons: that instead of being based on parental love, termite society at least was based on fatty dermal secretions. It was only bribery or perhaps blackmail that caused the addicted insects to fulfill their familial duties: "Our ancestors, like other solitary insects, originally set their offspring adrift to shift for themselves as soon as they hatched, but it was found that the fatty dermal secretions, or exudates of the young, were a delicious food and that the parents could reciprocate with similar exudates as well as with regurgitated, pre-digested cellulose. Thenceforth parents and offspring no longer lived apart."[19] Beyond these obvious intertextual connections, a deep debt to Pareto was shared by Aldous Huxley and Wheeler.[20] Both writers took from the sociologist a "top-down" view of behavior (the idea that individual behavior is determined by the group) and a cynical attitude toward human nature.

Julian Huxley's book *Ants* contained no original research. What it did instead was to refocus current knowledge according to a functional, economic model.

The book's central chapter was titled "Food Economics," which was the linchpin for the whole system. Julian Huxley compared the arrangement of ant and human societies thus:

> The members of a human civilized community are tied together by economic bonds ... With us, of course, there is a universal medium of exchange in the shape of money, and by the use of such a medium we raise our system of mutual exchange of services to a far greater level of flexibility than was possible by means of payment in kind or direct barter. The ants in an ant-colony are equally tied into a single economic whole; but the means by which this is accomplished are ... unlike and, indeed alien to those employed by man.[21]

Huxley, like Wheeler, saw exchange as the thing holding the nest together.[22]

Huxley's characterization of trophallaxis as an alien form of exchange was not entirely fair, as we can imagine his brother Aldous thinking. For what was soma but trophic exchange? It was precisely what King Wee-Wee the termite described in 1920: "an elaborate exchange of exudates, veritable social hormones ... which, continually circulating through the community, bound all its individuals together in one blissful, indissoluble, syntrophic whole, satisfied to make the comminution and digestion of wood and mud the serious occupation of existence, but the swapping of exudates the delight of every leisure moment."[23] The capacity of the shared honey-dew to pacify the individual, and thereby to maintain the greater harmony of the colony, meant that it performed exactly the same function as the ubiquitous drug of Aldous Huxley's novel. Although freely available from the State, characters in the book *gave* soma to one other, quoting one of the relevant phatic aphorisms they had been taught from childhood. "A gramme is better than a damn." "A gramme in time saves nine." "One cubic centimetre cures ten gloomy sentiments."[24] Soma was the all-purpose reward, consolation, and pacifier without which the inhabitants of the Brave New World could not stably exist.

Soma was even at the core of the "solidarity services" at the Community Singery. In a parody of the Eucharist, a cup of strawberry ice cream soma was passed around, with the twelve-times-repeated formulation, "I drink to my annihilation."[25] It was the individual that was annihilated in this ritual; he submerged himself into the community, losing even his sexual identity and individuality in the "orgy-porgy" that followed. The ideology of *Brave New World* was thus that of the superorganism: "the social body persists although the component cells may change."[26] Like the individual ant, the inhabitant of Mond's world was singly insignificant, and his role was constructed through the use of a social drug. The distribution of Soma thus fulfilled the same social function as the honey-dew of the ants. Its dispersal in *Brave New World* was trophallaxis by another name.

More generally, Julian Huxley's use of an economic system as the chief model of what holds together the insect society echoed the mandatory overproduction and overconsumption that bound and perpetuated the system of *Brave New World*. In Aldous Huxley's world, people were exhorted to use fantastical goods and services, which were constantly being "improved" and added to. They were never supposed to mend things but always to throw away and consume anew. This constant obsession with material goods gave everyone something literally and metaphorically in common—a version of socioeconomics very different from the liberal Smithian lesson that used to be provided by the beehive.[27] In that case, each bee faithfully labored in order to contribute to the common good. Now, the ants seemed to be unwitting slaves of the system instead of its authors. The processes associated with the colony's benefit were not their aim; rather, they were the addictions that happened to define it.

There was a precedent for looking at so-called alien exchange within human society. Here we take a step back into Victorian thought: the savage as mirror to civilized man in his baser moments. Bronislaw Malinowski devoted his 1922 book *Argonauts of the Western Pacific* to a form of exchange prevalent in the Trobriands named "Kula." He wanted to dispel certain misconceptions about savage life: first, that the savage was "happy-go-lucky" yet governed himself by rational, utilitarian motives, and second, that the savage was not capable of organized labor or its corollary, trade and economics. On the contrary, argued Malinowski, there existed a strict though nonutilitarian exchange that was a form of economics. He summarized: *"the whole tribal life [of the Trobriands people] is permeated by a constant give and take;* . . . *every ceremony, every legal and customary act is done to the accompaniment of material gift and counter gift;* . . . *wealth, given and taken, is one of the main instruments of social organization, of the power of the chief, of the bonds of kinship, and of relationship in law* [emphasis in the original]."[28] What we have here is the ant as primitive version of human: irrationally economic. It is a surprising comparison, for we have just seen the ant-people of *Brave New World* as the apex of modernity. Yet Wheeler too read the *Argonauts* and cited Malinowski in his writings about the ants.[29] His reading of Pareto had convinced him that there was not so very much to distinguish between the savage and the American—or, indeed, the ant.

It is almost tempting to write an alternative history of sociology and anthropology as myrmecological disciplines. Espinas' political philosophy was heavily based on myrmecological study and in turn influenced the human sociology of Durkheim. Durkheim's disciple Marcel Mauss took his master's method as an injunction to record the entire exchange system of a human culture to see how it was held together functionally in the absence of a "market" as such, by acts of giving.[30] Meanwhile, Malinowski was providing myrmecologists with a plausible model of nonrational economics for ants and termites, analogous to Pareto's. Wheeler's Paretian contemporaries in sociology, Homans and Curtis, moreover, discussed the importance of institutions and social exchange among "civilized" humans. Despite the protestations of Wheeler that one could not make direct comparisons between ant and human, this common, naturalized economic vision of nature created a social agenda for humans that was no less powerful for being covert.

Making Scientists Antisocial

> Look down the escalator at Piccadilly Tube Station during a crowded hour. A motley mass of humanity moving down into the bowels of the earth, a lesser one ascending. But for the lack of uniformity in their component parts, these swarms of densely packed humanity would be indistinguishable from two streams of Working Termites flowing through sealed and covered tunnels to and from their work.[31]

This, Herbert Noyes' 1937 vision of mass man as mass insect, is exemplary of the modernist condition: the loss of oneself within the crowd. His image of standardized men recalls those memorable scenes from the film *Metropolis* (1926) in which streams of identical workers shuffle forward in synchrony to their shifts of mechanical underground slavery.[32] This vision of ants and their kind bewitched intellectuals of the early twentieth-century: biologists, psychologists, educationalists, sociologists, philosophers, and novelists. Might one be overwhelmed by the termite-men on the London Underground? Worse still, might one even become one of them, perhaps without knowing it? This, ultimately, was the dilemma Wheeler had to resolve—another version of Hoover's tension between the individual and the state. Wheeler had made ants social through economics, thus making humans intrinsically social, too. But Wheeler himself did not want to be one of the termite-men.

Wheeler was always keen to point out how the scientist eluded Paretian generalizations and was a fundamentally antisocial creature. His family could only concur with this assessment. Morton's son Ralph found that Dora (Morton's wife) and Adaline (his daughter) misunderstood the behavior of the males of the family. They "ascribe to malice or impatience some of the things you and I do through oversight or shyness," Ralph explained to his father. Fortunately, the women's embarrassment over such insults and antisocial behavior was limited to those occasions when guests were present, since, he continued, "I believe they no longer expect those marks of kindness to themselves that they once did."[33]

Unlike some members of his family, Wheeler's male colleagues (at least those with whom he did not engage in poisonous feud) relished his humor and gusto for discussing all sorts of topics, irrespective of the taboos of nice society. Pitirim Sorokin, with whom Wheeler created his Harvard course on comparative sociology, was also impatient with etiquette, and complained about the trophallactic circuit that he had to complete upon arrival in Cambridge: "We had to 'eat our way through Cambridge and Boston' at many lunches and dinners given by Harvard professors, 'proper Bostonians' and various dignitaries of both cities. I have never much cared for the 'social life' of going from party to party; nevertheless, like all newcomers to Harvard I had to go through this ritual to comply with the established mores."[34]

Wheeler's friend David Fairchild was one of his greatest kindred spirits. Having been introduced to Pareto by Wheeler, Fairchild took up the cause with evangelistic fervor, giving away copies of his book to friends and colleagues.[35] Four years after first reading Pareto's *Treatise,* the impression was still great, putting his social interactions in a new light: "If I had not read Pareto perhaps I'd never have been so alert to these emotional things [inflicted by 'petty people'] but since I read his philosophy I cannot see people other than through his glasses."[36] Inspired by Wheeler's social-biological take on Pareto, Fairchild wrote to Wheeler in 1927 seeking further enlightenment:

> Why can't you find out experimentally what this social force is which made those Echitons [ants] . . . support the whole mass of their fellows in the heavy curtain of individuals which hung from the branch to the ground? . . . Why didn't they complain and make some of the others take their places? . . . Wasn't it the same kind of an emotional force which makes me afraid even to offend the chambermaid when she wants to come in and interrupt me and sweep this room? I've about come to the conclusion that I am as truly a slave as though I were bound by shackles . . .
>
> The emotional reflexes are stronger than any steel and wherever I turn I find them exercising an effect on whoever they touch. I used to think that one could get away from bondage by going off into the wilds but one cannot, for there one comes into touch with savage or low class personalities which are just as insistent as those of civilized society. Whenever a two legged two eyed thing called a person

crosses one's trail he attaches his tentacles to you and binds you with a force that is terrible ...

I think it is for you dear Morton to point out how this same SOCIAL FORCE acts in the societies of insects.[37]

Fairchild's outburst followed a description of how a "scientific crowd" with whom he had visited the West Coast of Africa had managed to avoid squabbling but had never truly integrated. With the benefit of two years' hindsight, Fairchild thought he knew why: "It has always struck me that so many of our scientists give nothing emotional for others to love or revere except an abstraction TRUTH which has too little of the force that binds people together in large groups."[38] Fairchild's conclusion that there was something about scientists that made them less susceptible to the ubiquitous "social force" was one with which Wheeler repeatedly and publicly concurred.

However beautiful, however fascinating the organization of the anthill, Wheeler and his fellow scientists resisted its force in their own lives. When asked to give a lecture on the organization of scientific research to the zoologists at the American Association for the Advancement of Science in 1920, Wheeler argued that there existed no such social force among scientists, and that no-one should attempt to induce it. He averred that science should be exempt from the organization required by society at large, tracing the modern "vogue" for organization to the reprehensible and unpleasant "mobilization of armies and resources for the World War."[39] In contrast to the politicians and their destructive organizing tendencies, claimed Wheeler, scientists operated as individuals. Indeed, some scientific individuals were geniuses and therefore not susceptible to organization at all. Wheeler compared science, which generally was not organized, to its despicable antithesis, religion, which invariably was. He even quoted the Marxist saw, "religion is the opiate of the people," to prove his point—a phrase that recalls Huxley's soma yet again. Religious activity appealed to the otiose residues lurking inside most of humanity, but the scientist knew better.

While Wheeler's friend L. J. Henderson was busy with his fatigue laboratory at Harvard, Wheeler took it upon himself to head off any suggestion that such analysis should ever be applied to intellectual workers:

> We are ... beginning to see that as civilization progresses it is necessary to maintain a certain number of our activities in a primitive, unorganized condition and for their exercise to set aside hours ... so that we can escape from the organized routine of our existence. And as the surface of the planet becomes more and more densely covered with its human populations, it becomes increasingly important to

retain portions of it in a wild state, i.e., free from the organizing mania of man . . . that stand for and return to a Nature that really understands the business of organization. Why may we not regard scientific research [and] artistic creation . . . as corresponding reservations of the mind, great world parks to which man must resort to escape the deadening, over-specializing routine of his habits, mores and occupations?[40]

If one tried to inflict "any suggestion of such things as punctuality, punching timeclocks and other efficiency aimed factory devices," then the precious individuals of scientific research would be unable to work.[41] Individuality was crucial to the accomplishment of good science.

Wheeler's mention of the "primitive" and of "reservations" indicates his attitude toward scientific individuality as an antidote to overcivilization. Here again he was in accord with Aldous Huxley:

Civilization is, among other things, the process by which primitive packs are transformed into an analogue, crude and mechanical, of the social insects' organic communities. At the present time the pressures of overpopulation and technological change are accelerating the process. The termitary has come to seem a realizable and even, in some eyes, a desirable ideal . . . A great gulf separates the social insect from the not too gregarious, big-brained mammal; and even though the mammal should do his best to imitate the insect, the gulf would remain. However hard they try, men cannot create a social organism, they can only create an organization.[42]

By no means all scientists exhibited the primitive gumption that Wheeler saw as essential to proper science. Laboratory-based scientists never saw nature red in tooth in claw; neither did they suffer the deprivations of fieldwork that might cause them to recognize related drives in themselves, drives whose management cultivated manly character. In 1917, Wheeler dismissed their researches:

After perusing during the last twenty years a small library of rose-water psychologies of the academic type and noticing how their authors ignore or merely hint at the existence of such stupendous and fundamental biological phenomena as those of hunger, sex and fear, I should not disagree with . . . an imaginary critic recently returned from Mars, who should express the opinion that many of these works read as if they had been composed by beings that had been born and bred in a belfry, castrated in early infancy and fed continually for fifty years through a tube with a stream of liquid nutriment.[43]

One of main targets of Wheeler's jibes was what he termed the "eunuchoidal professors," the "inhibited" "priests" who spoiled the joy of biology for their students. He called them the "celibate," "damned professors," and recommended that they all be subjected to "Freudian tests." Slyly alluding to scandal, Wheeler noted that the greatest achievement of the behaviorist John B. Watson had been to remind academics that they too were animals.[44] Clearly, sex is the main instinct that Wheeler has in mind here; in the popular mind, and even in the not-so-popular mind like Aldous Huxley's, sex *was* the Freudian instinct. It was also the trait that humans shared with animals, the atavism that made them bestial; in biology, sex was one of the three main instincts or goals of life, the other two generally being nutrition and protection.[45] Nor did all naturalists escape Wheeler's censure. Louis Agassiz was faintly praised as a "dear old, mellow, disinfected" professor, exemplary of a certain brand of academic natural history.

Wheeler's individuality was unmistakably constructed as male, pitted against a female mass. Since creativity was allied to the male traits of dominance and aggression, it was no coincidence that creative individuals, who were male, tended to be less sociable than the polite mass of merely reproductive woman. "For obvious biological reasons," he wrote, "the female is the social sex *par excellence*, whereas the male was originally and throughout the evolution of the Arthropod and Chordate phyla, except in a few fishes, amphibians and birds, the unsocial sex. In many animals, in fact, he might more properly be called the antisocial sex."[46] Moreover, in Wheeler's view, such men were in a minority. The majority of men lived "in collaboration with the women" and merely maintained the structure of society. There was only a "very small class" of "less social individuals whose dominance was manifested mainly in the . . . great cultural values (sciences, arts, technologies)" as well as in the "great cultural illusions (philosophies, theologies, social utopias)." Wheeler thought that Russia was probably headed down the same evolutionary cul-de-sac as the termites because it had so socialized all its males. Like all bisexual societies, it would be "peaceful and harmonious, but also stationary and incapable of further social evolution." If a society had no antisocial males, it had no creative members; "the matriarchal clans of primitive man advanced towards civilization only after they had become patriarchal."[47]

Wheeler's concerns fitted with a general concern in the 1920s that society was becoming feminized. In Britain there was widespread concern about the effect of the predominantly female teaching staff on the boys in their charge. In North America there was something of a backlash against the women who crusaded for the "more civilized" world of which they themselves were the representatives and forerunners. This was precisely the overcivilization, the overspecialization against which Wheeler inveighed.

One may go further and identify the element of femininity that Wheeler eschewed specifically as maternity. Wheeler's account of the genesis of insect societies, as we have already seen, focused on the role that was played by maternal bonds. Thus the formicary provided an antimaternal counterpoint to Wheeler's masculine, scientific subsection of society.[48]

A certain fascinated disgust with motherhood was a recurring theme in the literature on ants. The queen ant, after her beautiful nuptial flight, returned to earth, digested her own wing muscle, and became what was invariably described as an "egg-laying machine." Most authors made great play of how she was merely a big, pallid, fatty barrage balloon, continually popping eggs out of her rear end.[49] In *Brave New World*, *mother* is one of the most taboo words there is; the use of contraceptive "Malthusian Belts" is drilled into those thirty percent of women who are fertile from childhood. It is an important device for Huxley, for he could easily have made all the women in his world physiologically sterile. Women are still obliged to have Pregnancy Substitutes every so often despite the weakening and controlling of the sexual urge through a promiscuous multiplication of its outlets. This is essential to keep them in good health; certain instinctive requirements, it seemed, cannot be overridden.[50]

Wheeler had appealed to sociologists in his account of evolution and, conversely, had constructed a desirable image of the scientist. He had shown that the ants evolved through maternalistic interdependence to their hyperorganized state; although many human beings could be described in this manner in sociological terms, the true scientist or artist was shown to rise above such constraints. Scientists would sometimes collaborate voluntarily, but the impression Wheeler gave of this grouping recalls Herbert Spencer's facultative account of the evolution of ant society rather than his own.[51] The scientist individuated himself from the maternalistic bonds of society (or was congenitally antisocial) and was thus liberated to fulfill his creative vocation.[52] Wheeler's science was defined around a set of related signifiers: it was individual, not organized; primitive, not civilized; male, not female; creative, not reproductive.

CHAPTER FIVE

The Generic Contexts of Natural History

Despite having achieved academic success beyond that of his entomological peers, William Morton Wheeler found that he was dogged by old accusations of "mere" natural historicism. As some of his peers at the Woods Hole Marine Biological Laboratory redefined biology in terms of benchwork, Wheeler pondered over his worryingly stamplike collection of ants and gathered what seemed to some anecdotal accounts of ant behavior.[1] In 1922, Wheeler exchanged letters with his friend T. H. Morgan on the topic of saving naturalists, whom the bigger organizations—specifically the zoological societies—were trying to drive to the wall. Morgan unsurprisingly suggested incorporating genetics into natural history in order to make it more professionally palatable, and facetiously recommended holding the naturalists' meetings as far out of the way as possible, since this absence was bound to make the zoologists' hearts grow fonder.[2] Wheeler felt under threat.

> In the scientific literature of the present time . . . natural history is so rarely mentioned that it seems to be the name of some extinct science, like alchemy or astrology. The term "naturalist" has also passed out of use. A few years ago, I was introduced to an audience by an eminent paleontologist as one of the last surviving naturalists, and, of course, the audience eyed me as if it were catching its last glimpse of a living Brontosaurus . . . I felt like the curator who overheard a little girl say, while she was being conducted through his zoological museum "Why, mother, this is a dead circus!"[3]

Certainly there were negative connotations of natural history from which Wheeler wished to distance himself: amateurism, lack of theory, antievolutionism, deism, anecdotalism, anthropomorphism, sentimentalism.

Notwithstanding these dangers, Wheeler was nevertheless keen to be counted as a natural historian. A committed supporter of the Boston Society of Natural History, he also retained a special loyalty to the American Society of Naturalists throughout his career, using its meetings as a platform to present some of his most important ideas.[4] Wheeler defended natural history in general scientific circles, speaking in its favor on a number of occasions at the American Association for the Advancement of Science. From very early in the twentieth century until his death, Wheeler consistently used his publications to reinforce the value of natural history, provoked by feelings of professional threat from the successful academic laboratory biologists. In response to negative attitudes, Wheeler hoped to defend "natural history" as a distinctive alternative program for biological research, for there were many positive aspects of natural history practice that he wished to emphasize in his own work and its written expression. These aspects included nonutility, fieldwork, the use of live animals, a philosophical inclination toward Aristotelianism, and a raft of positive cultural connotations, including childlikeness, manliness, primitiveness, and perhaps a commitment to speaking to the common man.

American natural history had a strong claim to fulfilling these desiderata; its pedigree was uniquely and intimately bound up with the discovery, or rather construction, of the continent on which its practitioners actually lived. Exemplars of the frontier naturalist were relatively easy to construct (especially without the historiographical inconvenience of their continued existence), but what might such a figure be in Wheeler's day? William Beebe, a professional scientist turned successful popular writer, struggled to define "naturalist" and "literary natural history" in his 1944 collection of natural history writings. He concluded that the ideal literary natural historian possessed the following qualities:

> Supreme enthusiasm, tempered with infinite patience and a complete devotion to truth; the broadest possible education; keen eyes, ears, and nose; the finest instruments; opportunity for observation; thorough training in laboratory technique; comprehension of known facts and theories, and the habit of giving full credit for these in the proper place; awareness of what is not known; ability to put oneself in the subject's place; interpretation and integration of observations; a sense of humor; facility in writing; an eternal sense of humbleness and wonder.[5]

This was truly a daunting set of characteristics. Notably, it included professional factors relating to the laboratory and its instruments, and a connection to the extant literature. It also included factors that were not commonly considered essential to science, such as humor, and even those that were downright suspicious, such as the "ability to put oneself in the subject's place."

As we might be led to expect by the technical, scientific desiderata, Beebe divided his collection chronologically into two parts: the earlier and merely amateur period (whose hero was its earliest writer, Aristotle) and the later, "proper" period of natural history. This period began with Darwin, but its apotheosis was Beebe's close friend and former co-expeditionary, W. M. Wheeler. Only Wheeler, the "ideal naturalist and scientist," met all the characteristics set out in Beebe's list."[6] Indeed, the piece of Wheeler's that Beebe chose to include in his collection was that odd eugenic plea for a sociopolitical role for biologists, "The Termitodoxa."

This chapter begins a discussion of the breadth of natural history writing that is continued in the next. In early twentieth-century North America, natural history was a fundamentally literary enterprise mounted on a cultural understanding of the nature of analogy. It encompassed a broad range of writers, each of whom defined individually his or her version of natural history. Children's natural history (or nature study) was an important part of the genre, and its key debates during Wheeler's career (c. 1900–1935) echoed and overlapped with "expert" issues of the day. When measured against the writing and reading of W. M. Wheeler, the genres of natural history reveal something about the problematic disciplinary identity of myrmecology.[7] Despite its pretensions to scientificity, natural history writing shared many characteristics with popular or even children's literature. As a would-be "elite naturalist," Wheeler needed to distance himself from "mere natural history" in order to retain the respect of his laboratory-based scientific colleagues. In particular, the rhetorical techniques of natural history writing offer insight into Wheeler's position with respect to amateur and semi-amateur entomologists.

Natural History: A Literary Enterprise

In the early twentieth century, natural history was above all mediated through the written word. The adult market for literature in this period grew considerably as expanding college curricula, correspondence courses, and reading clubs all contributed to the demand for books. As Joan Shelley Rubin has argued, the genteel values of late-nineteenth-century literature survived this transition.[8] Classic American texts, reconstructed as canonical, formed the basis of such institutions as the Book-of-the-Month Club and prospered in "chastened and redirected form" until the 1940s. Indeed, Rubin places literature at the center of the New Humanism, the early twentieth-century American search for selfhood. Models of democracy, culture, and education were all worked out in tandem with an intense discussion about the nature of literary criticism itself—in other words, how these concepts should be read from the classic texts.

Nature writing was where the frontier spirit and the American search for selfhood through literature came together. The themes of survival, spirituality, and political identity were enshrined above all in the writing of Henry David Thoreau. Around the turn of the twentieth century, the American reading public made space for more writers in this mold. Their appetite for nature writing expanded to include magazines, recreational nature guides (such as bird-spotting guides), practical nature guides (such as fruit-growing guides), and fiction.

Material changes in publishing also shaped the literary market in natural history. In 1881, trichromatic half-tone plates were developed, making cheap reproductions in color possible for the first time.[9] The technique, which was used extensively for cover illustrations of dime novels during the first few decades of the twentieth century, also enabled the production of field guides that allowed naturalists to identify specimens on the basis of coloration. W. J. Holland was one of the first authors to make use of the technology. His *Butterfly Book* attracted praise from the long-established entomologist Samuel Scudder and found a large market amongst enthusiastic adults and children alike; later editions were dedicated "to the boy scouts of America." Similar guides were produced on birds, and Ellen Eddy Shaw contributed a large number to the Pocket Garden Library, brought out by the New York City-based Country Life Press. The books were pocket-sized and beautifully produced, often bound in imitation leather and boxed, with as many as 300 color illustrations. They were items to treasure. Such books often assumed considerable expertise on the part of readers. E. O. Wilson's copy of Holland's *Butterfly Book* showed only one side of wing patterns, the side that was not visible when the insects were at rest. The user was expected to know the inner wing coloration corresponding to each illustrated outer pattern.[10]

One particularly prominent natural historical author was Liberty Hyde Bailey (1858–1954), a botanist, educator, and founder of the nature study movement at Cornell University. Besides achieving publishing success in his own right, his connections to the publishing industry embedded him deeply in the business of nature writing. He edited numerous practical nature guides, such as Macmillan's Rural Science Series, intended to give useful hints to the farmer. Bailey was a prolific author; his bibliography contains some 700 titles, including 63 single-authored books. Bailey's most celebrated publication was *The Holy Earth* (1915), in which, as the title suggests, he spoke to the religious instincts of readers. It created an agreeably spiritual account of nature that could be interpreted as conventionally Christian or vaguely deist—even mystically pantheist—just as one pleased. It spoke to the city-dweller's fear that he had lost touch with nature, creating a literary door through which the urban resident could step into that imagined lost world. Bailey also published quantities of poetry that appealed to similar tastes. His practical books, meanwhile, generally ran through ten or more editions, selling 15,000 to 30,000 copies; a botany book topped 80,000 sales. In total, Bailey's royalties consistently reached $10,000 per annum in the 1910s and 1920s. During the Depression they dropped, but not as much as one might have expected—to around $8,000 a year.

Magazines—the *Rural New Yorker, Country Gentleman, Country Life in America*—catered to similar audiences, from the practical rural dweller to the romantic urbanite.[11] The Rural Publishing Company was, in its early days in the 1890s, successful in such markets. Its magazines and book series were explicitly aimed at florists, amateurs, country gentleman, fruit growers, estate gardeners, and market gardeners. Despite efforts to expand, the company had failed by 1893 and was bought up by De La Mare. Small nature publishing ventures often suffered financial crises, at least as much because of poor management as because of poor sales, and were bought up by major publishing houses that aimed at a wider, more urban audience than the rather genteel and rural one intended by the Rural Publishing Company.[12] *American Garden* became part of the Garden Publishing Company, and Country Life Press was incorporated into Doubleday, Page. The shift in the market was physically symbolized by Doubleday's move out of urban New York and into Garden City, Long Island, in 1911.[13] A move to a genuine rural setting was in reality no more desirable to the company and its 1,000 employees than it was for its readers. A more practical maneuver was to bring something of the countryside into urban life, whether this meant bringing greenery into Garden City or one of the company's books into a sixth-floor apartment. Doubleday's press release about the move spoke ambiguously in these terms, stating that the editors aimed to "draw people from crowded cities into open spaces . . . foster a love of the wide outdoors, the home of health and broad horizons."

There was thus a well-developed market for nature writing in early twentieth-century North America. Initially differentiated for a variety of readers, from the practical to the escapist, its various streams were somewhat channeled together as the twentieth century wore on. Written natural history retained, however, a unique American flavor; apart from Wordsworth and Emerson, there was remarkably little overlap between British and American publications. And Wordsworth, as we shall see, was a doubtful choice for the healthy Stateside male reader. W. H. Hudson and Richard Jefferies predominated in the United Kingdom, while Bailey, along with John Burroughs and Theodore Roosevelt, were the most popular modern American nature essayists. In part the unique character of written natural history that emerged in North America derived from Americans' relationship to the continent. American land was at once home and the frontier of exploration, whereas Europeans split their exotic, colonial nature writing and their domestic natural history writing into separate genres. The American nature essay offered much; it carried a great deal of weight when it came to affirming the natural, political, and personal order.[14] If Paley's "happy world" was Genesis inscribed in England's green hills, then the U.S. Constitution was written in the deep—but not measureless—waters of Walden Pond.

The metaphor of the book of nature is, to state the exceedingly obvious, an ancient one.[15] A tradition discussing the relation between the proper knowledge of nature and proper reading or writing was explored particularly during the Romantic and post-Romantic eras. The creativity of writing was profoundly linked to the act of Creation, even for those who did not ascribe that act (or process) of origin to God. The literary-natural comparison was often a symmetrical one: just as literature was used to enhance the understanding of nature, so some commentators applied natural history techniques to poetry. The Victorian engineer and writer Henry Dircks created a "classification" of nature poetry, ordering it according to that which focused on minute or single objects and that which embraced varied and complex views; the imaginative versus the fanciful; poems that inspired association, reflection, comparison, or meditation; and so on.[16]

Analogy was the key question. Properly formed analogies were central to understanding the universe and to the composition of good poetry. As Dircks put it, his method was to "trac[e] the influences of external nature [in shaping metaphor] to their true sources, and their consequent operations on our mental faculties."[17] Coleridge's taxonomy of metaphor was an important point of departure for all in this discussion, and Emerson's celebrated essay *Nature* (1836) was a universal reference for Americans in particular.

The contemporary literary historian Gillian Beer urges us to treat scientific and literary analogy as cut from a single cloth. Darwin, she writes, used the multifariousness of the world as both "material and idea," delighting both in substantiating metaphor and in restoring the wonder of everyday fact. According to Beer, analogizing is about finding identity between apparently unlike things, and then using the shifting energy of their relationship to reach into unstable realms such as the future. The analogy is itself open to change; its pleasure and its power both reside in its precariousness. Beer highlights the connection between analogy and magical or religious revelation suggested by these traits; analogical discourse is implicitly transubstantiative, changing the homely to the transcendent. It claims a "living, not simply an imputed, relation between unlikes."[18] This characterization is a good one, and in Wheeler's period it is given a historical substantiation. Evolutionary theory provided a marvelous mechanism for the old idea of real, transubstantiative analogy. One could easily remove God the designer from nature and replace him with some other form of the demiurge. Or, less anachronistically, one can see Darwinian and post-Darwinian biology as particular kinds of response to the tradition of analogy. The important notion is simply that *nature can speak to us*.

Immediately before and during Wheeler's career, the most important writer on analogy was John Burroughs (1837–1921). Burroughs, "the Sage of Slabsides" (the name of his cabin in the hills), was a national figure, regarded as priest and prophet, in addition to progenitor of the American nature essay. Beyond the realm of literature, his influence was partly responsible for the nationwide impulse toward nature study both within and without the school curriculum. In the four decades prior to his death in 1921, Burroughs was said to have had "more of a personal following, more contacts with his readers, both through correspondence and in person, than any other American author has had, and, probably, more than any other author of modern times."[19] He was also recognized by a number of scholarly institutions, receiving honorary degrees from Yale University, Colgate College, and the University of Georgia. His iconic status gave him an unparalleled mandate to speak on nature and its analogical significance for the nation.

Burroughs examined this matter of naturalized understanding in his collection of essays, *Literary Values* (1903), and particularly in the essay "Analogy—True and False." Burroughs traced back to Emerson the intuitive truth that, thanks to the natural order, something beautiful was in a deep and fundamental sense also true: "The method of the universe is intelligible to us because it is akin to our own minds. Our minds are rather akin to it and are derived from it . . . The truth here indicated is undoubtedly the basis of all true analogy—this unity, this one-ness of all creation."[20]

Burroughs criticized some of the analogies produced by an unquestioning faith in this fact, but was very clear that true analogy, springing from a "unity of law," *was* possible, originating in the very material patterns of the universe. The metaphysical status of true analogy had an epistemological implication. Because human minds were a part of that universe, true analogies "ha[d] the force of logic; they shed a steady light."[21] Essential analogies not only appealed to the fancy, they also bred true understanding. The literary response to nature was thus all-important, because appreciation was based on kinship. Evolution was both the source of analogy and the grounding for its human comprehension, as well as being the best example of analogy in its own right. This was essentially one of Wheeler's arguments in favor of natural history: it was "the perennial rootstock . . . of biological science . . . because it satisfies some of our most fundamental and vital interests in organisms as living individuals more or less like ourselves."[22]

In evaluating Burroughs' contributions, it is difficult to determine how far his philosophy as opposed to his image permeated culture. It is very rare that one finds, for instance, a proponent of nature study using the argument of natural kinship to explicate a child's understanding of nature. On the other hand, Liberty Hyde Bailey's successful poetry, notably the frequently reprinted "Brotherhood," sprang from exactly the same intuition of interconnectedness. Henry Wadsworth Longfellow (1807–1882) was frequently quoted in a similar vein:

> It was his faith,—perhaps is mine,—
> That life in all its forms is one
> And that its secret conduits run
> Unseen, but in unbroken line,
> From the great fountain-head divine,
> Through man and beast, through grain and grass.

Similarly, Walt Whitman (1819–1892) held that "a vast similitude interlocks all, spans the universe and compactly holds and encloses them."[23] Whitman's vast similitude, like Burroughs', included the mind of man, or his experience and the way in which he dealt with the greater whole.

Native Americans were to many the missing mystical-natural link, priestly intermediaries between the colonists and the land. Ernest Thompson Seton went on expeditions deep into the American wild, wrote about the nature he encountered, and encouraged boys in his Woodcraft movement to develop (or perhaps recover) survival skills, the connection with the landscape that came naturally to the "Red Man."[24] The red man's knowledge of nature was intuitive; his rituals were natural analogies with practical power to help him survive, a form of spiritual power that yielded reverence for the land from which he had sprung and by which he was sustained. The Native Americans' totemic animals were also adopted as icons by European Americans; both Seton and Jack London identified themselves as "wolves."[25] So perhaps we should see Burroughs as articulating a philosophy of knowledge and literature that was part of a general urge toward nature mysticism (transcendentalism is the wrong word in this context). Whether or not readers followed his arguments in every detail, he clearly connected with the things that mattered to them.

It is a little easier to measure Burroughs' belief in analogy as trustworthy explanation against his scholarly contemporaries' philosophy of biology, which was frequently well articulated. In particular, Burroughs' epistemological intuitionism bears comparison with Henri Bergson's insistence that one must submit to one's intuition in order to understand the dynamic processes of the developing universe, processes that bear some causal relation to the mind, since man is himself a product of that universe. Bergson, of course, inspired professional American biology during the first two decades of the twentieth century; W. M. Wheeler was one of his greatest advocates.[26] No matter that the cultural origins of Bergsonism were different from so-called transcendental naturalism; expert and popular culture appeared for this brief period to be aligned in their understanding of the human relationship to nature.

In the 1930s and 1940s, a century after America's golden age of natural history, a number of authors began to revisit the question of what constituted American nature writing.[27] Turning away from philosophical questions about the nature of analogy, they began to trace histories of how "proper" nature writing—scientifically informed, yet rooted in American experience and identity—had arisen. There was a rush of histories of entomology as the first generation of professionals began to retire and take stock.[28] Needless to say, these narratives all told similar directional stories about the professionalization of naturalism. They also covered similar ground in their discussion of founding figures, notably Thomas Say, and the relation of these figures to Jeffersonian ideals of American science, asserting independence both from Europe's scholarship and from its landscape. In this era a new group of writers of (rather than about) natural history also came to the fore, notably William Beebe, Julian Huxley, Gustav Eckstein, and Donald Culross Peattie. Their observations and style were compatible with the new natural history ideals.

Telling one such history of progress in 1950, Joseph Wood Krutch posited that nature writing was a distinctively modern phenomenon that steered a careful course between the precise observation of science and the unreined subjectivity of literature.[29] Its first key feature was a kind of species humility: a sympathetic and humane realization that humans had much in common with animals and that animals had their own lives and identities independent of the uses humans had for them. This realization, Krutch argued, could be derived from Darwin's evolution just as easily as from a natural theology. In this sense, the Europeans Bacon and Descartes came off poorly as natural historians. But where Americans really triumphed was in the second quality of nature writing, a sense of the sublime.[30] The problem with European nature writing was that it saw beauty in tame nature. The Romantic enthusiasm for wildness was all too often expressed through the "minor and somewhat dandiacal" form of landscape gardening.

In pinning the uniqueness of American nature writing to this reconstructed aesthetic of the sublime, Krutch followed numerous other commentators. Notions of a Spencerian sublime—a harshly beautiful landscape that only the strong can survive to appreciate—pervade the writing of Jack London, for instance. Other writers explicitly formulated the concept in Krutch's mid-century era, notably Donald Culross Peattie, a purely popular writer. Peattie gave Darwin his place at the forefront of European nature study, but it was in the second part of his book *Green Laurels,* set in the New World, that the true pioneers could come forward. The American landscape was splendid, unyielding; it could swallow up children and even hunters. When explorers came looking for the exotic, erotic East, what they found instead was an innately masculine, puritan continent: "Instead of a voluptuous and adaptable Nature, America proffered a hard, clean and somehow intractable biota."[31]

Peattie was ambivalent about the European expansion in America. In one sense he descried natural law in the triumph of one species over another; more profoundly, he saw degenerative processes at work, whether through violence or decadence, and lamented them:

> Europe has evolved a modern, aggressive biota, a pushing, compromising fauna and flora so subservient to man and so much cleverer than he, that it makes its way around the earth ... European man has entered the other continents ... he hoes out what he finds growing to make way for that symbol of chivalry and eroticism, the rose, his queen of flowers ... But there was in Michaux's America an aristocracy of primordially ancient lineage that would make no compromise with the invader ... American biota was great without prettiness, strong but not elastic, proud to die, but not able to bend the knee.[32]

Certainly, Peattie feared the effects of teaching this European attitude to American children; if they studied Wordsworth they would develop an aesthetic that was "delicate, subtle, subdued to live with men"; they would learn to watch the skylark and not the flapping eagle.[33] And while Peattie had no sympathy for the "atavistic pleasure" of shooting, he emphasized that the skills of tracking nature to kill were required in "an actual struggle for survival" not so long ago; they were Emersonian qualities of self-reliance. Darwin himself was refracted through an Emersonian lens; his chief virtue was imagination, his principal achievement to "infect the world ... with belief in the worth of effort."

William Martin, chair of zoology at Syracuse University, told a progressive story of natural history much like Beebe's and Peattie's. Its pioneers' key qualities were a spirit of adventure and imagination. Because of his own position, Martin was also keen to identify and explain a trend of professionalization in natural history. Here too his story was much like Beebe's, for he concluded that though the naturalist had at the turn of the twentieth century fallen for a time to the status of amateur, he could now be confidently characterized as "a man who studies animals and plants in their natural environment and seeks to apply to his investigation all the techniques which modern science has invented."[34] Beebe ranked his contributors as "scientists of the first rank" (such as Wheeler), "zoologists of note," and "articulate naturalists" (including popularizers), but all were included in his *Book of Naturalists*.

Peattie chose to highlight an entomologist as the best representative of the last generation of naturalists. C. V. Riley was "a Gulliver in the Lilliput of the insect world"; the insects were that remaining part of the American landscape that had "yet to be reckoned with." Insects perfectly fulfilled the qualities for nature study, requiring both imagination to comprehend them and vigor to combat them: "When men march out to slay them, they turn into some other form . . . [they] are bizarre, fantastic, thoughtless but knowing, differently motivated than we are . . . In fact, theirs is a way of life so unlike ours that it is astounding to find them on the same planet . . ."[35] Perhaps thinking of Wheeler, with whom he had briefly studied in 1921, Peattie finished by concluding that ants were the most fascinating of all.[36] Peattie's ants, autochthonous representatives of a tough, masculine American landscape, were ideally suited to the cloth of the naturalist cut by Wheeler.

Natural History for Children: The Nature Study Movement

One cannot give a proper account of the genres of natural history in early twentieth-century North America without paying significant attention to its composition for, and consumption by, the juvenile market. For men like Wheeler, writing for children was often the most dubious form of natural history. Stories like *Betty and the Little Folk* or *Mother Nature and her Fairies* were the nadir of the genre so far as they were concerned.[37] Others were more problematic: presenting themselves as serious educational texts, they might nevertheless stumble in their treatment of sentiment or in their implied methodology. Their intended audience might in itself devalue the natural history that Wheeler sought to protect and promote, for many of the values embodied in children's natural history literature were indeed, with certain caveats, acceptable to experts. As with adult literature, the relationship of expert writing to this genre was a complex matter, and the comparisons drawn are not always to be taken at face value.

Cornell University was one important birthplace of nature study, as natural history for children was generally called in the United States. Liberty Hyde Bailey used his time as director of the Cornell College of Agriculture to encourage the movement.[38] The mission was intimately commingled with Cornell's Extension project to educate and assist the general public, providing reading courses for farmers and their wives and improving agricultural and domestic techniques. Bailey fostered links with public schools, encouraged the production of leaflets for teachers, and in 1898 hired a nature study instructor to attend teachers' institutions and present the case for nature study. In that year the instructor spoke to 14,400 teachers at seventy-two institutions. Bailey's leaflets, meanwhile, were printed in batches of 25,000 to satisfy the demand from New York State teachers alone.[39] Bailey's trick in promoting nature study was to combine an appeal to the popular romanticism of the urban back-to-nature cult with a commitment to science and education as means to achieve practical agricultural improvement.[40]

Bailey's colleagues Anna and John Comstock were also very important in the development of nature study for children, most notably through their books.[41] The Comstocks were entomologists; though entomology was comparatively poorly funded among Cornell's extension projects,[42] it had a disproportionate influence. John Comstock taught Vernon Kellogg and Ephraim Porter Felt, who went on to exercise considerable influence; he also helped David Starr Jordan set up a department of entomology at Stanford University during the years 1891–1900. Cornell operated as significant center of collection for economic entomology, through which it was able to develop (although more slowly than Wheeler at Harvard) a more purely scholarly reputation. It is still an important center of entomology today. At the turn of the twentieth century, insects were considered an ideal topic for nature study because children allegedly found their life stories fascinating, while they also had great economic relevance.[43] Insofar as nature study was utilitarian, insects were a central topic of study.

Another member of the Cornell nature study network was the self-dubbed "Uncle John." John W. Spencer was a fruit grower by background, but by 1900 he had remade himself fully in his new persona, cultivating children's study of nature through his publications, communications with teachers, camps, and Junior Naturalists' Clubs. In 1899 there were around 1,500 clubs with 30,000 members. The initial subscription declined somewhat, but in 1902 there were still about 17,000 members, and a higher number of Junior Gardeners (another of his ventures with the more practical aim of improving the physical condition of school grounds). Spencer requested that teachers let him know how they were following his suggestions, and even the names of the children involved, so that he could write to them individually. Uncle John is a difficult figure to assess. His adopted identity sits uneasily with contemporary sensibilities—an impression that is only increased by the discovery that he corresponded with adult colleagues in his Uncle John persona—and much of his self-publicity was pure bluff. One teacher reported, "The name 'Uncle John' carries with it a special charm for [my pupils]," but the singularity of the letter rather undermines its ostensive content.[44] Certainly Bailey was sufficiently impressed with his achievements to make him supervisor of the Bureau of Nature Study after himself. Cornell supported Uncle John in his ceaseless quest for feedback, as it provided a means to know and keep tabs on interested parties (teachers, mothers' clubs) who encouraged nature study in all its guises; it was a way to sustain funds for all aspects of their work.

The promoters of nature study could have done very little without the enthusiastic collaboration of teachers from urban areas, and those of New York State in particular.[45] Together these teachers created a network of training, discussion, and publication dedicated to the furtherance of nature study in schools, and a healthy reflexive debate as to the character and aims of the movement itself.[46] The *Nature-Study Review*, devoted to nature study in an elementary school setting, was started at the New York City end of the nature study network in 1905. Its founder was Maurice A. Bigelow, a trained biologist who had by then moved to Teachers College, Columbia (the institution where Dora Emerson Wheeler received her second degree in 1898). The *Review* was issued every two months but soon proved so successful that in its second year it moved to monthly publication during the school year. The *Review* was privately published until 1908, when it was adopted as the official organ of the newly founded American Nature Study Society (ANSS), based at Cornell.[47] From a relatively small distribution in 1910 it grew massively; receipts from subscriptions in the school year 1912–1913 totaled around $1,200, at which point publication was taken over by the Comstock Publishing Co., right at the center of the Cornell network. The ANSS and its journal drew together elementary schoolteachers, educationalists, and biologists, including a significant number of women at all levels except university professors.

From the outset, the *Review* treated nature study as a stable and specific entity, not the same thing as natural history. Unlike the latter, nature study was not an organized body of knowledge but "an outlook, a point of view, a method of studying nature."[48] Its subject matter might include agriculture, elementary science, or "popular picnics in the woods," but above all it was about the cultivation of sympathy with nature. Although the *Review* was at that time based in New York, it took its lead from Bailey at Cornell: "[W]e are not to teach Nature as science, we are not to teach it primarily for method or for drill: we are to teach it for loving ... On these points I make no compromise."[49] The discussion as to what nature study was went on through successive issues, and its proponents were sufficiently confident in their opinions to allow a fairly robust exchange on the topic, reprinting in full some very critical comments.

From its origins in the New York metropolitan-upstate nexus, nature study was promoted and taken up farther afield.[50] A survey conducted in 1915 revealed that nature study was required in the school curriculum by law in fifteen states and encouraged in a further fifteen. By way of comparison, agriculture was legally required by twenty-five states and unofficially promoted in seventeen states. Twenty-three states produced their own course for teaching nature study, only two fewer than those producing courses on agriculture.[51] The compilers of the study claimed "a definite increase" in the use of nature study, although there were no comparative statistics from previous years. In 1925 an ANSS- and Rockefeller-sponsored Coordinating Council on Nature Activities, based at the American Museum of Natural History in New York, counted the ANSS as one of dozens of nature-related societies, from outdoors clubs like the Boy Scouts to astronomical organizations. By 1928 the list had more than doubled. Although not all the organizations can be counted as groups dedicated to nature study, the list does illuminate the diverse concerns with nature to which the ANSS answered and which it hoped to influence through the production and dissemination of nature study material.

The most bizarre indication of the nature study movement's success was that it attracted criminal attention: a man traveling through New York and New Jersey offered fraudulent subscriptions to the (real) magazine *Birds and all Nature*, said to include sixty color plates, via the fictional "Nature Study Co., 1135 Broadway, NY." This elderly con man apparently succeeded in this endeavor for some time, despite his memorable appearance and poor disguise: always using the same name, consistently professing Quakerism, and lacking two fingers on his left hand.[52]

The literary manifestation of the nature study movement was considerable. Some nature study books were published on both sides of the Atlantic, but to a large extent the oeuvres were independent of one another. This separation can be largely explained on the basis of the differing biota of the two continents; a species-specific guide was simply no use on the wrong side of the Atlantic. Differences in approach, however, cannot be explained on these grounds. The British catalogue of serious children's nature study literature was largely made up of rather dated Victorian titles by Lubbock, Henslow, Kingsley, and the like. Though most of the books were not overtly religious, a large proportion were brought out by religious publishers, notably SPCK, the Religious Tract Society, and Hodder and Stoughton. Educational publishers such as Nisbet and Macmillan were on the rise, but a look at their complete catalogue shows that nature study, and science in general, formed a minor part of their largely literary offerings.

In North America, an extensive literature for or about teaching nature study to children was produced. A 1906 survey of teachers and educationalists produced the following list of favored authors and titles:[53] Bailey's *Nature Study Idea*, Blanchan's *Bird Neighbors* and *Nature's Garden*, Frank M. Chapman's bird books, Clifton F. Hodge's *Nature Study and Life*, Stanley Coulter's books on plants, and David S. Jordan and Vernon Kellogg's *Animal Life*. In 1911, Anna Comstock's *Handbook of Nature Study* gained its place as bible of the movement. There were books on the theory of nature study for teachers to read, books about nature to inspire teachers, books and pamphlets containing lesson plans for teachers to use (with or without additional notes for the teacher), popularizations of nature study for children that might be employed in the classroom or read at home, and nature storybooks for children. Writers in the *Nature-Study Review* came from a variety of backgrounds and included professional scientists (notably W. C. Allee), naturalists, educationalists, and teachers.

By 1913 some fairly serious books were being reviewed by the journal, especially those on genetics and evolution. The topics covered ranged from "baking buns as nature study" to the study of stars, via hygiene, but most were on nature as traditionally conceived. The dividing line between acceptable and unacceptable nature literature was a fine one. A review of April 1914 wrote off *Mother Nature and her Fairies* by Hugh Findlay but praised Fabre's *Life of the Fly*, so often condemned by critics of nature faking. "Simple and forceful writing ... accurate and worth-while observations ... The book is not only interesting but there are many suggestions of ways and means of studying insects ... As a matter of fact the specialist will read it with as keen interest as the layman." Those concerned with nature study *soi-disant* thus constituted a large and often self-aware market for nature writing. Bailey, revered as a teacher as well as a writer, received more than $3,000 of his total royalties from Macmillan educational publications at the height of the Depression, giving some indication of the continued success of nature study literature for children in the early twentieth century.[54]

As the twentieth century moved along, the character of nature study and the cultural needs it was shaped to fulfill changed somewhat. At its inception, there was widespread concern that urban children were losing touch with the natural world. This nostalgia took a gentle (feminine?) Emersonian form; for others, such as promoters of scouting, it was fashioned in a Spencerian form of frontier ideology, often with the red man as priestly mediator of the continent's spirit. A utilitarian aspect of nature study, though present in its early days, grew over time. School gardens were a key feature of the movement in the early twentieth century, and by the time of the Depression, nature study activities had become a way to give children healthful recreation. Thus, the argument went, they were educated in matters of hygiene, parsimony, and work, and correspondingly less likely to become delinquent.[55]

One striking feature of the ANSS's evolution was its increasing claim to scientificity. Bailey, arguably its foundational figure, stated clearly in *The Nature Study Idea* of 1903, "Nature-study . . . is not science. It is not knowledge. It is not facts. It is spirit. It is an attitude of mind."[56] Quite quickly, however, other proponents of the movement began to assert that it was a precursor to science, or even a kind of science (if an elementary one) in itself.[57] Staking this claim, the ANSS affiliated itself with the American Association for the Advancement of Science (AAAS). By the mid-1920s there was an AAAS representative in the ANSS, and in 1929 the ANSS held its annual meeting in conjunction with the large scientific society. This collaboration resulted in the development of *Science on the March* (1938), a radio program aimed at high school students and adults. Partly because of the age range at which it was aimed, nature study was subsumed into more traditional scientific topics; according to listener responses, astronomy was far and away the most popular topic. The following year the ANSS collaborated with the AAAS on another radio program, *Science Everywhere*. Before NBC dropped it because of personal feuds, this program was projected to feature four pupils, a teacher, and a scientist, and to appeal more to the ANSS's traditional constituency. Notwithstanding the difficulty in practice of raising the national profile of scientific nature study, the aspiration of the ANSS just before the Second World War was to ensure that "no distinction is drawn between nature study and science, since their materials are often identical."[58]

Throughout the evolution of the natural study movement, its proponents continued to raise questions about their identity. The issues they raised, and the answers they negotiated, can largely be explained with reference to the professional aspirations of these people, as validated by their scientific superiors (as they perceived them). Natural historians such as Wheeler, it turns out, considered very similar issues, reflecting their own particular disciplinary issues. The danger of this similarity was that Wheeler et al. might be lumped together with the teachers of nature study. Strategies to avoid this outcome are examined in the following chapter.

One lively topic of conversation among educators concerned the question of whether children were or were not natural naturalists. Plenty of people thought they were, or at least that nature was a natural topic of study for children—a formulation that had more to do with the didactic moral potential of nature.[59] Benjamin Franklin's "Proposals for Education of Youth in Pennsylvania" had commended nature as a topic falling "within the capacity . . . of children," suited to their natural curiosity, and a "pleasant and agreeable" study, more like recreation than "painful and tedious" forms of schooling.[60] In the late nineteenth century, John Burroughs' writing was commended for its quality of "childlike fun" and its healthy, quiet enthusiasm, suited for teaching city children.[61] An early figure in the nature study movement argued that "nature study should lead the child back into his natural intimacy with nature."[62]

An enthusiasm for linking children with nature was also present in the realm of experts. W. M. Wheeler and his friend David Fairchild spent a lot of time encouraging their sons in entomology. Fairchild's son, Graham, was the optimal kind of child naturalist who apparently outdid many an expert in his curiosity and commitment to the subject. When Graham Fairchild was made a member of the Cambridge Entomological Society, in Massachusetts, David wrote rather creepily of his son's pleasure, "He was so delighted . . . I think it puts the last pin through him which will fasten him to an entomological box for the rest of his life. He has been mounted."[63] Preserved like a butterfly in a state of boyhood, Graham would be the perfect naturalist: "the . . . enthusiastic boy we hope he will always remain." Even advanced biology was considered by many to be theoretically within the grasp of older children. Wheeler considered that "no facts or theories in entomology—or for that matter in any biological science— . . . transcend the understanding of any fairly intelligent lad of fourteen."[64] In his old age he came to believe that the young were actually better at theory than mature adults; their neotenous minds produced more important ideas than those of the hidebound.[65]

But in a deliberately provocative article published in 1907, Maurice Bigelow slashed away at fond assumptions regarding children and nature. In doing nature study, he suggested, children were generally following parental prompts without much enthusiasm, or else they tended to view specimen collection in a competitive spirit rather than for its own sake. No matter how much they appeared to enjoy nature study, children would always do something else if given the choice. Even their so-called "sympathetic relation with animals" rarely prevented an active enjoyment of cruelty to them, a natural tendency that could only be stopped by a threat of spanking or, unscientifically enough, bad luck.[66]

Bigelow's article provoked some strong reactions, many positive. These correspondents expressed relief that at last someone was being frank and unromantic.[67] Naturalists, many of these writers emphasized, were made and not born. The editor of *American Botanist* wrote in to explain, "it seems to me that if nature study is ever to get anywhere, it will have to be used as a drill on observation, not as a stimulus to the child's interest in nature for which most children have no abiding interest. The child is very much like an electric motor. It keeps going and interested so long as you turn on the current."[68]

The child-nature discussion was intimately connected with teachers' aspiration toward scholarly professionalism.[69] In particular, nature study provided a rare opportunity for women, who made up the majority of elementary schoolteachers, to claim scientific and intellectual respectability. Denying the child's natural affinity with nature meant that his learning must be stimulated by a good teacher and countered the suspicion that there was something easy or childish about nature study. Debates in the *Nature-Study Review* revolved around the premise that the teacher, not books—as Anna Comstock would have it—was the intermediary between children and nature. The discussion about children's innate predisposition, or lack thereof, toward nature study came about in 1907 and 1908, just after the foundation of the ANSS, when proponents of nature study were already moving toward calling it a science. Indeed, at the first ANSS meeting, held at the University of Chicago, nature study advocates were anxious for approval from the university's academic scientists.[70] Thus the discussion was, in its timing, a useful source of evidence for professional recognition.

Despite his belief in the abilities of children, Wheeler made his excuses when asked to speak in schools; there was a difference between the natural historical rhetoric of childlikeness and the indignity of speaking to actual children. To one such correspondent he parapractically complained, "I am a very poor lecturer for children as I have an eradicable habit of talking over their heads."[71] Wheeler's Freudian slip reveals the gap between the rhetoric and reality of children's affinity with nature. It was one thing to mythologize the connection if one published nostalgic country romance but quite another if one were a professional whose position was at all uncertain, whether "expert naturalist" or "female science schoolteacher." Affirming the importance of nature study would also have meant affirming the feminine influence on children and a feminized version of natural history, since most teachers were women.

A second question for debate among the nature study movement concerned its educational aims: Should the skills cultivated be observational or problem-solving? Many educators were obsessed with the importance of training children in observation, an ability that was supposed to cultivate moral values of honesty, persistence, and accuracy.[72] Karl Pearson's then-recent *Grammar of Science* recommended science first and foremost for the value of "the efficient mental training it provides for the citizen."[73] For many, observation had less of a utilitarian, and more of a spiritual, rural-romantic dimension: "Its peculiar function is to develop the perceptive powers, and through this development bring the child into an intimate and sympathetic relationship with his surroundings."[74] Uncle John's writings often began with a photograph and the question, "Can you see . . . ?" and the Comstocks' insect guides emphasized the challenging matter of wing venation as the key to advanced observation. Observational beehives and formicaria were especially recommended for the schoolroom as convenient, interesting, and morally appropriate items for observation over an extended period. Even the president of Harvard University, Charles Eliot, wrote in to encourage the *Nature-Study Review* in promotion of observation as a supplement to book-learning.[75]

Indeed, many educators were utterly unbending on the exclusive primacy of observation. "The child must be taught to see before he is taught to explain . . . for this reason many of the observations the pupil is called upon to make in these lessons bear upon no conclusion."[76] Children were not allowed to become etiolated in character through false encouragement that they were making new discoveries: only observations were possible. An emphasis on observation could be a way of cautiously limiting nature study's pretences vis-à-vis proper science. Maurice Bigelow emphasized that nature study "deals with facts primarily for their own sake without particular regard to organization into a system; on the contrary modern natural science deals with facts primarily as they stand related to generalizations."[77] But mostly the arguments were positive ones regarding the moral and practical benefits of developing observational skills in children.

Again, even Wheeler echoed these aims, writing while director of the Milwaukee Museum, "The Museum should be a place where everyone . . . [learns] something about the material objects of the wonderful planet on which we live. But the extracting of knowledge from objects requires skill in observing. So important is such skill that it is no exaggeration to say that the man who observes the things about him with greatest accuracy will lead the most successful life, whether he be artisan or scientist . . . Such training must be one of the prime objects of our common schools."[78] His attitude had been shaped by schooling that was ahead of the nature study trend; the principal of the "Englemann Schule" that he attended had come direct from Fröbel's Germany, and created a museum (later developed by Henry Ward) so that the children might be given object lessons in natural history.

Other educationalists under the influence of Pestalozzi and Fröbel disparaged pure observation as a dry and pointless activity.[79] A New York professor of education typified the useless detail thus gathered in the following, apparently genuine fifth-grade description of an oak leaf: "Size, 7½ inches long; 4 inches widest part; shape, somewhat oval—widest at top; lobes, alternate, long pointed, 10 lobes on leaf; indentation, 10 indentations, rounded, deep, alternate; petiole, short, thick, dark brown, mid-vein thinner near top of leaf; veins, alternate, thin; color, dark brown, near mid-vein."[80] The description is more than a little reminiscent of Gradgrind's "gramnivorous quadruped." Besides, argued others, children are already natural observers; the challenge is to teach them something else. "The teacher of the first standard of a public elementary school . . . asked whether a cat has feathers, scales or fur on its back, and was rewarded by 'Lor blimey, miss! Ain't you never seen a cat?' from a boy in the front row."[81]

Like the question of the child's empathy with nature, the question of observational skills was related to the scientificity of nature study. Three biologists connected with the ANSS concluded that both science and nature study were essentially concerned with problem solving.[82] This view was in line with the majority conclusion once the ANSS was well established: that nature study was the infant form of science. Nature study was for the child, science for the adult; the skills developed in the former could be expressed to their full potential in the latter.

Whether advocating pure observation or self-directed discovery, the common point of agreement between commentators was that didacticism was not the way to teach nature study. It was not a set of facts to be conveyed but a method to be absorbed, whether utilitarian or sympathetic, observational or problem-solving. The role of the teacher was key in either case, and the potential to develop true science by the elementary teacher was asserted. Some scientists also saw their interests served by guiding and protecting this process and the definition of science that it incorporated. W. M. Wheeler was likely among the audience that heard a talk at the American Society of Naturalists in 1907 praising the wave of nature study and its promotion of science and biology but decrying the "hurricane of devastation" it was causing through an imbalance of enthusiasm and training. The speaker praised the potential of nature study to develop a scientific spirit in the child and highlighted the role of scientists and naturalists in helping the movement steer a course between the extremes of utilitarianism and sentimentalism.[83]

At another extreme, the harshest critics of nature study implicitly argued that there was not enough observation, or that the observation was of the wrong sort, and that imagination was given too free a reign. A biology professor at Chicago complained that nature studies trained only poorly in observation, and that from this children were encouraged to make a hasty leap into a sentimental love of nature. Imagination was encouraged too much and too sloppily. "Nature study, imbedded as it is in conventional education, is the one chance for exact and independent observation, for cultivating the ideas that between cause and effect there can be no hiatus . . . and that there should be no playing fast and loose with the truth."[84] Of all the issues in teaching nature study, the place of sentiment was the thorniest, perhaps because questions about its role tapped into deep conflicts about the nature of nature and its connection to humans; it was still being discussed by the ANSS in 1936.[85]

At the disreputable end of the sentimentality spectrum came such vocal and memorable figures as Mrs. Suckling of the Humanitarian League. Mrs. Suckling, by contributing financially to a conference on nature study in schools, was able to buy time at the podium. She explained her methods of inculcating kindness in children—"Nature songs," "the Pledge of Kindness"—which, if Mrs. Suckling was to be believed, apparently had remarkable success. Even the toughest of boys supposedly learned to love the "dear, delightful creatures" of nature. The majority of the conference, teachers and educationalists, objected to such talk and gave that well-intentioned lady decidedly short shrift.[86]

Sympathy was but a short step from anthropomorphism, another besetting sin of natural history (Wheeler called it the "eighth mortal sin").[87] Yet here too the dividing lines were not as clear as some might have liked to think. William J. Long's tales of animal life, and in particular his claim that woodcocks could improvise leg plasters out of mud, provoked a major and acrimonious "nature-fake" debate in 1903, drawing in the naturalists John Burroughs and Theodore Roosevelt (both sponsors of the nature study movement) and W. M. Wheeler, among others.[88] Burroughs' and Wheeler's objections to the wood-folk pieces were twofold. They found the sentimentality of their anthropomorphism distasteful, and they criticized their psychological stance. The problem was that Seton, Long, et al. attributed conscious thought processes to their subjects. Long wrote some rather sophisticated responses to such charges. In *Mother Nature* he cleverly suggested that the cruelty and competitiveness commonly attributed to nature were themselves anthropomorphic, and that in their place he merely presented an equivalent, cooperative account of nature's order—something that Wheeler himself advocated. What is interesting about Wheeler's criticisms of Long is that they appeared in specialist scientific journals, not popular or nature study–related publications. Likewise his antifaking essay, "The Obligations of the Student of Animal Behavior," was published in the ornithological journal *The Auk*, most likely to get at his then colleague at the American Museum of Natural History, Frank Chapman, who had been involved in the nature study movement.

Wheeler was concerned to distance himself from anthropomorphism in the eyes of fellow scientists; he was not concerned with putting elementary schoolteachers straight. In the eyes of the former his field methodology might appear rather too similar to anecdotalism; his insect psychology and cooperative view of nature might look rather too like Long's.

Many writers in the *Nature-Study Review* were concerned to avoid charges of anthropomorphism and undue sentimentality leaking into the nature study movement and adversely affecting its teachers' claims to professionalism. Inaccuracies in books pretending to present valid nature study were lambasted in its pages.[89] Even professional biologists could be accused of nature faking. C. F. Holder and David Starr Jordan's *Fish Stories* was dismissed as containing "masterpieces of nature-faking."[90] Vernon Kellogg, a respected entomologist in scholarly circles, was anxious to dissociate his support of the nature study movement from childish sentimentality. His letters to John Comstock bemoaned the lack of microscopes at the recently founded Stanford University. Worse, his associate, Mrs. Mailland, had been using this equipment unscientifically to look at spiders. "She finds suggestions for child study in them!" he complained. "Thinks their little 'spinning tails' are cute. 'Spinning tails,' Oh Lord!"[91] The desire to combine sympathetic characterization of animals with factual accuracy sometimes produced unlikely styles. A California supervising principal of schools wrote about children developing a friendship with the ants in their garden. Though the ants' ability to speak caused him no qualms of realism, he was apparently scrupulous to avoid the accusation that the ants were better educated than the average insect. When asked difficult questions, the queen ant hedged with answers like "it may be" and, with a remarkable display of intellectual rigor for an invertebrate, directed the children to further reading.[92]

Some nature study writers and teachers were positive about certain sentimental values in nature stories for very young children. At times the *Nature-Study Review* commented favorably on Long and Seton as a means to encourage sympathy in children.[93] A Canadian teacher concluded, "if any teachers of animal life have been distressed of late regarding the value of stories by Thompson-Seton, W. J. Long, Roberts, and others because of criticisms from men who speak from the standpoint of the hunter of animals, not from the standpoint of the sympathizer with animal life, let them rest assured that neither fact nor fiction is the aim in nature study, but sympathy for all that is good in life."[94] Another educationalist assured readers regarding his sentimental *Mother Nature's Children*: "Our purpose is to teach only the actual facts about nature. But as every fact is a fairy tale in the mind of the child, these facts will take 'form and limb' in a way that would make them untrue for us."⁹⁵ Anna Botsford Comstock's *Ways of the Six-Footed* followed an apparently teleological narrative, appealing to the "wisdom" of insects' actions, and was certainly anthropomorphic. Yet Comstock was an earnest advocate of evolution by natural selection of the fittest; for her, anthropomorphism and purpose were merely convenient literary tropes.⁹⁶ For some, then, a certain amount of sentiment was a way of distinguishing nature study from utilitarian agricultural education in the university extension programs.⁹⁷

Uncle John somewhat belatedly joined the nature-fake fray, criticizing the "yellow magazine writing" of the *Atlantic Monthly* and Doubleday's *Country Life in America* and *The Garden Magazine*. Yet he himself wrote such gems as "Sunny Pete and Mrs. Pete the Squirrels," in which Mrs. Pete was revealed to be, unusually for scientifically described rodent, a keen whist player. The difference, perhaps, between Uncle John's work and the yellow writing was that John's, though occasionally anthropomorphic, was very unsentimental. His nature study was about the inculcation of manly perseverance, industry, profit, and self-sufficiency, in both an economic and a political sense. His "Instructions for Organizing a Junior Naturalist Club" (1900) were less about nature than they were about an exercise in democracy; their main emphasis was not on the nature to be studied but on the importance of electing officers and having a constitution.

Various professional interests therefore shaped the debate over sentiment in much the same way as they shaped the debates concerning the place of observation and children's sympathetic relationship with nature. Some popular writers, counting dollars rather than citations, were perfectly happy to write whatever sentiment would sell, cashing in on romantic notions of childhood, nature, and morality. Some readers, including some teachers, were happy to accept this material. Some educationalists, seeking to gain ground in terms of education methodology rather than science, were also willing to accept the "unscientific" if it encouraged qualities that they wanted to see in the classroom—whether sympathy, imagination, self-improvement, or democracy. Intellectually ambitious teachers and naturalists had most to lose, and in their different ways and on their respective platforms they distanced themselves from sentimental and anthropomorphic writing. W. M. Wheeler fell into the category of intellectually ambitious naturalists. Eschewing the practical responsibilities of the economic entomologist, he aimed to carve out an unheard-of niche for himself, that of professional natural historian. In essaying this task, he too would have to respond to issues of observation, sentiment, and anthropomorphism, in particular delimiting the value of amateur entomologists in these respects.

CHAPTER SIX

Writing Elite Natural History

"Naturalists may attempt to achieve a scientific objectivity toward the creatures they study, but fortunately for editors they invariably fail." With this sneaky compliment Alan Ternes, editor of *Natural History* magazine, introduced a collection of essays by the staff of the American Museum of Natural History.[1] Scientific objectivity was an issue pertinent to an area, popular natural history, with which Wheeler was well acquainted. A number of his students went on to write popular natural history, and of those that did not, a significant proportion wrote unusually readable book versions of their research.[2] Wheeler's early amateur associates included the showman Henry A. Ward; later friendships with popularizers included William Beebe, David Fairchild, and Thomas Barbour. He often praised the work of amateurs, notably the British Army major R. W. G. Hingston. Wheeler even confessed to enjoying the forbidden aspect of natural history invoked by Ternes, ruefully categorizing anthropomorphism as "the eighth mortal sin."[3]

While it was relatively easy for the professional entomologist to differentiate himself from children's writing and nature study, adult writers or readers of sentimental nature presented more of a problem. Besides the inherent appeal of sentiment in all its guises, entomology still, even in the twentieth century, depended on the observations of adult amateurs for large quantities of raw data. In other words, it was not only the readers of myrmecological literature but also its contributors that blurred the boundaries of "mere natural history" and professional science. Having laid out in the previous chapter the generic background against which the entomologists were judged, I examine here their very careful intertextual placement with respect to that background. Wheeler in particular attempted to create a literary-methodological genre of elite natural history through a variety of literary maneuvers: delimiting the usefulness of amateurs, conflating critiques of their work, and praising them in carefully constrained ways that suited his own agenda.

The Problem of Fabre

The problem of generic placement was presented most forcefully by Jean-Henri Fabre (1823–1915), a phenomenally popular writer of insect observations and meditations, or, to use his preferred term, *souvenirs*. Fabre was in his later years a recluse living in the south of France, and made his living by writing popular books of science. His *Souvenirs entomologiques* (published in ten volumes in 1879–1907) were wildly popular around the globe, reaching even larger audiences in the ten years after the author's death. They were read by adults and children alike, though publishers also produced bowdlerized versions specifically for children. If an ordinary American knew anything about insects and their ways (apart from the need to eradicate them, which he would have learned in agricultural classes), chances were the knowledge came from Fabre. Yet despite Fabre's refusal to be incorporated into the disciplined networks of science, much about his method, both scientific and literary, resonated with Wheeler.

Wheeler had a good deal of sympathy for Fabre's approach and his popular advocacy of the study of living subjects. At the Woods Hole Marine Biological Laboratory in the summer of 1910, among morphological and physiological contemporaries—including, probably, the arch-reductionist Jacques Loeb—Wheeler gave his "superorganism" talk emphasizing the value of natural history as the field study of living organisms. He boldly claimed that such was in fact the future of biology, not its past:

> Twenty years ago we were captivated by the morphology of the organism, now its behavior occupies the foreground of our attention. Once we thought we were seriously studying biology when we were scrutinizing paraffine sections of animals and plants or dried specimens mounted on pins or pressed between blotting paper; now we are sure that we were studying merely the exuviae of organisms, the effete residua of the life-process . . . It is certain that whatever changes may overtake biology in the future, we must henceforth grapple with the organism as a dynamic agency . . . In using the term organism, therefore, I shall drop the adjective "living," since I do not regard pickled animals or dried plants as organisms.[4]

In the strength of this claim, Wheeler echoed Fabre. In one extended piece of polemic, Fabre inveighed against marine laboratories, founded at great expense, where sea animals were cut up, and which "scorn[ed] the little land animal which lives in constant touch with us, which provides universal psychology with documents of inestimable value." Fabre asked, "When shall we have an entomological laboratory for the study not of the dead insect, steeped in alcohol, but of the living insect; a laboratory having for its object the instinct, the habits, the manner of living, the work, the struggles, the propagation of that little world with which agriculture and philosophy have most seriously to reckon?"[5]

Although Wheeler never disputed the value of places like the *Stazione* (indeed, he was a regular at Woods Hole), he agreed with Fabre's defense of the study of living insects. And, like Fabre, he did cultivate something like a personal relationship with his insects, based on nostalgia:

> Every specimen has its history, written or remembered, oftenest the latter. This history is made up of a vivid mental landscape of the spot where the specimen was found, all the little incidents connected with its finding . . . If transferred to another merely its anatomical or structural value is left, the subtler poetry or lively incident and the many pleasant emotions which trooped into the collector's mind at the sight of a specimen, long sought and at last obtained only with difficulty, have forever vanished like an odor in the wind.[6]

Wheeler's challenge, therefore, was to transform the interest in live observation that he shared with Fabre into some kind of science.[7] In 1902, Wheeler wrote a piece entitled "'Natural History,' 'Œcology,' or 'Ethology?'" in which he pondered what name should be given to his area of study, animal behavior and related subjects. He concluded that "natural history" was rather too broad and ended up settling on "ethology," listing Fabre as among those in favor of the live observation that would be the basis of this "upsurging" science.[8] Wheeler's obituary of Fabre went further:

> [Fabre] alone realized the great significance of the study of animal behavior at a time when other biologists were involved in purely morphological work . . . Not only was Fabre to realize the full important study of animal behavior but he was the first consistently to apply the experimental method to the investigation of the animal mind. The "Souvenirs" abound in accounts of experiments, performed for the purpose of elucidating the nature of instinct, not the less illuminating and conclusive because they were carried out with crude, home-made apparatus. It is as instructive as it is humiliating to read his results and to reflect on the mountains of complicated apparatus of modern laboratories and the ridiculous mice in the form of results which only too frequently issue from the travail of "research."[9]

Another of Fabre's laudable characteristics was that he possessed to "an extraordinary degree the gifts of a virile and penetrating observer." Wheeler's praise in this respect echoed the agreement of the nature study movement that observation was a crucial—probably *the* crucial—skill of the naturalist. His emphasis on the gendered aspect of this assessment was reaffirmed by his positive review of Peattie's book, *Green Laurels*. He praised its astute characterization of the practitioners of natural history, which, as we have seen, were decidedly manly. He relished *Green Laurels* for its definition of natural history as a historically justified form of investigation in virtue of its masculinity.[10] Wheeler contrasted masculine exploration and collection with the feminized "collecting-mania" that afflicted some so-called naturalists. This trait was no better than that occurring in "little girls passionately fond of collecting useless buttons and business cards and in the fashionable lady who stores her house with old crockery."[11] Wheeler was, to put it mildly, no feminist. Compared with the Comstocks, Wheeler had very few female entomological correspondents, and it was with reluctance that he fielded inquiries from schoolmarms and entomologizing ladies. Few women were encouraged in his laboratories. For Wheeler, promoting natural history as a science entailed protecting it against the threat of femininity, embodied in the undisciplined practice of collection.

Fabre's deist spirituality was eminently suited to an American public that so eagerly consumed the poetry of Liberty Hyde Bailey and the philosophical musings of John Burroughs. Fabre laid out his natural theological stance at the start of the eighth volume of the *Souvenirs entomologiques:* a row of lilacs formed his chapel, his daily observations were his rosary, and an "oh!" of admiration was his only prayer. Fabre was condemned by professional biologists as a flat-earther, blinded by religion, because of his rejection of the transmutation of species. But his writing contained no orthodox religiosity whatever. Ironically, his retreat to Serignan was precipitated in part by a teaching scandal, a dismissal from his post on account of the shockingly secular nature of his curriculum.[12] Wheeler himself, though fiercely anticlerical, was not above applauding a kind of spirituality resulting from contact with nature.[13] After his mental breakdown he grew increasingly interested in the meaning of life's experience, instigating a kind of heterodox "Sunday School" with some of his Harvard pals.[14] Wheeler was not so far from urbanite consumers of Thoreau and Emerson in the pleasure he took in his escapes to the wild on collecting trips, where he reveled in "Nature, Nature everywhere, unobstructed, unobscured!"[15] And it was more than an enjoyment of landscape; spiritual qualities were developed through this process of collection: "One of the most important . . . mental benefits is the development of that grand deep feeling of sympathy, which is taught to extend itself to the smallest and most insignificant living object, yes, even to the inanimate stone. The feeling that one is merely a fragment of the universe, a very tiny fragment, co-existent with countless other such tiny fragments, all struggling to live, needs cultivating."[16]

Notwithstanding these points of sympathy, Wheeler was aware that he trod a dangerous line in praising Fabre, for the Frenchman had his embarrassing characteristics too: reclusive, antievolutionist, and a master of purple prose. As he prepared to write *Demons of the Dust* (1931), Wheeler reported: "some of my councilors suggested, with a just perceptible curl of the lip, that Fabre and his great popular success might be worth emulating."[17] Wheeler's remarks take on an additional significance because *Demons* was a book that explicitly aimed to engage the nonspecialist reader. Even when an entomologist tried to contrive some public appeal, it seemed, Fabre was an unworthy model. Thus Wheeler's partially favorable obituary of Fabre in the *Journal of Animal Behavior* also made sure to smear his name, portraying him as mistaken in most of his theories and personally unattractive to boot.

One problem was that Fabre identified himself with the apparent insignificance of his subject matter, presenting himself as a humble student of nature's humblest creations. Fabre used this trope to calculated naivist effect: "whoever humbles himself like this child is the greatest in the kingdom of heaven."[18] His use of supposedly juvenile language was defended as "the right way for the simple to understand one another," a turn of phrase that covered the insect and the observer, as well as the author and the reader. Although the childlike qualities of naturalists were in some cases desirable, this identification of the bug hunter with his bugs was a troublesome stereotype afflicting insect specialists. Even the economic entomologists still had to struggle with the ridicule that was attracted by this perception.[19] Noneconomic entomologists, lacking the utility argument to justify their labors, had to work even harder to distance themselves from their objects of study and so achieve credibility amongst their biologist peers.

Fabre's claims to "juvenile" language were debatable. He tended to write prose of the purplest variety, a tendency that was only exacerbated for English-language readers by his translators. Here he describes a mantis in typically florid terms: "A pointed face, with walrus moustaches furnished by the palpi; large goggle eyes; between them, a dirk, a halberd blade; and, on the forehead, an extravagant head-dress that juts forward, spreading right and left into peaked wings and cleft along the top. What does the Devilkin want with that monstrous pointed cap, than which no wise man of the East, no astrologer of old ever wore a more splendiferous?"[20]

Most commentators, however, were convinced by Fabre's claims of childlike truth-telling. The implication was that the reader would observe all the same things as Fabre, were he in Fabre's place: "[T]he most precious of [Fabre's] natural gifts was certainly that . . . [h]e excelled in persuading his readers to participate in the lucidity of his analyses. His descriptions are quite devoid of the subjective element . . . [he] puts himself in the reader's place and retouches his text until it becomes, so to speak, a faithful and concise image."[21] This manufacture of authoriality in order to make the writer disappear from the text, and then let the reader read nature directly, has been discussed by Simon Ryan in *The Cartographic Eye* (1996) as the mark of travel writing.[22] For Fabre and many nature study writers, the author was not an explorer but an observer or an experimenter; nevertheless, the integration of actor and narrator was essential to make the book seem as though it interpreted nature directly to the reader.[23] Here were more undesirable traits for the professional entomologist: overdone descriptions, or, on the other hand, textual self-effacement of the author. In the former case, he would not appear serious, and in the latter he might not appear at all.

Fabre rejected all the institutions of science and considered that his work was morally superior to the output of professional scientists. He and his critics were at least united in hoping that people would not read his books and think they were science. Nature study writers and educators, on the other hand, were generally very well disposed toward Fabre, finding that his approach and presentation meshed very well with their own. Insofar as this group was by the 1910s attempting to present itself as scientific, there was a danger that they would downgrade the seriousness of entomology by presenting it to the next generation in this way. In response to this dangerously solitary and eccentric science, Wheeler came to pay great attention to the practice of citation. William Beebe identified Wheeler's practice of almost line-specific references as being ahead of normal protocol in scientific writing, and he highlighted Fabre's failure to refer to others' research as Fabre's greatest sin.[24]

Other theoretical entomologists shared Wheeler's ambivalent reaction to Fabre. When the Société Entomologique de France published its centenary book in 1932, a brief, anonymous biography was inserted that safely celebrated Fabre's *literary* genius only, while in another the writer Jean Rostand steered carefully between acknowledging this and downplaying Fabre's scientific credentials.[25] In general, then, Fabre was acknowledged as a master of observation and its literary retelling and dismissed as a biologist. All in all, it would not do for the entomologists to be too much like Fabre, the object of polite scorn from the scientific community despite (and partly because of) his popular success. True, he brought fame to their otherwise overlooked subject matter, but he also attracted ridicule by virtue of his antievolutionism and his florid naivist style. The challenge for the noneconomic entomologists was to resolve the tension between popularization as a means to promote their field and the threat that populism posed to their professional ambitions vis-à-vis their university biologist peers.

Dealing with Fabre: The Conflation of Sentiment, Teleology, and Anthropomorphism

Of course, the biggest problem with Fabre was that his natural theological accounts of insect "wisdom" were theoretically risible among biologists. At least three critiques were interwoven in this complaint: sentiment, natural-theological teleology, and anthropomorphism. Fabre's solitary wasp had a God-given wisdom—instinct—according to which she knew and planned the care of her beloved offspring. Or so it seemed.

Fabre's alleged sentimentality cannot be accepted without question. The charge implies a particular set of sentiments coloring the description of animals: endeared, tender, soft-hearted. Nothing could be farther from the truth in Fabre's descriptions of praying mantises, for example. Having described their duels to the death ("the swelling of the ovaries perverted my flock, and infected them with an insane desire to devour one another"), he went on to a "yet more revolting extreme." After their "lengthy embrace" they were "made one flesh in a much more intimate fashion" when the female "methodically devour[ed] [her husband], mouthful by mouthful, leaving only the wings." Fabre's analysis of the action was that "here we have no case of jealousy, but simply a depraved taste." Next, he wondered to himself about the reception of a second male: "The result of my inquiry was scandalous. The Mantis in only too many cases is never sated with embraces and conjugal feasts . . . in the course of two weeks I have seen the same Mantis treat seven husbands in this fashion. She admitted all to her embraces, and all paid for the nuptial ecstasy with their lives . . . insects can hardly be accused of sentimentality; but to devour [the husband] during the act surpasses anything that the most morbid mind could imagine. I have seen the thing with my own eyes, and I have not yet recovered from the surprise."[26]

Sentimental anthropomorphism was casually conflated with the Victorian fashion for drawing morals from the animal world. Here too the charge was unfounded when it came to Fabre. Typically, the example of maternal care in the animal kingdom was used to reinforce its importance for the human family or its value to human society. Though Fabre at times came close to making points like these, at others it is hard to imagine any but the most perverse moral lessons being extracted from his insects. Fabre's fascination with insects, which so many found infectious, was a fascination with them as "other" to himself. This fact stands in opposition to any unsophisticated charge of moral anthropomorphism. Beasts were to Fabre unalterably bestial, as may be seen in his version of one theme that entomologists took up in the early twentieth century, namely, the effect of the stomach on social life. Auguste Forel wrote that the mutual feeding of worker ants was the mark of their utopian communality, but for Fabre, "the taint of the belly" was quite literally the "mark of the beast."[27] While Forel believed that emulation of these ways of ants would bring about "true universal fraternity," Fabre wrote that human progress was limited by the tyranny of the gut. "The intestine rules the world," Fabre declared, since there must always be war to defend food; Forel considered that armies could be relegated to the realm of history. Fabre's frontispiece and cover illustration to *Social Life in the Insect World* (1912) are visually very similar to Forel's *Social Life of the Ants Compared with that of Man* (1921). However, where Forel has mutual feeding in his design (the social stomach), Fabre has two mantises locked in mortal combat. In an era when social insects—the most "human" insects—grew so much in importance, it is noteworthy that Fabre resolutely ignored them. This, like his use of the mantis image, was in keeping with Fabre's reluctance to draw strong analogies between the nature of humans and the nature of insects.

Fabre's insects are quite alien to human life, and any points of apparent similarity serve only to remind the reader of his minor place in the vastness of nature and of the animalistic elements within his own, supposedly civilized, self. Thus Fabre sets up one passage describing the cruel "massacre" of the bird-trapper in such a manner as to make the reader identify this man as abhorrent. He then compares the *oiseleur* to the Epeira spider, making the latter also "other" to human sensibilities. But then Fabre hits home with the punchline: this otherish man *is* an animal. "*Il ya dans les veines de l'homme du sang de bête fauve*" ("The blood of wild beasts flows in the veins of man").[28] Elsewhere, Fabre concludes a chapter on his education by remarking "*l'instinct est le génie de la bête*" ("instinct is the genius of the beast").[29] This likens the insect to another otherish human—not impossibly cruel this time, but impossibly brilliant.

There were other ways to read the human-animal analogy of mind. Some writers produced anthropomorphic accounts that were more radical than mechanistic ones, in the sense that they were actually theriomorphic, showing how human psychology was just an extension of the insects' psychic faculties.[30] In France, J. C. W. Illiger and J. T. Lacordaire had seen reasoning faculties in the insects comparable with those of man.[31] Annie Besant's translation of Büchner's *Mind in Animals* reveled in its rebuke to the "educated:"[32] "[The ants'] cleverness in discovering [food sources] is so great, that as the instinct-mongers say, their instinct "borders on human reason;" in reality it is reason, and often surpasses the human acuteness which it is thus vainly sought to defend."[33]

These radical writers earned the ire of religious antievolutionists, for they raised the beasts dangerously close to man in the moral order of things. Jack London was another such writer; his *Call of the Wild* was on one level an anthropomorphic story about a dog, but on another an amoral, theriomorphic tale about how humans are subject to the laws of survival of the fittest. What was condemned as anthropomorphism might actually be politically challenging for reasons quite opposite to genteel Christian sentiment. Fabre's British publishers T. Fisher Unwin repackaged those of his essays that touched on the instinct question as a single volume, indicating interest in animal psychology to adjudicate on such theological questions.[34] Fabre himself, though not a proponent of orthodox Christianity, countered radicals' opinions that insects exercised intelligence, and dismissed those who tried to "find the origin of reason in the dregs of the animal kingdom."[35] In refusing to discuss ants, Fabre also stayed away from the issue, since these were the paradigmatic insects of the instinct-or-intelligence conundrum.[36] He thus distanced himself from the whole intelligence issue, though oddly enough because he thought this elevated the place of humankind in creation, and not because, as radicals intended, it would lower humans nearer the animals.

Anthropomorphism was a slippery charge. John Burroughs accused London and Seton of the vice in the nature-fake furor. Yet his 1887 notes on bees could not be described in any other way: "When a bee gets back from good food the others 'smell out the secret.' No doubt, also, there are plenty of good gossips about a hive that note and tell everything. 'Oh, did you see that? Peggy Mel came in a few moments ago in great haste, and one of the upstairs packers says she was loaded till she groaned with apple-blossom honey which she deposited.'"[37]

A teacher admiringly wrote: "Burroughs' way of investing beasts, birds, insects and inanimate things with human motives is very pleasing to children. They like to trace analogies between the human and the irrational."[38] Even while the furor was still fresh in his mind, Burroughs heartily commended the Peckhams' *Wasps Social and Solitary* (1905) in an unashamedly anthropomorphic style: "[This book] is a wonderful record of patient, exact, and loving observation, which has all the interest of a romance. It opens up a world of Lilliput right at our feet, wherein the little people amuse and delight us with their curious human foibles and whimsicalities, and surprise us with their intelligence and individuality."[39]

Charges that Fabre was "anthropomorphic" could actually concern other issues. The celebrated account of the mother *Philanthus* and her careful provision for her unhatched offspring was supposed to be flawed in this respect. It is important to notice, however, that the accusation of anthropomorphism was interwoven with one of teleology: not only did the *Philanthus* "know" something, she also "planned" her actions in the light of that knowledge. But Fabre was very careful to deny that insects "knew" anything of their actions, still less their reasons.[40]

Another irony of such debates was that mechanists could produce accounts empirically similar to Fabre's versions. Naturalists had no time for these mechanists, however, because they overlooked the wonder of the animal kingdom—precisely what was condemned as sentiment in Fabre. Natural theologians, radicals, mechanists, and "expert naturalists" each had accusations concerning each other's work; these did not, however, add up to a consistent negative image that could be reversed to produce a positive picture of their own work. Rather they were deployed in an ad hoc—often ad hominem—manner. The apparent division of writing into scientific (unsentimental, not anthropomorphic) and merely natural historical (sentimental, anthropomorphic) was, and is, simplistic. In the early twentieth century, this division reflected careful generic placement by expert entomologists.

Making Use of the Observers

The expert entomologists, or expert naturalists, did not want to erase the amateurs all together. The Entomological Society of America, for example, welcomed contributions to the science made by amateur collectors. Amateurs could also provide recruits to the field or patrons for new universities and their entomology departments. In the context of literary natural history, amateurs also constituted a significant market for the publications of the elite. Fabre in particular whetted the appetites of a general audience for insect psychology with his accounts of mysterious, prescient instinct and paved the way for other writers, including the Belgian *littérateur* Maurice Maeterlinck ("discoverer" of Fabre at the turn of the century), whose lives of the insects dealt with their respective psychologies at length.[41] There was even a play based on Fabre's idea of insect psychology, first performed in 1923.[42]

Since elite writers could not escape the psychological slant expected by their readers, most popularizing entomology books had at least one chapter on insect psychology, usually called by some variation of "Instinct and Intelligence." Nor did the elite wish to throw off the science's psychological connotations. Insect psychology was a prominent (perhaps the most prominent) feature of nonapplied entomology in the late nineteenth century and early twentieth century. Even if we cannot point to an intentional adoption of psychology in order to prop up theoretical entomology as a discipline in its own right, there is no doubt that between 1880 and 1930, nonapplied entomology was most strongly identified by its focus on insect psychology. This became its guiding program de facto and a means to distinguish itself from the amateurs, the "mere observers."

A considerable number of amateur entomologists participated in insect psychology during this period.[43] Two important students of insects were George William Peckham (1845–1914) and Elizabeth Gifford Peckham (1854–1940), who lived in Milwaukee.[44] George was principal of the public high school that hired Wheeler to teach physiology, and he studied insects with his wife in their free time. They published together prolifically between the years 1882 and 1909. The Peckhams conducted their research in the gardens of an acquaintance with the assistance of their children, all of which gave their accounts a rather homely feel. Acknowledging their amateurism, they accepted professional judgment in the production of their books. The specialists whom they cited in the course of their books were mentioned very respectfully, more so than in contemporary publications of disciplined science. Despite having a fairly adventurous evolutionary perspective on their studies, the Peckhams relied heavily on experts, implicitly defining themselves as amateurs, in order to produce their books of entomological natural history.

Philip Rau (1885–1948) and Nellie Lois Rau (1885–1972) were another well-known pair of amateurs who spent time researching and writing together.[45] Philip was forced out of formal education through the death of his parents at the age of ten, and in adulthood became a storekeeper; Nellie, however, was a university graduate. They published a number of papers together in the local *Transactions of the Academy of Science of St. Louis;* Philip published alone in various national journals. Nellie and Phil Rau spent the early 1910s tracking wasps in and around the city of St. Louis, Missouri, especially in the well-worn baseball diamond occupying a nearby vacant lot. An account of this work was published as *Wasp Studies Afield* (1918). Like the Peckhams, they produced entomological knowledge in homely locations. Though Philip Rau managed to join scientific entomologizing trips and had ambitions to articulate novel evolutionary theory, he remained better known in his domestic context.

Other "humble amateurs" of the era included an African American, Charles H. Turner, as well as a number of women, notably Adele M. Fielde and Edith Patch.[46] All researched insects in their backyards and gardens near their homes, publishing their observations in local journals and low- to middlebrow books. Collectively, they may be thought of as domestic entomologists. Some were associated with the ANSS; Patch was president of the society in 1937–1938, while Rau made the profits from his *Jungle Bees and Wasps* over to the ANSS scholarship fund.

Apart from Fabre, most of the domestic entomologists interested in behavior were North American. (By contrast, homely or amateur entomologists in England during this period tended to be collectors and amateur taxonomists rather than observers of behavior.[47]) Why should this have been so? The historian W. Conner Sorensen suggests that the North Americans turned to the study of insect behavior in the absence of the European literature that would enable them to keep up with systematics.[48] Although he intends this explanation for the nineteenth century, it also works for socially marginal groups in North America right into the twentieth century. Their study of insects was limited by the inaccessibility of specialist literature and constrained by what they could not do—economic and academic entomology—by reason of gender, race, and poor education.

It is easier to ask why the domestic naturalists chose to study insects in particular. For one thing, insects lay close at hand—right in the backyard. For socially marginal groups that lacked funds for travel or access to laboratory resources, insects presented a rare opportunity for original research. The activity was compatible with the ideology of the nature study movement. "It is the glory of nature study that it deals with commonplace things; the plants of the dooryard, the animals of the wayside, the weeds of the waste places are its appropriate subject matter," a 1911 editorial in the *Nature-Study Review* claimed.[49] Within a homely setting, insects had an appropriate moral value for their investigators; the domesticating activity of naturalism worked reflexively on the agent and on his or her subject. By studying insects at home, the investigator imbued her subjects with domesticity. By allying herself with her insect subjects, she took on their humble characteristics. Like the earlier traveling naturalist entomologists, then, the domestic naturalists found in insects a source of moral reflection. While some balked at the socially conservative lessons they were expected to draw, others embraced the reflexive humility that was supposed to result from studying William Petty's "vilest" creatures. The language of the ingenuous and trustworthy traveler marked their writings, indicating that anyone could do as they did; no superior skills were claimed by the domestic entomologists.

In this respect, the domestic entomologists again echoed the proponents of nature study. Part of the latter's rhetoric was to emphasize that they had no pretensions that children might make new discoveries. Indeed, some were very stern on this point. Marion H. Carter insisted that children be taught their place through nature study: the teacher should supervene with greater knowledge—not her own, but from experts—about whatever children thought they had discovered.[50] By contrast, T. D. A. Cockerell of the University of Colorado claimed that children could add much to the corpus of science in areas such as insect-plant relations. "With the work of the Peckhams as a guide," he suggested, "there is no reason why any intelligent person, young or old, should not discover numerous new facts."[51] Cockerell's relative generosity may perhaps be compared with Carter's as an example of gender-specific professional confidence.[52] Cockerell was sufficiently secure in his role as expert not to feel the need to protect it from children.

It is striking that many of the domestic entomologists chose the Hymenoptera—ants, bees and wasps—for their most detailed studies. On the negative side, they were the most promising group that remained for original investigation. Butterflies had long been collected by the social elite and patrons of the traveling naturalists. Moreover, butterflies had been well catalogued and the function of their patterns elucidated. Pestiferous bugs, the territory of the professionals, had been equally well documented. On the positive side, the economically irrelevant (in the pestilential sense) Hymenoptera were morally promising contenders for study. They contrasted with the distinctly unsuitable Isoptera, which included destructive termites and distasteful cockroaches. The Hymenoptera also exhibited a fascinating range of behaviors. They included solitary wasps, whose "maternal" instincts in protecting and provisioning their eggs were easily constructed within a domestic idiom. Meanwhile, ants and bees provided a mystery in their methods of food location and route-finding back to the nest, not to mention their social arrangements, which might be represented as familial or civic.[53]

The solitary Hymenoptera were reasonably large, moved relatively slowly, and did so on their own. Thus they could readily be followed singly as they went about their business, which made it possible to construct them as individuals, an essential part of the process of domestication. Indeed, the works of the domestic entomologists frequently addressed the insects themselves in familiar terms, as if they were domestic pets. By contrast, Fabre despised the nonsolitary ants, with which he could not cultivate the day-by-day observational relationship he so prized. It is notable too that Fielde, the domestic entomologist who was most obviously unhappy about being categorized in this way, was an ant specialist.

It was for Wheeler to translate the work of the domestic entomologists to his own advantage, making sure that the connotations of humble, backyard amateurism did not contaminate his own work. As part of this translation Wheeler used psychological theory—closely allied to issues of anthropomorphism in discredited naturalists—and used it as a lever to raise himself above the mere natural historians.

In technical terms, Turner, the Peckhams, and the Raus provided the perfect bridge between Fabre's popular natural theological account of instinct and Wheeler's elitist one. Fabre had always studied solitary insects, whereas ants existed only in social form. Fabre's most famous accounts of insect behavior described the solitary wasps *Philanthus* and *Ammophila,* and therein lay a transition between solitary and social insects.[54] Wasps were well-known to be the phylogenetic progenitors of ants. Furthermore, some species of wasp—species not studied by Fabre—were known to have semisocial lifestyles. Thus they linked Fabre's solitary insects with the myrmecologists' social ones in terms of behavior and phylogeny—the two things that Forel had united in his seminal ant studies.[55] The Peckhams, the Raus, and Turner all studied the behavior of semisocial wasps, and Wheeler was able to use their work as the foundation for his theories about the evolution of social behavior.

The switch to social insects also allowed a change in authorial perspective and authority. Fabre cultivated a profoundly personal relationship with each of his living insects. In doing so, he domesticated them for representation as nature's perfect creatures, and the relation was naturally framed by a first-person narrative. It would have been harder to do all this with insects that apparently had self-sufficient relations with one another. Burroughs' animal narratives switched from a neutral third-person account to the anecdotal first-person singular and the didactic plural, while the Raus and Peckhams used a narrative first-person plural (because they had actually worked together). Wheeler, of course, wrote in the third person. Although this was standard for science, it can also be read as a product of the difference between writing about solitary and social insects. Their less domesticable, less personal nature presented the possibility of an "objective" stance—the mark of proper science—to practitioners of elite natural history.[56]

The Peckhams and the Raus also made it easy for Wheeler to put some distance between their writing and his own in their emulation of the self-effacing faux naiveté that marked Fabre's oeuvre. Fabre had always claimed to have closer friends among the insects than among his peers, and he constantly emphasized his humility in the face of nature's six-legged marvels. The Peckhams and the Raus invoked Fabre in this mode as their constant source of inspiration. Like Fabre, both couples stressed the domestic and humble nature of their work, and they involved their children in their research, just as Fabre had.

Wheeler, who like his peers had solved the Fabre problem by praising his humble observations but deploring his theories and his rejection of the scientific community, now did a similar thing to the domestic naturalists of North America. He applauded their patient, humble observational skills, thereby implicitly limiting them to these activities and debarring them from theoretical work. And what better place to do this than in the introduction to one of their books? Although the Peckhams—friends of Wheeler—had an introduction composed for *Wasps* by John Burroughs, Wheeler got his chance courtesy of the Raus' *Studies Afield*. Wheeler's foreword began with a fairly sophisticated summary of the psychological and evolutionary significance of vespine studies. Next, Wheeler discussed the problems of such work. These difficulties included the necessity of conducting research in the field, since "the solitary wasps have so many and such intricate relations with their inorganic and living environment [that] their study in the laboratory [is] impossible or, at any rate, very inadequate."[57] A more serious drawback, however, was the difficulty of performing proper experiments in the field; accurate observation was the most one could realistically hope to achieve. It was in this specific sense that Wheeler singled out his exemplary naturalists for praise: "Still results of considerable value have been obtained by simple field experimentation as will be apparent to the reader of the ingenious studies of Fabre, the Peckhams and the Raus."[58] In his carefully worded summary, Wheeler faintly praised the Raus while damning the limited value of their results.

Charles Henry Turner (1867–1923) was also praised and constricted in the same breath. Although he held a chair in biology at Clark University between 1892 and 1905, most of his career was spent teaching biology at various grade schools and teachers' colleges, for Turner was excluded from high-flying professional circles because of his race. Throughout the turmoil of his teaching years, Turner continued to research the behavior of insects at home and in the local parks.[59] One speaker at his memorial remarked, "Nature lovers and scientists cannot but feel grateful for Dr. Turner's admirable contributions. In making his studies he did not venture on lengthy and costly voyages to far-away countries teeming with fascinating allurements ... he had the ability to take the material that was near at hand and make the most of it."[60]

One can only speculate to what extent Turner would have agreed with this analysis, since he was compelled to purchase all his own equipment and books and was denied laboratory access in all of the institutions to which he was attached.[61] Despite Turner's prolific publishing record, it was a particular set of *personal* qualities that his obituarists celebrated. These qualities were the same as those aped by Fabre: "the humble simplicity of the truly great man disarms us quite completely," stated one of Turner's obituarists, "and we crane the neck to overlook exactly that which we seek."[62] Indeed, one might well say that this modest overlookability was exactly the moral quality that Fabre ascribed to his insect subjects. The African-American Turner represented a peculiar combination of intellectual respectability and "humble simplicity" which enabled him to generate a palatably domestic and domesticated entomology ideologically allied to Fabre's.

Wheeler, it will be recalled, starkly contrasted the easily organized mass of humanity to the individualistic scientist and artist. The former resembled the ants, whose highly socialized origins lay in maternalistic society, while the latter were its direct opposite. Amateur myrmecologists were accordingly granted a status somewhere in between true worker ants and their antithesis, the workers of science: "[A]mateurs . . . constitute a very large and important 'bloc' of investigators . . . The many members of our numerous natural history . . . clubs, who hold monthly meetings and contribute modestly but effectively to the sum of our knowledge, regard themselves as anything but 'solitary' workers. That designation would seem to be more applicable to some of the professionals in our universities and research institutions."[63]

The domestic naturalists thus represented an intermediary stage between "amateurism" and "professionalism" because in their faux naiveté they were domesticated, just as Fabre claimed to be. Fabre created *himself* in the mold of humble domesticity—a model that his supporters were then pleased to celebrate—and through this tamed nature for his natural theological purposes. The domestic naturalists, on the other hand, were tamed or constructed by the academic entomologists, who accepted their observations while valuing their personal over their professional or intellectual qualities. In this way Wheeler, Fairchild, Jordan, and others achieved a vicarious domestication of nature, without making themselves appear too humble. Wheeler praised the amateurs' observational abilities in a way that elided with moral (rather than scientific) approbation; in effect, he translated their research through an elitist matrix. These amateurs provided raw materials for his theorization without threatening his "professional" or expert status. Although early twentieth-century entomologists shared a polemic of natural history, they had also to achieve a scholarly status comparable with that of laboratory-based biologists. The domestic naturalists were a critical intermediary in the network between the values of natural history and of professional noneconomic entomology, propped up on its psychological claims.

Scientific Sheep, Anthropomorphic Goats

In literary terms, different people meant different things by natural history, and the divisions between it and fanciful writing, on the one hand, and scientific writing on the other were not at all clear or undisputed. David Starr Jordan founded an entomology department at Stanford University yet was accused of writing "nature-fakery"; the Comstocks sponsored Uncle John yet promoted an advanced study of insects. The same book could contain "quaint" insect myths, extremely precise taxonomy, and a casual mention of ants' "spitefulness."[64]

In this context, the separate names for nature study and natural history gain obvious significance. The professional entomologist Vernon Kellogg was wary of the former term and tried, unsuccessfully, to scotch its use. In a letter to John Comstock, he reported that he was putting on an "introduction to nature study" course at Stanford, as Comstock had suggested. But, he added, "I call it in the catalog the natural history of insects, an introduction to the study of natural history based on the observation of insect structure and habits."[65] At the first meeting of ANSS a dozen years later, Stanley Coulter declared, "It is doubtful, indeed, if any modern educational movement [besides nature study] has been so hampered by definition, so obstructed by material, so deflected by sentimentalism."[66] Ongoing debates in the *Nature-Study Review* were testament to the truth of this statement; sentiment was the besetting sin of nature study.

The fact that key issues of nature study—natural sympathy, observation—were worked out so as to retain primacy for the teacher, not books, gave a curious status to the mass of nature literature in the earlier part of the twentieth century. It was supported most by the very people who wanted to delimit its validity. Notwithstanding the efforts of the Comstocks, most teachers of nature study echoed Agassiz's celebrated dictum, "study nature, not books." At best, nature literature was a supplement to a method that only they could inculcate. Perhaps this self-effacing character of nature literature played into the hands of the would-be elite naturalists.

Proponents of nature study and writers of respectable popular natural history were fighting for ground with the entomological naturalists, or, as one might put it in contemporary marketing jargon, they suffered from insufficient product differentiation in their respective brands of natural history. This produced a good deal of self-consciousness on the part of writers in both genres. Each was anxious to rebut and deflect charges of sentiment, anecdote, and anthropomorphism. Sentiment and anthropomorphism, though often tangled together, could theoretically be separated if one wished. (In 1950, Krutch made humaneness toward animals an important strand in his Whiggish story of the development of nonsentimental nature literature.) Others could deliberately and consciously affect anthropomorphism but avoid sentimentality, or vice versa. It was in Wheeler's interest to attempt to run them together and write them all off as "mere" natural history or, in Burroughs' phrase, "mere literature."

In this way, the nature-fake dispute was resolved fairly easily for the fault-finding writers. They restored a clear division between acceptable and unacceptable nature writing by conflating anthropomorphism and childish sentimentality. Roosevelt dismissed it as insufficiently manly, Wheeler as insufficiently scientific. In effect, a tacit understanding was in operation that some anthropomorphism was a knowing literary flourish and therefore acceptable (even desirable), while for others it was sincerely meant and hence beyond the pale. But there were no transparent hermeneutic methods for drawing this distinction between the scientific sheep and the anthropomorphic goats. Networks of those in the know, those who defined acceptable nature literature, were sustained along academic and political lines. Bailey could be included among the sheep because of his status at Cornell, and Burroughs by virtue of his political connections with Roosevelt. Fabre was excluded because of his determined isolation from scientific networks, Long by virtue of his small-town preacher identity. Rhetorical techniques were employed to retrench this divide, which was centered on a distinction between mere observers, reliable in what they saw yet naive in their interpretation, and expert theorizers.

Many of the same rhetorical ploys and professional accusations continued to be made well into the twentieth century. Von Frisch and his dancing bees were clearly, in the mind of the general reader, heirs to Fabre. Von Frisch described some of Fabre's observations but did not mention him by name, perhaps because he was tainted by amateurism. If Fabre's sin was not to cite, his punishment was, by scientific karma, not to be cited. The charge of anecdotalism was also to remain a potent one. In the 1960s, allies and students of the ant specialist T. C. Schneirla were to condemn the new ethologists, Morris and Lorenz, as amateurs, in part on the grounds that all they did was tell stories.[67]

How successful was Wheeler in demarcating elite natural history from popular natural history and nature study? A comparison of the fortunes of the Entomological Society of America (ESA), American Association of Economic Entomologists (AAEE), and the ANSS provides a helpful answer to this question. Wheeler's appointment at Harvard in 1908, a mark of his professional success, coincided with some important dates in the history of these institutions. The first annual meeting of the ESA, chaired by J. H. Comstock, had been held just two years earlier, in 1906. The ESA was an alternative to the AAEE (founded as the Association of Economic Entomologists in 1889), with the aim of promoting nonapplied entomology—in other words, expert natural history. Some of the aims of the ESA, as one might expect, were compatible with the aims of nature study: to consider questions pertaining to the educational side of entomology, and to encourage the formation of local entomological societies. Others indicated the problematic status of nonapplied entomology; it was also intended to give some much needed "dignity" to the field. Despite the disputed disciplinary status of nonapplied entomology, the ESA initially had twice as many members as the AAEE had that same year. But this amounted to only 216 members, and by 1918 there were very few affiliated with both the AAEE and the ESA. The AAEE had grown dramatically and had become the larger of the two societies. The ANSS hit the ground running, with a larger membership than the ESA and a greater impact on official and unofficial educational policy. The ESA had only moderate success in forming a credible complement—let alone alternative—to the AAEE. Members of both, most notably John Comstock, made more of a historical impact on the nature study side. Thus, although two historical natural history societies were founded around the time that Wheeler defined his new, expert natural history discipline of myrmecology, they did not succeed terribly well. The specifically entomological ESA was eclipsed by its rival in applied science, while the educational ANSS was never a serious contender for recognition among professional scientists. Wheeler's success did not reach much beyond his personal connections, a fact emphasized by his appointment not at the Harvard main campus, but well removed from this at the Bussey Institute.

In literary terms, things were complicated by the fact that Wheeler, though undoubtedly elitist, apparently disagreed with the literary elitism of Babbitt and More. The latter, like Wheeler, commented critically on popular writing in the early twentieth century, but Wheeler embraced the Romantic notions of untutored genius that they rejected.[68] In upholding Romantic concepts of nature and its explorers, Wheeler was in a position to be vindicated by Peattie and Beebe in the 1930s and 1940s. The nature study movement had reached its heyday between about 1890 and 1914; by the 1930s it was moribund as an educational movement (though many outdoor clubs were still going strong). At this time the spirit of gardening promoted by the nature study movement gave way to a new natural aesthetic of the sublime, at least for adults. Nature ceased to consist of apples and caterpillars and came instead to be signified by sequoias and wolves.[69] The old connection of nature and children might have appeared to conflict with the new adult nature writing that emphasized masculinity. However, primitiveness was used as mediating quality that could be constructed as both childlike and tough.[70]

Wheeler's "reservations of the mind, great world parks" suggest both this new desire to preserve native landscape and the congruent character traits of the men who would study its contents. Indeed, that sympathetic recipient of Wheeler's thoughts on the primitive antisociality of scientists, David Fairchild, went on to create Florida's Tropical Everglades Park Association in 1929. Such parks were a physical instantiation of naturalistically inclined scientists' social aspirations and desired identities.

By connecting with values of masculinity and the frontier spirit, natural history could again find an audience in visitors to parks and readers. Arguably, this success came at the cost of further popularizing natural history. But by this time the professional status of myrmecology was changing again. The Second World War was about to interrupt the activities of biologists and natural historians. After the war, readers of popular natural history were to be members of the Rockefeller Foundation and Carnegie Institute. They would help take myrmecology in still another direction.

CHAPTER SEVEN

Ants in the Library: An Interlude

At this point the story of myrmecology takes an unexpected turn. As a discipline, it lacked direction after death of Wheeler in 1937. William Creighton's massive taxonomy of North American ants (1950) was of little interest to anyone besides collectors, and T. C. Schneirla's ant psychology of the late 1940s and early 1950s was unconvincing to those very taxonomic experts who had welcomed Creighton's book. It was not until E. O. Wilson that anyone managed to integrate taxonomic credibility with theoretical innovation, as Wheeler and Forel had done previously, and drag myrmecology by the scruff of its neck into a new era. But before myrmecology was reworked as the communication-based science adopted and promoted by Wilson, some unlikely scientific genealogies shaped its identity in the interim period. In particular, linguistic theory had a remarkable effect on the development of "hard" science through the twentieth century, for it was a pair of British linguists, C. K. Ogden and I. A. Richards, who formed the intellectual and social bridge between Forel, Wheeler, and Wilson.

Ogden nurtured an international circle of scientific readers, promoting an interdisciplinary approach to science and encouraging entomologists and theoretical biologists who would later be important to the cybernetics set. Between them, Ogden and Richards introduced Forel to the nonspecialist English-speaking world, advancing his theories on international language. They befriended Wheeler and encouraged his philosophical excursions, and finally, they established communication as the key feature of comparative psychology, thus ultimately setting Wilson on his path. The pair's interest in linguistics would also come to the fore as an approach to communication that was important in a cold war context. As it happened, Ogden had modeled communication on ants, just as Wilson was to do in his sociobiology. Richards' awareness of the myrmecological aspect of the project, though it did not show up in his own work, filtered through to that of his student, T. H. White, in his natural history of human behavior. Richards himself wrestled with instinct, that exemplar of the insect mind, in the making of his literary criticism. This, as will be seen, connected with entomological concerns in its relation to the wider problem of the mind of the masses. Moreover, Richards' early interest in literature as a social question anticipated the social concerns of the cybernetics set; this was also the spin put on Ogden's Basic English, much to the latter's chagrin. Between them, Ogden and Richards covered issues of language and representation, and the linguistic and psychological aspects of control, all of which were of interest to the cyberneticians.

The background on language provided by this chapter is important because it sets up some important questions regarding Wilson's communicational myrmecology: Who else was interested in language in the postwar period, and why? To what extent was this interest connected with the biological study of language (especially in ants)? Can it help to explain or at least contextualize Wilson's decision to treat ants as primarily linguistic entities?

The International Languages of Ogden and Richards

Something about ants appealed to, or inspired, researchers' interests in language. Or, to put the matter in less essentialist terms, issues of language were suggested by the study of ants in many of its early twentieth-century contexts. Adele Fielde, for example, who studied ants as a model of civic, feminist, international socialism, simultaneously promoted the study of Esperanto. Esperanto was to be a world language that permitted communication without misunderstanding. Just as Fielde's ants "spoke" to their nestmates with their specific odor, proving their amity, so humans would be able to avoid the kind of misunderstandings that led to war if they spoke the same language. In the summer of 1886, Auguste Forel removed the antennae from several species of ant that were normally mutually antagonistic and placed the ants all together. The results astonished him: "It might have been . . . Paradise, where cats, mice, foxes, lions and hens licked each other and drank from the same platter." Removing the antennae, the means to olfaction, prevented the expression of bellicosity, or rather, as Forel assumed, its generation, through communication of alien odor.[1] Later in life, Forel came to the same linguistic conclusion as Fielde; the slaughter of 1914–1918 convinced him of the necessity of Esperanto to "promote mutual understanding."[2] For Fielde and Forel, then, issues of language mapped onto issues of international understanding.

The childish fascination with watching ants and wondering what they were "saying" to each other as they waggled their antennae was, for Forel at least, also connected to the satisfaction of having a secret language of his own. In order to maintain a private world safe from his mother's prying eyes, he developed a code for his notes on ants and his private thoughts. At some point before leaving home, he shared the code with his favorite sister Blanche, permitting them to exchange secret messages in letters that were seen by the whole family.[3] There is, I think, no tension between the solipsism of Forel's early, private code and the international nature of his latter-day language, Esperanto. In both cases ants were the key to desirable communication. Forel kept the childhood satisfaction of secrecy by positioning himself in later life as the only person who could explain what the ants were really saying, and hence how humans should speak.[4]

The international language of ants brings us to Forel's translator and Wheeler's friend, Charles Kay Ogden (1889–1957). Ogden acted as an important node in the network of those interested in issues of science, the mind, and society, and in doing so he ended up advancing the projects of Forel and Wheeler in ways they never would have imagined (or, arguably, recognized).

Ants scurried around the periphery of Ogden's scope in various ways. He was friends with the preeminent British myrmecologist Horace St. J. K. Donisthorpe, and it may well have been conversations with Donisthorpe that inspired Ogden to translate Forel's massive five-volume *Le Monde Social des Fourmis* and to launch his publication series *Psyche Miniatures* with a book (also translated by himself) on ants.[5] And in his review of the supplementary volumes to the 13th edition of *Encyclopaedia Britannica,* Ogden bemoaned the omission of Forel, Pareto, and especially Wheeler from the canon of modern philosophical "greats."[6]

Of all Ogden's ant connections, the most mutually pleasant was his friendship with William Morton Wheeler. Ogden first wrote to Wheeler early in 1926, when he was beginning to translate Forel's *Social World of the Ants.* Wheeler, however, was hospitalized at the time with a nervous breakdown and was barely aware of receiving the letter. Having regained his mental health later in the year, he replied, mentioning that "I have enjoyed your books on Meaning and Psychology more than almost anything I have read within recent years."[7] Wheeler was not noted for his *politesse,* and we may take his comment seriously (although having a mental breakdown right after reading the rebarbatively labyrinthine *Meaning of Meaning* was not exactly the kind of recommendation that every author might care to advertise). Ogden visited the United States a couple of months later, met Wheeler, and immediately hit it off with the myrmecologist. Besides their shared interests in ants, philosophy, and general science, they were bonded by a number of curious little coincidences that cannot have escaped their attention.[8] Ogden was editor of the British psychological (and parapsychological) journal *Psyche;* this had the same title as the Harvard entomological journal in which Wheeler published most of his papers.⁹ Ogden's long-lived pseudonymous "collaborator," the punningly named Adalyne More, shared her moniker with Wheeler's daughter, also called Adaline. To complete the connections between the linguist and the myrmecologist, one batch of Ogden's letterheads accidentally replaced "etymology" with "entomology." Symbolized by these quirky connections, their friendship was based in a shared enjoyment of playful intellectual speculation. They continued to correspond for some time thereafter, meeting again in 1933.

Ogden had laid out the foundations for disseminating the philosophical, renaissance science beloved of Wheeler long before the two ever met. During his time at Cambridge before the Great War, Ogden was at the center of an extraordinary set who would go on to dominate their respective intellectual fields, if they were not famous already: J. M. Keynes, William McDougall, G. E. Moore, G. H. Hardy, Bertrand Russell, George Bernard Shaw, and G. M. Trevelyan, to name but a few. Norbert Wiener briefly joined the circle when he came to study with Russell at Trinity College in 1913–1914. Their main meeting point was the Heretics, a society founded by Ogden in 1909 and devoted to the free discussion of religion, to which he succeeded in attracting some notable speakers. Remaining in Cambridge as a sort of perpetual undergraduate, Ogden prevailed upon his friends to contribute to his next venture, the *Cambridge Magazine*.¹⁰ He took this on in 1912, the year after his graduation, and saved it from financial collapse through his tireless work in writing (under several pseudonyms), commissioning, editing, and promotion.

Ogden's colleagues were to shape Britain's intellectual landscape in the interwar period and beyond; together, their influence was so great that the economist P. Sargant Florence, when asked which college he had attended, was always tempted to reply, "Ogden College."¹¹ The time following the Great War was a period of hope, and Ogden's friends felt they had come or returned to Cambridge for important reasons; between them, they were ready to shape the new world.¹² Ogden continued to prevail upon the same elite group to contribute to his various publishing ventures. Through these, he was to exert a considerable influence that deserves reassessment in the light of recent "history of the book" historiography.

Ogden's chief means of influence in terms of publishing were four: the *International Library of Psychology, Philosophy and Scientific Method*, the *Psyche Miniatures*, the *To-day and To-morrow* series, and the *History of Civilisation* series. Additionally, he published a journal (also named *Psyche*) and *Science for You*, a smaller series that was eclipsed by *To-day and To-morrow*. All were published in London by Kegan Paul, Trench and Trubner and in New York by Harcourt Brace.

The journal *Psyche*, of which C. K. Ogden became editor in 1920, was intended for a fairly general audience but did not meet with much success. It fell between the stools of the truly popular and the "properly scientific." Ogden paid for articles, but so little—a tenth as much as newspapers, or even less—that many authors withdrew their papers when they discovered the rates. Conversely, professional scientists questioned the value of a journal that paid its contributors; Joseph Needham cautioned him, "I am sure [that] if . . . journals to which I contribute paid for papers, there would be a lot of skillfully disguised rubbish sent in, perhaps even researches which had never been done at all!"[13] Subscriptions staggered along at the 200–250 mark, and the journal was subsidized by the more successful book series or, according to some, Ogden's own mysterious independent means. The *Psyche Miniatures* were similarly unsuccessful; only Ogden's own *Basic English* and I. A. Richards' *Science and Poetry* generated any significant profits.[14]

Of the book series, the *International Library of Psychology, Philosophy and Scientific Method* was the most successful as well as the most influential. It grew rapidly from 1921, when Ogden assumed editorship, reaching 80 titles by 1929 and over 130 by 1944. By the time of Ogden's death in 1957, more than 150 titles had been published, and the series had made a profit overall, particularly during the peak period of the 1920s. The Library promoted the kind of interdisciplinary, philosophically oriented science that would soon be attempted by the Macy conferences on feedback systems, though without the military-industrial roots of the latter. Contributors to (and readers of) the series were British, Continental European, and American. Many of its publications quickly became famous, such as Wittgenstein's *Tractatus Logico-Philosophicus* and various works by Carl Jung, Jean Piaget, and Bronislaw Malinowski, to name but a few. Others were quirky, obscure, and little-known. The Library held together various forms of psychology, from the experimental to the philosophical, the psychoanalytic to the empirical, the parapsychological to the pragmatic, and published them side by side with accounts of anthropology, history, philosophy, biology, and statistics. Together, the Library's "virtual members" constituted a geographically dispersed but intellectually connected network committed to exploring metascientific as well as scientific questions, meditating on the nature of scientific method, its role in society, and the connections between the disciplines.[15] The Cambridge zoologist Solly Zuckerman declared that the Library's titles were all considered "required reading" for scientists.[16]

Looking backward, the Library was perhaps the last ever coherent instantiation of a generalized, widespread scientific clerisy. As such it developed and promoted the interdisciplinary, Aristotelian style of science practiced by Wheeler; as "required reading" it did a great deal to advance the pragmatist philosophy underpinning such science, notably that of C. S. Peirce. In a more specific sense the Library disseminated some otherwise marginal areas of science connected with myrmecology, including books by Bugnion, Forel, and Wheeler, and comparative psychology in general. Looking forward, many of the Library's contributors worked on or were cited in the postwar interdisciplinary science of cybernetics; again, the pragmatist philosophy of Peirce et al. was to play a role in both semiotics and biology and their cybernetic amalgamation. Other Library authors have a fair claim to the ancestry of theoretical biology in the form that emerged during the same period, most obviously Jacob von Uexküll, whose book of the same name was published by Ogden.

Of all Ogden's collaborations, the most important was with a young academic named Ivor Armstrong Richards (1893–1979). Richards was educated at Magdalene College and remained in Cambridge as lecturer in English and moral sciences (1922) and fellow of his alma mater (1926) until his move to Harvard at the beginning of the war. During his time at Ogden's college, Magdalene, the two collaborated extensively, becoming firm friends until relations soured some fifteen years later. Eventually, Richards was to become a "vicarious colleague" of Wheeler's at Harvard; Wheeler died two years before Richards arrived, but connections with his erstwhile confreres such as Talcott Parsons and the rest of the "Pareto Circle" remained.

Richards' early career reverberated with the shifting disciplinary lines of early twentieth-century British academia. He first made contact with Ogden as a disgruntled history undergraduate; the older man persuaded him to switch triposes. Richards' father was a professional scientist, and after graduation Richards himself dabbled in chemistry and physiology in the years 1918–1925, hoping variously to enter medicine or psychoanalysis. The study of English literature in its own right, such as Richards eventually pursued, was a recent innovation for academic study, and Richards was one of its chief advocates, trying to stake out some territory for it between the falling kingdom of religion and the rising kingdom of science.[17]

Together, Ogden and Richards wrote *The Meaning of Meaning* (1923), a radically nominalist account of language.[18] Layers of meaning were accrued in the contexts of sign formation and sign reception, so that words did not refer to things but rather to the context of thoughts, feelings, and emotions in which they were generated and used. At best, references were reliable, functional machinery; at worst, they were mere "word magic." The power of objective representation was removed from all (or nearly all) humans. Their theory was resolutely particularist, in that it was impossible to give a general account of how meanings were formed; one always had to look into the specific context. The book had a mythical genesis on Armistice night, when Ogden's bookshop was subject to an arson attack because of the *Cambridge Magazine*'s supposedly pacifist outlook (it had translated and published press articles from a variety of perspectives and countries, including, fatally, Germany). Ogden is supposed to have visited Richards in the aftermath of the event, and in the course of a two-hour conversation in the latter's stairwell thrashed out the entire structure of the book. Like all such myths it tells us more about how Ogden and Richards saw their mission—to create a trustworthy form of language rising from the ashes of war and distrust—than it does about actual historical events.

But what did the connection between linguistics and ants mean in practice at the height of the Library's output?

Instinctive Linguistics: Richards' Literary Criticism

Undergraduates spilled out from the overflowing lecture theatre and into the street as Richards, a gripping lecturer, wrestled with aesthetic and moral problems of the scientific age. At times he portrayed science as the main aggressor in the cultural chaos that he perceived to be threatening society. At others he employed concepts from psychology to give literature a scientific methodology, or the appearance thereof. Poetry, he claimed, revealed nothing less than "what may be called the natural history of human opinions and feelings."[19] Richards' intellectual biographer makes plain his debt to the sciences in this regard. Although Richards was familiar with the late nineteenth-century argument about aesthetics and rejected it, his work was nevertheless profoundly evolutionary in its construction: "[Richards'] own criticism would contain an evolutionary element; mental well-being under aesthetic contemplation may overflow onto the physical side; his psychological language frequently points up to the higher functions of the mind and down into the body; his model begins in neurophysiology. He would be accused of being the last of the Darwinians."[20]

In summary, Richards used (and was criticized for using) science heavily, yet did not do so in order to destroy the intuitive, gut response to literature. On the contrary, he used it to shore up this response, reconstructed as instinct. Although this application might seem paradoxical on first examination, there was plenty of room for such an approach among the ambiguous and sometimes contradictory early twentieth-century notions of instinct.

Between the world wars, the nature of instinct constituted the most important connection between linguistic theory and science. Just as comparative psychologists wrestled with the nature of instinct in human and animal, so literary critics struggled to define the instinctual and intellectual elements involved in the appreciation of literature, a relative newcomer to the academic humanities. The debate took up from a more general nineteenth-century one about the question of evolution and art; the general opinion, expressed by Herbert Spencer, was that the creation and enjoyment of art was made possible in the higher (or highest) animals by an excess of energies and faculties. It was a superfluous, if enjoyable, exercise of the higher feelings and capacities.[21] After Spencer, and on into the era of Freud, the discourses of instinct were marked by a tension between the concept of instinct as dangerous inner residuum of the soul and as an urge whose natural satisfaction would bring liberation.[22]

Two extreme literary critical responses to the place of instinct in literature in the 1920s and 1930s are exemplified by D. H. Lawrence and Aldous Huxley. In 1923, Lawrence published two books on what he termed the "unconscious"—*Fantasia of the Unconscious* and *Psychoanalysis and the Unconscious*. In them, he praised the unconscious, "the spontaneous life-motive in every organism," which, needless to say, was extremely close to contemporary ideas of instinct.[23] The *Fantasia* was the second and fuller of the two works. In it Lawrence identified the instinct/unconscious as biological, elaborating a theory of four bodily "dynamic centers" as the origin of the deeper, dynamic consciousness (or unconscious, in his terminology) and functions. Lawrence called this polar organization the "biological psyche."[24] Lawrence's antipathy toward science has often been discussed, so at first glance it is somewhat surprising to see him using the word "biological" in connection with something he regarded so positively. Although the phrase only appeared in the chapter titled "The Birth of Sex," it had a wider significance because of the importance that Lawrence placed on sex in human life. Indeed, throughout *Fantasia* the sexual self was strongly identified with the self as individual in the broadest and most important sense.

Fantasia was largely about the matter of children's education, and the main thrust of Lawrence's argument was that children should not be educated, at least not in an intellectual manner. Instead, their dynamic-biological psyches—instincts—should be allowed to develop in freedom: "To introduce mental activity is to arrest the dynamic activity and stultify true dynamic development[, creating] helpless, selfless, floundering mental entities."[25] The middle classes, Lawrence wrote, were worst affected by thinking, the most impoverished in their vitality, though even the proletariat was beginning to be corrupted. His remedy was a startling one: "*The great mass of humanity should never learn to read and write—never.*"[26] The irony was that Lawrence *wrote* all this, although he expected that few would really be able to understand it. In doing so he distinguished himself from those who had lost touch with their instinctual selves; he also implied that those who had the right sort of instinct or character or qualities could handle intellect.

For D. H. Lawrence, the unconscious, or deeper conscious, self played a part clearly identifiable with instinct in a Bergsonian, life-urge sense. Lawrence's instinct was a force to be nurtured, not ruined by intellectualization and idealization. It was a force to be respected and to be relied on for guidance. For all this, Lawrence's instinct was not like William James' inner light, which found its expression in the northeastern American culture of Unitarian Christian religious morals.[27] Rather, it was formed in reaction to the stale bourgeois morality that Lawrence (and for that matter the Huxleys) perceived in English life, and that Wheeler and H. L. Mencken discerned in American culture.[28] It was also formed in reaction to science and the mechanization of the self.

Unlike Lawrence, Aldous Huxley did not have any faith in a return to the instinctual self of humanity, particularly when instinct was identified with sex. Sex in *Brave New World* was used to *prevent* the development of individuality, exactly the opposite of the use that Lawrence had for it. Although instinctual, sex in the Brave New World was dull, automatic, and machine-like. Neither was the "savage" of the novel really savage in the expected instinctual sense; he does not return to nature but to extreme religious practices and Shakespeare. Huxley's solution to the world's ills was more intellect, not more instinct: "Clearly, the remedy must be homeopathic. The only cure for too much art and too much mind is not more matter and more nature (which would almost certainly destroy our complicated modern world) but more art and more mind . . . [T]he art of coordinating and regulating remains to be invented."[29]

Huxley was also outraged by the use of psychoanalysis in criticism of the arts. He hated the notion that what was depicted in art, or the motivation that lay behind it, could be reduced to the base impulses described by Freud. In fact, his obsessive composition of "ironic" psychoanalytic critiques would give plenty of grist to a psychoanalyst; he did protest too much, perhaps.[30] Huxley considered riddle solving, not the identification of sublimation, to be one of the principal delights of poetry reading. He even compared it to "the crossword puzzler's delight in working out a problem," an analogy that would no doubt have provoked Lawrence's disgust.[31]

Thus Lawrence celebrated vital instinct as the antidote to too much rational thought (à la Bergson), while Huxley feared and loathed the expression of instinctive, base automatism in the masses (à la Pareto). Richards, both on his own and with Ogden, steered a middle course between these two responses to instinct. In going over the same ground as Huxley and Lawrence, Richards identified two separate but related questions. The first was epistemological and concerned the role of the sciences in understanding the instinctual process of interpretation. The second was metaphysical and concerned the role of instinct in the act of interpretation.

Ogden and Richards put communication to the fore. To them, literature was above all about communicating the experience of the poet to the reader. The words, "which seemed to be the effect of the [writer's] experience," were to become "the cause of a similar experience in the [reader]."[32] However, experiences cannot be transferred directly between the interiors of heads. The same poetic image, Richards pointed out, would be interpreted differently in the specific contexts of sensibility and memory unique to each reader. Even seemingly objective stimuli, he claimed, were received in a context of "need"; a puff of wind unnoticed by a landlubber would register instantly with a sailor, while the odor of food would seem quite different to a hungry and a replete sniffer. Understanding the process of interpretation came down to psychology. Psychology explained what predisposed a certain reader to react in a certain way to particular images and helped to predict what reactions would be broadly shared, and by whom. Quoting the psychologist Titchener, Richards averred: "You cannot show the observer a wall-paper pattern without by that very fact disturbing his respiration and circulation."[33]

This epistemic framework produced the ontological question, what was the nature of this experience, and of these reactive predispositions? They were mostly (though not entirely) innate; they were biological; in short, they were instinctual. Richards used his own synonyms for the term instinct, including "appetency," "intuition," "emotion," "interest," and "impulse;" all were clearly counterpoised to "rationality" or "intellect" and were biological in origin (whether this was described in ultimate, developmental, or proximal terms). Richards identified a recent trend to read poetry too intellectually: the "average educated man [wa]s growing more conscious" and was missing the true, instinctual purpose of poetry. In short, people thought too much. There was a need for the "average educated man" to understand his psychology and so to recover his "place in Nature"—a project that amounted to much the same thing as learning to read poetry properly. Arid intellect was not enough. One needed that natural instinctive spark in order to enjoy literature, and to be at home in the world generally.

Having the correct psychological predisposition to replicate the poet's experience was a value-laden issue for Richards; the interpretation of literature was relative to the reader *provided that it was the right kind of reader to begin with*, thus restricting in practice the range of possible valid interpretations. Beauty was not inherent in the inspirational experience or in the words themselves but in the writer's or reader's ability to appreciate them. And just as it was a moral ability that enabled the poet to appreciate the experience inspiring his poem, so a moral ability had to exist in the reader if he was truly to appreciate it.[34] It was as impossible to fake the authentic reading experience as it was to write a good poem by "cunning and study, by craft and contrivance."[35] Truly instinctive artistry contrasted with an ability merely to reproduce art mechanically, robotically. One could learn art, but one could not fake artistry. Like the good poet, the good reader needed that special something within—that irreproducible, instinctual quality. Being educated was such an ineffable thing. Public schoolboys memorized reams of Latin, yet unlike Gissing's Cockney, they mysteriously turned out to be truly cultured rather than conjugating machines.[36] Ultimately, Richards' theory was obscurantist. Either one had the right kind of instinct or one had not.

Numerous literary historians have commented on Richards' and others' natural historical hierarchy of mental faculties involved in the appreciation of literature.[37] LeMahieu's *A Culture for Democracy* uses Richards, among other figures, to argue that intellectuals reasserted a "cultural hierarchy" after the First World War.[38] He claims that their rationale for this was a construction of the popular press, cinema, and jazz as media that appealed to the "lower instincts." LeMahieu locates the origin of this construction in the social psychology of crowds—a collocation of instinct and the mass that is entirely consonant with my own entomological argument in chapter 4. Richards' *Practical Criticism* especially dealt with the question of the small community versus mass society and the place of literature. Similarly, *The Meaning of Meaning* mulled over the mass "sinister potentialities of the cinema and the loud-speaker."[39] Meanwhile, in North America, Irving Babbitt and Paul Elmer More had already begun to articulate a similar socioliterary criticism.[40] Their chief assumption was that human nature was divided into a higher and a lower self. The lower self consisted of impulse, instinct, and passion, while the upper self, comprising restraint and spirituality, was the source of art. Explicitly counterpoising themselves to the romantic notion of the natural untutored genius, they championed an "antinatural" imagination that would lead to progress in civilization.

Besides betraying a certain amount of snobbishness, Richards' position reveals the inherently ambiguous nature of biological instinct and intelligence during the 1920s—an ambiguity that Richards may have communicated to, or absorbed in part from, Wheeler.⁴¹ An apparently supreme act of the intellect, namely, the appreciation of art, required some instinctual reaction. But the bestial kind of instinct would not do.

And yet while Richards emphasized the role of intuition in the appreciation of poetry, a certain amount of intellectual ability (or educational background) was the *sine qua non* to his scheme. The world was much more complex and diverse than in Chaucer's day, and there were correspondingly more specialized forms of experience that the poet might try to communicate. There was homework for the reader to do before he could understand the experiences, or allusions thereto, in poetry. That an emotional reading was displayed by "the right kind of reader" presumed that such a man was already able critically to appreciate a poem. Instinct in literary appreciation was qualified by its moral typology and balanced with the right intellect in the form of education.

Richards naturally assumed that he was writing for readers who understood (and possessed) both these qualifications of instinct. If one read a poem and did not experience any intuitional appreciation there were, strictly speaking, two possible explanations: either it was a bad poem or else one was a bad reader. Yet Richards contemplated only the former problem, writing, without irony, "the test is this . . . only genuine poetry will give to the reader *who approaches it in the proper manner* a response which is as passionate, noble and serene as the experience of the poet [emphasis added]."⁴² Needless to say, there is a certain circularity to these criteria for good reading and writing. Happily for Richards, his abilities as a critic were confirmed by his appreciation of the era's best poet, T. S. Eliot, whose abilities were accredited in turn by that able critic, Mr. Richards. Eliot's arcane allusions did not, according to Richards, obscure from the instinctual reader the "radical naturalism" of his work.⁴³ Eliot sat, like Richards himself, somewhere in the middle of the Huxley-Lawrence spectrum of opinion on instinct, a position that is perhaps best seen in his *Poems* of 1920, in particular "Mr. Eliot's Sunday Morning Service." As it happens, the poem meditates on the instinct problem with the aid of two insects, caterpillars and bees, beginning with an epigraph from Marlowe's *Jew of Malta*: "Look, look, master, here comes two religious caterpillars."

Taking sex as the biological-Freudian exemplar of instinct, the poem becomes an ambivalent meditation on Richards' subject, as well as its more obvious set of theological questions. The extraordinary first word of the poem, polyphiloprogenitive, is caterpillar-like in itself with its improbable train of syllables, and it introduces the reader to the nature of sex as one of Eliot's prime aesthetic problems.⁴⁴ "Polyphiloprogenitive" is a term Eliot almost certainly borrowed from Matthew Arnold, on whom he lectured in the year that he wrote "Sunday Morning Service." In an essay refuting the ideas of Arnold, a critic had characterized God as possessing uniquely the quality of "divine philoprogenitiveness." Arnold wrote a scathing rejoinder, noting how uplifting it was to observe how the Divinity shared this quality with "the British Philistine, and the poorer class of Irish."[45] Eliot's use of the word has a clearly Malthusian tinge to it, implicating sex as mass reproduction, the repulsive activity of the Brave New World or the spotty, awkward youth clutching their penitential offerings: "The young are red and pustular / Clutching piaculative pence."

The insects—bees and caterpillars—that appear in the poem are also representatives of automatic instinct. The former are first shadowed as the "sapient sutlers of the Lord"—ironic invocation of wisdom—that "drift across the windowpanes" in the first stanza. The pollinating bees figure both as merchants following an army on the march and also as the religious scholars flogging their wares to an undiscerning audience. The year of the poem's composition was 1917: religion, like war, held blind men in its thrall and profited from them. The hideous proliferation of this doctrine and can(n)on is reinforced by another scientific word in the second stanza, *superfetation*. The term is usually a medical-zoological one, referring to a second fetus conceived and grown within the gestation of another. As a botanical term it describes the situation where a single ovule is fertilized by different pollen grains. It captures a rather repugnant notion of an accretion of theological tradition, or, to use the biological metaphor, the acquisition of habit. All in all, the first stanza is a powerful reversal of natural theology: a flat denial of man's being at home in nature.

Other images in the poem call to mind the second kind of instinct; the Lawrentian sort that is counterpoised to dry academic intellect. "Polyphiloprogenitive" achieves this effect too; it is a ridiculous quasi-scientific mouthful for a concept that was sometimes expressed more simply by Lawrence in Anglo-Saxon. (French serves the same purpose as science for Eliot in "Mélange Adultère de Tout.") Aldous Huxley discussed such use of language in his essay, "To the Puritan All Things Are Impure." He wrote:

> Lawrence concerned himself primarily with . . . psychological reforms. The problem, for him, was to bring the animal and the thinker together again . . . [although] the conscious mind has taken extraordinary precautions to keep itself out of contact with the body and the instincts. Very significant in this context are the tabooed words which describe in the directest possible manner the characteristic functions of bodily life . . . [to] the normal bourgeois and his wife . . . The circumlocutions and the scientific polysyllables do not bring the mind into this direct contact.[46]

The second stanza's phrase "enervate Origen" is a rather complex one that seems also to criticize the lack of vital instinct. "Enervate" generally means weakened, but it is a peculiarly biological metaphor, implying a literal loss of nerve. It also has resonance with the nervous conditions epidemic among soldiers (clients of the sutlers) during the Great War. The inclusion of Origen, a biblical scholar whose work was diffuse, lengthy, and allegorical, expands interpretation of the reference. He was literally enervate, having castrated himself as an extreme means of spiritual discipline. The first stanza ends and the second begins with the same line: "In the beginning was the Word," and in this context Origen is inevitably read as a pun on "Origin"—a biological plea, or at least a plea for a more definite beginning than those provided by the subtle schools. (The phonemes of polyphiloprogenitive's three [pairs of] feet anticipate almost exactly the syllables in the word Origen.) The effeteness of the church seems to desex even the fertilization performed by the hairy bees on the flowers of the seventh stanza, where they finally appear by name:

> Along the garden-wall the bees
> With hairy bellies pass between
> The staminate and pistillate,
> Blest office of the epicene.

There is little or no respect from Eliot for the "blest office of the epicene"—the latter word denoting something of either sex, or neither.

Reading the final stanza of the poem reinforces the ambivalence with which Eliot approached his aesthetic problems, sex and literary criticism:

> Sweeney shifts from ham to ham
> Stirring the water in his bath.
> The masters of the subtle schools
> Are controversial, polymath.

The bathetic image of Sweeney shifting "from ham to ham" in his tub is a crude, fleshly parody of the unearthly, inoffensive baptismal scene that is portrayed in the third and fourth stanzas ("through the water pale and thin / Still shine the unoffending feet"). It is also a world away from the dialectical choppings and changings of the theologians. Sweeney is more real than the enervate religiosity encountered earlier in the poem—and yet one would not wish to say that his image held any attraction, either. Eliot is left in the end rejecting the asexuality of intellectual aesthetics and unable to reconcile himself to the alternative he perceives: brutish, mass carnality.

Difficult though it was—perhaps impossible, so far as Eliot was concerned—Ogden and Richards had rationalized to their own satisfaction how it was that the right sort of reader utilized instinct in order to appreciate literature in its deepest, realest sense. They were, however, left with a profound unease about the power of words in the hands (or ears) of the insufficiently moral or educated. Pareto, Ogden, Richards, and Wheeler all talked of how the undiscerning man, or improper reader, confused emotional conviction for proper reasoned persuasion.[47] Richards, for example, described the almost magical powers of words to move, sway, and convince the primitive man, or else the plain stupid. Yet a nonrationalized experience was exactly what Richards recommended in *Principles of Literary Criticism* and *Science and Poetry*. So what about those who claimed to be genuinely moved by trite doggerel, or worse, political sloganeering?[48] How could one tell this apart from the right sort of reaction? Richards was convinced that since instinctual psychology was at the root of literary appreciation, an engagement with the life sciences was the only way forward: "[If t]he next age but two ... is to be ... as Mr. [J. B. S.] Haldane supposes ... an Age of Biology; [it] will be introduced by a recognition on the part of many minds of their own nature, a recognition which is certain to change their behavior and their outlook considerably."[49] Only a scientific literary critic, or a literary scientist, could judge whether a person had the right sort of instinct.

Wheeler concluded that his social mission in entomology was the same as Richards' in literature:

> Perhaps one's attitude towards words should be that of the observer in the tropics towards insect fauna. Some words are like gorgeous butterflies and harmless, others (especially those so frequently used by the philosophers, theologians etc.) like the blood-sucking Diptera and Hemiptera which are vectors of subtle viruses. Are there not also parasitic[,] symbiotic and predatory words like the corresponding groups of insects[?] Also mimetic, warningly and protectively colored words? ... The entomologist resembles the ... philologist. Like insects words lack meaning except as their behavior in connection with their ... environment is brought into the picture to form a context.[50]

Wheeler had arrived back at Ogden's "accidental" typographical conflation of entomology and etymology. Using a mysterious blend of instinct and intelligence, Wheeler and Richards would both elucidate the true worth of their subjects, a meaning that was inaccessible to the insects—both two-legged and six-legged—themselves. This social control was, in part, what the Macy Conferences on Circular Causal and Feedback Mechanisms in Biological and Social Systems (latterly known as cybernetics) sought to address.

Between Ant War and Ant Utopia: Ogden's Basic English

Ogden's take on international language combined a somewhat idealistic view—resulting in the attack on his shop—with intellectual curiosity for its own sake. Such tensions shaped the various constructions of linguistics, myrmecology, and related sciences during the period of the Second World War period and beyond. Ogden was in no way as utopian as Forel, and he never preached through his writing as the elder man was doing during the same traumatic interwar period. Teaching the world Esperanto (Forel's preferred language), moreover, seemed like an uphill task to Ogden, especially when so many people already had at least a smattering of English for contingent historical reasons. Therefore he proposed a simplified "Basic English," an idea that grew out of *The Meaning of Meaning* and its highlighting of the frequent redundancy of everyday language. By contrast, Basic English had a vocabulary of 850 words and, crucially, only eight verbs. This obviated the need to learn numerous irregular conjugations, replacing them with simple verb-preposition compounds. In 1918 Ogden had founded an Orthological Institute for the "systematic study of language on thought"; this shortly became in practice an organization for the promotion of Basic English.

Basic English was conceived as more than just a useful tool. For Ogden, linguistics was at the base of all science and philosophy, illuminating their epistemologies and providing them with subject matter. It could provide metaphysical clarity in setting the boundaries and aims of investigation; it could facilitate clear and unambiguous communication along these lines between scientists, thus speeding the progress of science; finally, these processes of learning, meaning, naming, and understanding were themselves amenable to scientific study in the form of psychology. What is more, ants provided an ideal case study in linguistics. Animal language had been "naturally in the forefront" of the program of Ogden's Orthological Institute from the time of its foundation; moreover he regarded myrmecology as being of special significance in this project.[51] "With the subsequent widening of the scientific field to include all forms of symbolism and interpretation—signs, codes and notations—," he wrote, "the ant [has become] hardly less important than the chimpanzee for those in search of new light on old mysteries."[52] Ogden noted previous research in the area; Pierre Huber, early in the nineteenth century, had identified an "antennal language" based on touch. The Jesuit priest Erich Wasmann had elucidated the language further around the turn of the twentieth century, identifying signals for "follow me," "please regurgitate food for me," and so on. Since the discovery that antennae were organs of smell, the ants' "vocabulary" had been extended to include odors.[53] The ants' language was a model of economy, clarity, and usefulness.

At its high point, Basic English was a moderate success. It was presented to the world in 1930 in the form of a Psyche Miniature,[54] and by 1935 there were representatives of Basic English in thirty countries. Neville Chamberlain appointed a committee of the Economic Advisory Council to consider the potential of Basic English in 1939, but this fizzled out as war issues took center-stage. Winston Churchill took up the cause again in 1943, and a specially appointed War Cabinet committee concluded that "definite encouragement should be given to the development of Basic English as an auxiliary international and administrative language." The outcome of this conclusion was a white paper on the subject published on March 9, 1944, and a sudden public interest in this novel linguistic notion presented by the master of English oration.[55] Ogden did not enjoy the fame concomitant with the bureaucratic rebirth of Basic English. His occupation, as he stated in *Who's Who*, came to consist almost solely of bedevilment by officialdom. Meanwhile, the Axis powers hastened to condemn Basic English as propaganda and a tool for world domination, thereby setting back the cause in many countries, notably Japan, where it had been promoted most assiduously prior to the war. The American foundations withdrew their support, and Ogden's reconstituted and much reduced Basic English Foundation tottered along until its founder's death in 1957.

There was a tension between Basic English as constructed in a wartime context—that is, as a tool of ergonomics—and Ogden's vision for it as a tool of mutual understanding and of philosophical inquiry into the nature of communication. Such differences ended up wrecking the friendship between Ogden and Richards, for Richards was more comfortable with the utilitarian approach to Basic English than his one-time mentor.[56] Ogden never forgave Richards for his politic attitude of compromise. In 1947, for example, Richards conceded to the French demand that theirs should be an auxiliary tongue for Europe, in return for their support of English as world language. Besides possessing tact in negotiation, Richards was also a master of financial diplomacy and extracted considerable support from the Rockefeller Foundation for the promotion of Basic English, particularly in China and Japan.[57] Ogden, by contrast, opted for a more old-fashioned approach to its dissemination, relying largely on personal contacts and institutions that were in fact of his own creation. The Orthological Institute, for example, was under Ogden only, not Ogden and Richards. Though it notionally existed "for the promotion of research on the science of language" (a very Ogden goal), it was in practice mostly a center for training teachers of Basic English. For a time in the early 1930s, the institute was an engine in which the energies of Ogden's and Richards' interests were yoked together productively, with Richards drumming up financial support for the institute from the United States. However, the academic assessors for the institute during this period remained an eclectic bunch of intellectuals: friends and acquaintances of Ogden's with no particular practical axe to grind between them, including besides Richards such figures as the psychologist Adelbert Ames, medic and eugenist F. G. Crookshank, economist P. Sargant Florence, zoologist H. Munro-Fox, anthropologist Bronislaw Malinowski, and William Morton Wheeler.

Similarly, Ogden's personal efforts to spread Basic English in the Soviet Union were largely dependent on one woman, Ivy Litvinov, the colorful British-born wife of the Soviet Union's foreign minister during the early 1930s. Her tireless promotion and teaching of the language achieved, among other things, an interest on the part of Sergei Eisenstein and an official decision to teach it to Red Army troops.[58] Litvinov and Ogden sustained a long correspondence in which Ogden directed her efforts at teaching and gave advice and encouragement on the curricula and translations that she planned.

Through their letters, Ogden came to see more clearly than ever the emotive and ideological power of language. The promotion of Basic English in the Stalinist Soviet Union forced him to reassess the exact purpose and use of the language in order to stay on the right side of Soviet ideology, as well as that of the numerous passionately eccentric fans of Esperanto. (For this reason, H. L. Mencken explained to Ogden that he was usually loath to publish anything on international language, as a deluge of angry letters from advocates of alternative systems invariably followed; he was, however, happy to make an exception for such a fine writer as Ogden.[59]) Litvinov struggled to re-form the ideological content of sample sentences in standard books or sent specially by Ogden without damaging the grammar (on whose finer points she remained uncertain). She counseled him in no uncertain terms: "NOTE—avoid words RIGHT WRONG in connection with WAR CRIME, etc."[60]

In 1934, Ogden made the mistake of appending Litvinov's name to a list of those desiring an "auxiliary international language." This immediately provoked an angry letter from W. G. Keble of the British Labor Esperanto Association, chiding her for propagandizing a language of British imperialism; as a Marxist, she ought to have known better. Had she not noticed the dubious definitions of "republic," "socialism" and "proletariat" in the Basic dictionary? Worryingly for

Litvinov, Keble's tip-off had come from a Soviet Esperanto group, to whom Ogden had sent a copy of his petition; not only this, but her husband had already spoken with Roosevelt about the language as a means of learning English, potentially tarring himself with the same unpatriotic brush. Litvinov hastened to assure her Soviet comrade that she only cared about Basic English as a shortcut to learning English, and held no aspirations for its establishment as an international lingua franca. Angrily, she told Ogden, "I remember distinctly always emphasizing my apathy on the international side, & telling you again & again that BASIC had to be fun for me, or nothing. Without wishing to be offensive I have to say that I have always considered the international-pacifist aspect as the seamy side of BASIC." And, she added, using Basic as a shortcut to standard English was the only basis on which one would be permitted to promote it in the Soviet Union.[61]

In a public debate in the Soviet Union, Litvinov was accused of promoting a language designed for doubly-underhand imperialist purposes. Her attacker claimed to have written proof that Ogden believed that "one of the chief advantages of Basic" was that, by training colonial workers only in this simplified language, "natives—coolies, sepoys, etc.,—wouldn't be able to understand what the white man were talking about" when speaking normal English. When Litvinov leapt to her feet to challenge this slur, the speaker was unable to substantiate his claim, and reluctantly backed down. The event might seem trivial, not least because one has the impression that Litvinov would have conducted herself with the same dudgeon at a Mothers' Union meeting. But in the Stalinist era, more was at stake than personal pride.

A couple of years later, Ogden was learning to be more diplomatic, or at least more carefully reasoned, in the expression of his hopes for Basic English. He explained to the wife of the Soviet ambassador that even the most anti-English Indians insisted on English as their lingua franca: that Basic was a "simple way of exchanging ideas among 900,000,000 persons [in China, the Soviet Union, and India] who now have more than 300 different languages" and therefore "more important when viewed as the probable second language of the 400 millions of China . . . than as a form of English or American."[62] The Rockefeller approach to Basic, on the other hand, was more amenable to quasi-imperialist approaches.[63] Richards, perhaps uniquely for a Harvard professor, spent time at the Disney studios learning to draw cartoons, the better to teach the masses.

The advancement of science was one area where Ogden and Richards were able to agree: a place where intellectual-clerical and utilitarian concerns coincided. Basic English offered a way for scientists around the globe to disseminate their research, stimulating further efforts and preventing the needless duplication of work. The Soviets, with their national impulsion to progress and their Cyrillic handicap, were particularly keen on this aspect of the language. "The carrying out of the 5-year plan in the USSR has made it necessary for the technicians to keep a close watch on European and especially American technical literature," commented one scientist in the early 1930s, explaining his enthusiasm for Basic English.[64] In the same period, French and English speakers alike were contemplating similar advantages. The *Annales Guébhard-Séverine* decided to use Basic English for its abstracts, and the editor of *Genetics* attempted to persuade his editorial board to do the same.[65] Ogden tried to convince Raymond Pearl to publish his *Biology and Human Trends* in Basic, as Haldane had agreed with his own manuscripts *Science and Well-Being* and *The Outlook of Science*. Pearl responded positively to this possibility of dissemination beyond native English speakers.[66] Ogden began compiling a scientific dictionary in Basic English before the Second World War; when it was published, in 1942, it defined more than 25,000 specialist scientific words.[67] A certain amount of enthusiasm for the language as a medium of science lasted into the 1960s, with papers on its benefits given as late as the meeting of the British Association for the Advancement of Science in 1962.

There is no evidence that Wheeler jumped aboard the Basic-for-science bandwagon, despite the enthusiasm of his friends Ogden and Pearl, and indeed he is an unlikely candidate to have done so. For one thing he was a talented linguist and had no need of such a scheme himself (nor, perhaps, with characteristic bombast, saw why anybody else should). Moreover, like Mencken, that reluctant publisher on international language (whom he greatly admired), he had an instinctive scorn for anything that smacked of worthiness. But Wheeler aside, Basic English had proved that a human language based on the ants' communicative economy was appropriate to the era and could be central to politics, science, or society in general.

Basic English or Newspeak?

Though there was a certain amount of agreement on Basic English as a possible strategy for improving communication among scientists—and hence the progress of science—other tensions in the fields of linguistics, myrmecology, and related sciences remained evident. These tensions, and the negotiations between them, structured linguistics and its use in the period of the Second World War and would continue to do so into the cold war era. Crucially, the Axis critique of Basic English, echoing the earlier doubts of the Soviets, highlights the mid-century obsession with propaganda in relation to language.[68] Propaganda may, in this context, be regarded as a spectrum of linguistic strategies ranging from the dissemination of outright lies to more subtle uses of language in order to promote military-industrial goals by means of maximizing efficiency, or the construction of linguistic frameworks intended to shape and constrain modes of thought.[69]

A fear of propaganda's potential underpinned the individualist instinctual critique originally propounded by Ogden and Richards. Their division of language into the emotive and the reliably referential was, among other things, a way of keeping the dangerous powers of "word magic" away from those too uneducated (or perhaps too stupid) to recognize and resist it. As numerous historians and critics have noted, a similarly gloomy picture of the "average man" propelled other writers to oppose the mere reproduction of art by and for the masses. Ortega y Gasset welcomed the thought that modern art was "antisocial," dividing humanity into two castes: the appreciators and the uncomprehending masses. The work of the literary critics F. R. and Q. D. Leavis, their "Minority Press," and their protégé D. H. Lawrence did even more to foster the view that the mass lacked the individualist potential—possibly instinctual—that was necessary to respond to art.[70] Together with Wheeler, these writers were part the same cultural response to the phenomenon of mass, insectan, reproductive civilization.

However, the danger was that by constraining linguistic options within Basic English one might end up with an even more insidious form of propaganda that worked by constraining the form of possible thought itself. Ogden was resistant to this interpretation of his precious creation, but in vain; others—particularly those of a younger generation—were bound to read it this way. T. S. Eliot warned Richards in 1944 of the potential misuse of Basic English and the danger that it would take on a "technological" momentum of its own.[71] It was just that form of culture feared by Leavis et al.: not creative but mechanistic, reproductive.

By the cold war, Esperanto—a close cultural cousin of Basic—had become the model of the "aggressor language" for U.S. troops, representing everything that was wrong with the communist foe. A 1962 field training manual in Esperanto honed the linguistic skills of troops while warning of the language's dangers. As "Colonel Alexander" explains in the manual's illustrative conversation, Esperanto is intended "to instill an awareness in exercise participants of the basic differences between United States and potential enemy forces." Among these tricksy and pathological differences is the enemy's lack of caste; "Oficiroj kaj soldatoj portas identajn uniformojn," explains Major Hubert ("Officers and enlisted men wear identical uniforms").[72] Esperanto had become the language of classless, communist worker ants.

I. A. Richards' student, T. H. White, was considerably exercised by the propaganda question before and during the Second World War, as his diary for 1939 reveals:

> There don't seem to be many people being killed yet—no hideous slaughters of gas and bacteria.
> But the truth is going.
> We are suffocating in propaganda instead of gas, slowly feeling our minds go dead.[73]

White, like Richards, decided that the only way to understand the troubling aspects of human existence—why were humans the only animal to make war on its own species?—was through a return to natural history. He wrote to his former tutor in 1940: "You see, I have suddenly discovered that . . . the best way to examine the politics of man is to observe him, with Aristotle, as a political animal . . . I have been thinking a great deal . . . about man as an animal among animals—his cerebrum, etc. I think I can really make a comment on all these futile isms (communism, fascism, conservatism, etc.) by stepping back—right back into the real world, in which man is only one of the innumerable other animals."[74]

In a flushed state of excitement about the potential of science to answer questions of human nature, White wrote to all sorts of scientists requesting recommendations of books that would tell him about brains of "animals, fish, insects, etc." In the light of his reading, both Hitler and Marx came off as, fundamentally, "bad naturalists." In White's own books, the quasi-scientist figure of Merlyn guided the boy Arthur, ruler-to-be, through the animal kingdom, teaching him about man as a political animal so that he would not make the same mistakes as those fallible idealists.

Just as one might suspect, White's debt to Richards, and through him to Ogden, went beyond a generalized interest in natural history and extended to a specific and significant enthusiasm for ants. White, who listed his recreation in *Who's Who* as "animals," observed ant colonies during his years in Ireland; in 1942 he instigated a series of experiments in which he tried to incite nests of *Messor barbus* to war, so that he could better understand the natural forces then at work in the human realm.[75] Like von Frisch's bees (chapter 8), ants were a convenient experimental organism during wartime, requiring fewer laboratory resources than bigger animals. White had read his Forel carefully but was not persuaded by the old man's naturalized justification of socialism based on trophallaxis; Arthur, when magicked into an ant's body, found to his dismay that he was "a dumb-waiter from which dumb-diners fed . . . even his stomach was not his own."⁷⁶ This episode, along with a visit to an exemplary natural community, that of the geese, was originally projected as the capstone and conclusion to a series of five Arthurian books on the natural history of man. In the event, *The Book of Merlyn* was not published until 1977 after White's death, and the episode was reworked for *The Sword in the Stone*.

White's two accounts of ants provide a useful insight into the development of his thought in the context of the political constraints of the publishing world.⁷⁷ The later version of the ant episode plays down the political aspects of the former; the ants greet one another with "hail" rather than "heil," as in *The Book of Merlyn*. *The Sword in the Stone* ants are also noticeably working class, where no class portrayal was evident before. The earlier emphasis on the enslavement of the ants by an evil system gave way to a condemnation of the ants themselves for the degree of ease with which they allowed themselves to be led. The later book reminds one much more of *Nineteen Eighty-Four*, where Orwell places similar emphasis on banal propaganda and prole songs. The ants' songs turn from *mammy-mammy-mammy* (Big Sister?) to *Antland, Antland over all* whenever a foreign ant is spotted, just as Orwell's humans switch from veneration of Big Brother to vituperative xenophobia in the two minutes' hate.

In fact, the language of White's ants was strikingly like Newspeak; value-laden and decorative words were nonexistent. Where Newspeak had the words "good," "bad," and all their derivations ("doubleplusgood"), the ants had only "done" and "not done," corresponding to the totalized life regulation "everything not forbidden is compulsory." The only possible word for "mad" was therefore "not done"; it was functionally defined as either doing what one should not, or not doing that which one should. Otherwise, it was literally unthinkable.

Orwell shared with Ogden and Richards a strong interest in how language could limit and shape thought. He started out by liking Basic English; its ruthless discarding of vapid expressions matched his own critique of polemic in "Politics and the English Language." However, he came to lampoon it after the war in *Nineteen Eighty-Four*:⁷⁸ "The purpose of Newspeak was not only to provide a medium of expression for the world-view and mental habits proper to the devotees of Ingsoc, but to make all other modes of thought impossible."⁷⁹

Just as Litvinov pressed Ogden to reconsider his use of the words "right" and "wrong" in the context of war and crime, so Newspeak had rendered meaningless any reference to virtue unless it was in the sense of party orthodoxy. Yet White's version of language went even further than Orwell's, in that it presented a strikingly digital notion of communication. One imagines that Richards might have noticed this subsequent to his engagement with digital theorists; beyond noting Richards' status as connective node between White and Shannon, however, it would be dubious in the extreme to claim that White in any sense anticipated digitalism. What one can do, however, is note the Second World War and its propaganda as a most important context—shared with Richards et al.—in which the cultural meaning of digitalism itself was established. White's ant system provided a totalized web of context or possible meaning in which only one-bit messages were necessary, or indeed possible. Thus one can read the digitalism later endorsed by Richards and the cyberneticians as a political definition dealing with choice, or the delimitation thereof. It is to this definition that we now turn.

PART III

COMMUNICATIONAL ANTS

In 1909 a young and eager would-be scientist received notification of his first ever publication. The paper, a comparative discussion of ant colonies, was to be published in *The Guide to Nature;* the young scientist's name was Norbert Wiener. That same year (at the remarkable age of sixteen) he began graduate school at Harvard, studying biology. Given his interest in ants, he may well have come into contact with Wheeler during the brief period before his father, using the excuse of Norbert's poor manual dexterity, forbade him to study such an unworthy subject as animals.[1]

This little-known factor of Wiener's scientific development hints at a surprising history of cybernetics, involving not only engineering and neurophysiology but also natural historical, whole organismic, and ecological study. It is my contention that ants in their then-favored forms of representation helped to create cybernetic science: that they provided a repository of disciplinary and natural historical metaphor from which it was convenient to draw. In particular, the experimental continuum posited by the cyberneticians from the biological to the social meant that ants were an obvious subject of scientific interest, thanks to their liminal status as organism and superorganism. The chairman of the Macy cybernetics conferences, Walter McCulloch, stated in the very first year of meeting "we have made, consistently, attempts to see to what sociological data, on ants and men, these [cybernetic] notions are applicable."[2]

The language of ants became an increasingly important feature of their representation in the cybernetic context. Just as Ogden's development of international language was inspired by his reading of Forel, so Karl von Frisch's bees and the ants of New York psychologist T. C. Schneirla helped cybernetics to model the nature and use of language. For one thing, Wheeler's biological functionalism fitted nicely with the new linguistic functionalism.[3] Moreover, the long-held neo-Lamarckism of the myrmecologists suddenly came in handy, reconstrued as purposiveness within the mechanical system. This was most palatable to a generation rediscovering Peircian pragmatics, and herein lay a great irony. The biology labeled "vitalist" by its detractors in the early twentieth century actually turned out to be the most important of all in the information age since it enabled the modeling of purposive systems. Forel's roots in—and Wheeler's latter-day interest in—psychological analysis had a part to play in the construction of ant colonies as functionally purposive systems. Forel pursued psychological disorder to the unconscious, where a dynamic, not physico-energetic, rewiring effected a cure.[4] In psychoanalysis, as in cold war myrmecology, there was also a particular focus on language. Freud explored the id's presentation of clues as word puns in dreams; both he and Forel emphasized the untrustworthiness of the spoken word. Pragmatic control for the cyberneticians came through language, in the form of both direct instructions and propaganda—a double-edged sword that nurtured the gullibility of those it sought to "educate."

In discussing this period one is acutely aware of the problematic definitions for various key terms, as employed both by historical actors and by contemporary commentators. Information,[5] communication, representation are all actors' categories, but with different connotations in the mouth of each. For some, representation was an intrinsically mentalist concept; for others, its psychological import was simply irrelevant (though if forced to define it in such terms, they would have adopted the language of behaviorism). Functionalism, though less of an actor's category, also reveals its limitations when applied to the period. Sebeok, proclaiming himself indebted to Lévi-Strauss's structuralism, was actually remarkably functionalist in his outlook. Perhaps the most problematic term of all is "cybernetics" or "the cyberneticians." It denotes an interest in systems, in organization, and in quasi-purposiveness; I also, however, use it as a short-hand for a variety of social and political interests (technocratic social control, anticommunism) while well aware that this in no way captures the interests of many of the Macy participants—quite the reverse, in some cases. Nevertheless, when considered in the context of their funding and patronage, the term is a useful, if imperfect, contraction.

The transformation of ants into communicational or informational entities can be seen most clearly by following the career of Edward O. Wilson (1929–), whose early taxonomic work on the Formicidae quickly switched to a focus on their chemical trails. As Wilson hooked up with mathematicians, this pheromonal language was incorporated into a wider set of questions about communication and society. For Wilson himself, ants and their pheromones formed the core of what he proposed as a whole new discipline investigating the biology of all social phenomena, humans' included. *Sociobiology*, the statement and blueprint of this new discipline, was a cultural as well as a scientific landmark whose publication in 1975 was attended by accusations of sexism and racism. These fights and their reverberations are still felt in current debates about "evolutionary psychology." An impartial historical account of these events has yet to be written, but it is an area so large that it is not my intention in this chapter to deal with it except in so far as it impacted upon entomology.[6] My preferred way to put the question is this: Was it mere coincidence that the author of *Sociobiology*, with all its controversies, was a myrmecologist by avocation? Besides answering this question with a definitive no, I argue that the institutional uncertainty experienced (and affected) by Wilson at Harvard was another reason why he was so keen to view his work as constitutive of a new discipline.

A delegation of ants thanks Forel. Presumably they are grateful for all the positive press; Forel recommended that his human comrades should live like ants. Cartoon gift from Charles Bach and his wife upon Forel's departure from the Burghölzli asylum, 1898. (Fonds du Département des Manuscrits de la Bibliothèque Cantonal-Universitaire, Lausanne)

An interior view of nest architecture showing the spatial arrangement of the brood. Engineers approached myrmecology and the evolution of social living with a focus on the energetics of building and other activities. (From Forel, *Social World of the Ants*, vol. 1, 451)

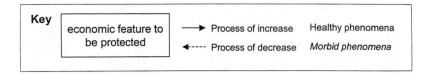

Human

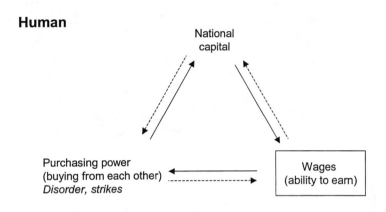

Ant

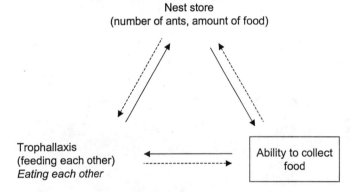

Diagram comparing Hoover's and Wheeler's economic models. An analogous hunger drove workers in both the human and formic cycles.

A fourth-grade nature study lesson in progress. A classroom ants' nest provided children with valuable lessons in observation and civics. (From *Nature-Study Review* 2 [1906]: 268)

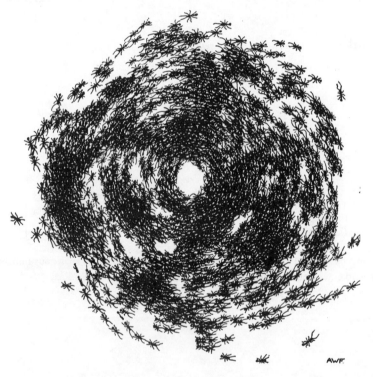

Schneirla's suicidal ant-mill. Caught in an unusual environment, ants circled hopelessly until they died; for Schneirla, this proved the ants' lack of purpose or intelligence. American Museum of Natural History, image #322190. (From Aronson et al., eds., *Selected Writings of T. C. Schneirla*, 766)

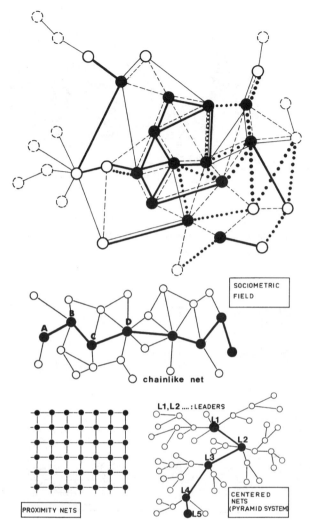

Sociometric nets as drawn by psychologist and information theorist Abraham Moles. This cybernetic approach to communication applied equally to ant or human, or indeed machine, as Moles's original caption made clear: "A sociometric set of communications is expressed by a sociogram, more or less developed according to the number of connections between individuals, related to some convenient unit. In fact, these communications can be made at various levels or with various channels, and one may be led to distinguish basic patterns of specialized sets according to the nature of the communication, e.g., food, war and love." (From Sebeok, ed., *Animal Communication*, 633)

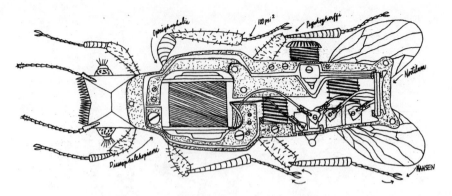

A machine-like insect, as conceived by the illustrator of a review of *Sociobiology* by Lewis Thomas. This illustration underlines how ants, as cybernetic models, mediated between representations of human and machine. (From *Harper's*, November 1975, 94)

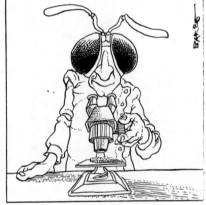

Cartoon from article "Les biologistes vont-ils prendre le pouvoir?" ("Do biologists want to seize power?") Some reviewers of *Sociobiology* concluded that Wilson's project, by treating humans as essentially no different from ants, would place scientists in a position to control the "nest." (Source unknown; Wilson papers)

CHAPTER EIGHT

The Macy Meanings of Meaning

Dancing Bees and Bumbling Ants

In the sweet heat of postwar Munich in summer, the drone of bombers gave way to the softer, more pastoral hum of bees zigzagging their way between hive and flower. As it happened, the bees' population had also been decimated across Europe during the war, in an epidemic apparently unrelated except by irony to the human carnage. The fifty-year-old man who now followed their paths so intently had coaxed the survivors into pollinating crops more effectively, perhaps forestalling, as he liked to think, still greater human deprivation.[1] Now, Karl von Frisch found himself wondering whether the returning bees were communicating to their nestmates the location of their nectar sources. He had believed for a long time that they imparted, through odor, a generalized signal of success. Recently, however, he had noticed that when foragers were trained to a food source, their nestmates would selectively go to it even if a second source were placed nearer the hive. One would not expect these results if the bees were responding to a generalized signal exciting random foraging behavior. "Was it conceivable," von Frisch asked, "that their 'language' should have a 'word' for 'distance'?"[2] The answer to this question was, needless to recount, yes. Von Frisch quickly elaborated a decoding of the bees' "dance" in the hive: the shape, speed, and direction of the dance all communicated specific details about the distance and direction (with respect to the sun) of the discovered source of pollen or nectar.[3] After some time, he further announced that the bees were using ultraviolet rays in their navigation with respect to the sun.

The dancing bees almost immediately became an animal icon, their astounding powers of communication accepted by almost all, and von Frisch was courted by Americans in particular both before and after the war. His prewar trip to the United States, arranged by the biologist Marcella Boveri, included a visit to the centers of myrmecology, Harvard and Cornell, and a trip to Indiana University, where he was hosted by Wheeler's former student, Alfred Kinsey. In 1949 he returned to Harvard, renewing his friendship with G. H. Parker (whom Wheeler counted as his best friend of later years,) and took trips to Florida and Chicago, taking in recent work on primates and termites. In New York, von Frisch delivered three lectures with films at the American Museum of Natural History and lunched with Warren Weaver, director of the Rockefeller Foundation, and half a dozen senior members of Weaver's staff. The foundation had previously funded the construction of a zoological institute in Munich; now they were very keen to hear more from the middle-aged apiarist.

The funders' interest in the bees concerned their analogical relationship (whether or not this yielded practical possibilities) with human communication. By a neat twist of fate, the challenges of human communication were highlighted by von Frisch's visit. Besides the linguistic barriers to surmount, von Frisch was having considerable hearing difficulties by 1949. He could not be expected to converse in cocktail party contexts; in his lectures his hearing aid had to be wired in, and questions from the audience had to be repeated by an assistant on stage. Von Frisch's words, produced by collective labor, were self-evidently precious. It all added weight to his stance on international communication. Von Frisch recollected: "As they all seemed to have the cause of goodwill and understanding between nations very much at heart, I talked to them, at Weaver's request, about communication among the bees. For in certain respects the bees' organization and mutual collaboration even in artificially mixed communities is indeed superior to our own efforts." His talk went down well, and following a private conversation with Weaver von Frisch received a grant for his projected research, an outcome that he reported with evident satisfaction as "very fine."[4]

Shortly after this, von Frisch's assistant Martin Lindauer also received Rockefeller money to visit India and Ceylon to do a "comparative philology" of bee language. The plan was to look at the subsocial or incompletely social species found there and see whether their language was likewise more primitive.[5] Lindauer found what he was looking for in *Apis florea,* which, unlike the domesticated honeybee *Apis mellifera,* could only perform its dance on a horizontal plane, and could not correct for the vertical orientation of a *mellifera*-type hive. Lindauer went on to play an important role in developing and promoting von Frisch's program of research to a scientific world that never seemed to tire of its wonders.[6]

In retrospect, scientists seem to have been extraordinarily credulous about the bees' ability to communicate abstract facts. By all accounts it was a "truly revolutionary" discovery, amounting to nothing less than "the use of symbolic . . . flexible communication" among insects.[7] As a consequence, ants were adopted by the cyberneticians in a manner that was unanticipated by previous formic specialists, and moreover was resisted by at least one myrmecologist, T. C. Schneirla. The cyberneticians' formic phenomena were not psychological or sociological but rather organizational, communicational, and navigational. Though these themes had been present in ant studies from the nineteenth century, they were now brought to the fore in the military context of the era and, in particular, the scientific context of military funding for research. The themes were also reinterpreted from their earlier forms; they were seen as purposive, effectively functional phenomena. As such, they were also freely amenable to interspecific analogical application, humans included. Analogies of such functions could even be drawn between the organic and the inorganic realm; indeed, the design of machines with colony-like properties of problem solving was a major aim of cybernetics. In doing this, however, I argue that the cyberneticians inadvertently picked up on an old tradition of holist biology that would have been palatable to Wheeler. It is with this historical footnote—the revival of vitalism—that the chapter closes.

An obvious reason for the ready belief in talking bees was the pragmatic commonality perceived between the topics of communication and orientation.[8] Historically, the two types of study were closely interwoven. The Macy meetings on communication mixed in questions of orientation without any acknowledgment that this required justification. The common link between the two was the sense of olfaction. In 1963, Wilson, staking out the territory of pheromones for himself, claimed that this sense had been neglected by previous researchers. This statement was unwarranted; in the nineteenth century John Lubbock had speculated on the possibilities of entirely new sensory experiences amongst insects: "There may be fifty other senses as different from ours as sound is from sight; and even within the boundaries of our own senses there may be endless sounds which we cannot hear, and colors . . . of which we have no perception . . . The familiar world which surrounds us may be a totally different place to other animals."[9] Questions about insect orientation had been pursued by French psychologists in the early twentieth century; Forel, and to an even greater extent Fielde, had looked at olfaction as a feature of communication, though not orientation (chapter 2). The French psychologists had also considered smell in insects. Von Frisch's work departed from all this in that it looked at orientation as a social phenomenon induced by other insects, unlike their experiments on the senses and faculties of individual insects.

These key insectan themes of orientation and communication had also gained a military significance since the Second World War. When communication was considered in connection with orientation, the links with military applications in the form of radar and sonar, missile guidance,[10] and human reconnaissance were clear. For example, Wheeler's old colleague Harold J. Coolidge spent the war developing a signaling mirror for downed airmen and a chemical repellent for sharks.[11] Arthur D. Hasler, a zoologist at the University of Wisconsin, was involved in similar naval ambitions. He advised on a project to synthesize odors that would repel fish so that the Navy could clear them from harbors, where their schools interfered with sonar detectors looking for enemy bombs or submarines.[12] Funding from the Office of Naval Research (ONR; founded in 1946) for such work was free-flowing.[13] The experience of the animal behaviorist J. P. Scott was typical in this respect. In 1948, he recalled, his Jackson Laboratory received a couple of visitors from Washington "who said, we have some money for research, why don't you apply?"[14]

Shortly after the war, in 1951, the entomologist Vincent Dethier organized a symposium on chemoreception at Johns Hopkins University with ONR sponsorship, bringing together academic and military interests, and military zoology continued through to the 1960s. Marine bioacoustics, navigation, and communication were especially well supported; of the contributors to a 1968 book, *Animal Communication*, Thomas C. Poulter (marine mammals) and William N. Tavolga (fishes) both consulted for the U.S. Navy and had their research supported by its funds; Adrian M. Wenner (honeybees) studied electronics in the navy and continued to draw on the funds of the ONR for his work; even Gregory Bateson's theoretical piece, "Redundancy and Coding," was aided by a US Naval Ordnance Test Station Contract (he had been working on communication in cetaceans).

Naval money paid for concrete results. The pressure was on to find instances of effective orientation and meaningful communication in the animal realm. This pressure sometimes led to an overhyping of results or the potential thereof. Poulter, a technical consultant to the U.S. Navy and winner of awards from Congress, contributed a pedestrian, anecdotal account of different whale species' noises to *Animal Communication*. Yet he boldly concluded, "Yes! We believe that marine mammals talk and that what they talk about makes sense to other marine mammals of the same species."[15] As psychologist and information theorist Abraham Moles put it, the aim was to use "the study of the animal world [to] help the communication physicist," and not the reverse.[16]

One perhaps unlikely recipient of military largesse was Theodore C. Schneirla (1902–1968), a psychologist by training who had come to regard ants as presenting the most interesting phenomena of his field.[17] From a modest social background, Schneirla wrote his doctoral thesis on learning and orientation in ants while teaching at New York University. In 1930 he spent a year at Karl Lashley's laboratory in Chicago, after which he returned to New York University, remaining on the faculty there until his premature death. From 1943 on he also held an appointment as curator at the American Museum of Natural History (AMNH), for which he designed, planned, and executed the behavioral alcove in the new Hall of Invertebrates in the 1960s.[18]

Schneirla was an outsider in science for his whole career.[19] His work with ants was not fashionable within psychology, and his disinterest in taxonomy disqualified him from being regarded as a myrmecologist (he was not, for example, on Creighton's list of automatic recipients of reprints). Meanwhile, his interest in live animals made him an odd staff member at the AMNH, whose main concerns were to build specimen collections for taxonomists and to educate museumgoers through its popular dioramic displays of taxidermy. Moreover Schneirla's own communicational style was, ironically, notoriously diffuse and difficult to follow. Alfred Emerson was perhaps his only dependable source of professional entomological support.[20] Yet Schneirla was a force to be reckoned with by all these groups. His name crops up in the correspondence of all professional myrmecologists in the period from the 1940s to the 1960s. They may have regarded him as an odd fish, but he was someone whose opinion mattered and needed to be factored in when questions about ant behavior required answering during the 1950s.

In 1947, Schneirla received a grant from the ONR to investigate the behavior of army ants, specifically "to investigate analytically a complex behavior pattern in a social insect in relation to environmental conditions and to underlying biological processes."[21] The work was done at the long-established Barro Colorado Island laboratory in Panama (Wheeler, Fairchild, and Beebe had all been earlier visitors to the site, and David Fairchild's son Graham was present on the island at the same time as Schneirla and provided him with companionship). Schneirla was delighted with the ONR's support and felt that the army ants were his best and most important research of his career thus far—a period of almost twenty years.[22]

The year 1947 was also the year that Schneirla's name first appeared in papers on un-American activities.[23] His "Marxism" and "pacifism" had chiefly manifested themselves in a vocal prewar commitment to the republican side in the Spanish Civil War and in his activity for the Teachers' Union, which resulted in some frustration with the left.[24] Meanwhile, Wheeler's old Harvard pal, the astronomer Harlow Shapley, was trying to promote the idea of an apolitical

National Science Foundation, much to the paranoiac alarm of J. Edgar Hoover and the FBI; ironically, Wheeler's sociological colleague Talcott Parsons, also from Harvard, assisted Hoover in investigating Schneirla's case.[25] Even the apolitical was political. Despite these investigations, Schneirla's funding from the ONR was renewed at intervals through to 1954. None of his work was rated as especially secret; Schneirla's boss Aronson surmised that the ONR's interest in the Panama project amounted to nothing more complex than in having its personnel trained in a tropical location.[26] Despite (or perhaps because of) the question mark over his politics, Schneirla cultivated a warm relationship with Sidney Galler, the chief of the biology branch at the ONR.

During the late 1940s, Schneirla set about publishing in various semipopular contexts what was to be his most memorable piece of work. This described an observation that had occurred in Barro Colorado Island and was by his own account an accident.[27] After a violent rainstorm, he came upon about a thousand *Ecitons* (army ants) behaving oddly on the sidewalk outside the laboratories. Most were gathered in an aimless cluster, but a few had begun milling around the outside of the group in an anticlockwise direction. Returning to the group at intervals, Schneirla found that "by noon all of the ants had joined the mill, which had now attained the diameter of a phonograph record and was rotating somewhat eccentrically at fair speed." By dawn of the following day, the ants had exhausted themselves and had not fed for 24 hours; "the scene of action was strewn with dead and dying Ecitons."

The irony, for Schneirla, was that the tragedy arose "like Nemesis, out of the very aspects of the ant's nature which most plainly characterize[d] its otherwise successful behavior."[28] Under the unusual conditions of a featureless terrain washed clear of odorific cues, the ants were caught between two simple, innate impulses: a "centrifugal" one, to resume the march, and a "centripetal" one, to stick with the group. In this unnatural situation, there was a perfect balance between the two vectors, resulting in ceaseless circular motion.

An account of the observation was first published in 1944, but Schneirla rewrote it and allowed it to be further bowdlerized and reprinted several times in 1948. It was his way of trying to establish a paradigmatic example of social insectan behavior. Schneirla's bumbling ants could not have been more different from their contemporary cousins, the dancing bees. The ants were unintelligent, uncommunicative, caught between simple impulses; the bees were intelligent, communicated symbolic propositions, and were able to coordinate complex navigational cues. This contrast reflected Schneirla's disinclination to participate in military-inspired projects and his lack of belief that human-insect analogies were politically or scientifically viable.

Schneirla's skepticism about the validity of analogy was expressed overtly in his review of von Frisch's book, *Bees—Their Vision, Chemical Senses, and Language*. He praised the efforts of his fellow worker in insect behavior, but without wishing to disparage his valuable allies (von Frisch himself, and his sponsor in American publication, Donald Griffin), Schneirla was compelled to draw attention to the anthropomorphic interpretation that was at least invited by von Frisch's words and those of Griffin in the book's introduction. "It is not correct," Schneirla commented regarding the latter, "that complex behavior (as this certainly is) is necessarily 'intelligent' . . . What the finder actually transmits to the secondary bees in the wagging run is still an unsolved problem; however, as a psychologist, this reviewer doubts that, when known, the transmission processes will meet the accepted psychological criteria of 'symbolic communication.'"[29] Schneirla was in real life apparently less measured in his criticism of von Frisch's anthropomorphism, using it in his teaching as an example of "what not to do."[30]

Schneirla's role in animal behavior and communication was generally, in intellectual terms, a critical one. He is best known for his denunciation of Lorenz's work on aggression, and he was always ready to act as a philosophical scourge to those too eager to generalize between species, especially when one of those species was *Homo sapiens*. Schneirla placed a continued emphasis on the different "levels" of various animals and their behavior, and the irreducibility or extrapolation of one to another. (The concept was more than a little reminiscent of Wheeler's hierarchies.) Within his insect work, and also in some mammalian experiments, Schneirla consistently focused on simple bipolar responses, not complex symbolic communication. Schneirla also insisted on seeing communication and psychological responses as part of a developmental process that could not be broken up in the laboratory and given achronic significance. Though Schneirla was to play a major role in bringing ants into cybernetics, he was politically and methodologically an unexpected candidate to do so.

Insects in New York

Over the past fifteen years, historical scholarship has emphasized the role of ergonomics expertise, initially developed in the context of the Second World War and personified in large part by Norbert Wiener, in shaping postwar culture. Historians of physics have pointed to the emergence of "Big Science," a term first coined in the early 1960s that invokes the large-scale nature of research in its number of researchers, the size of its equipment, and, most important, the amount (and sources) of funding.[31] As this mode of science was pursued from the 1950s onward, the "military-industrial-academic complex" crossed institutional and disciplinary boundaries, and began to exercise notable power in the social sciences.[32] Steven J. Heims has explored this latter change, explaining the emergence of "human engineering" in a context of "post-war circumstances... conducive to a ready acceptance of the political status quo and to a technological or technocratic optimism."[33]

These scientific questions and sociopolitical interests came together most famously in a series of conferences that ran from 1946 to 1953, undergoing various name changes until the name stabilized in 1950 as "Cybernetics: Circular Causal and Feedback Mechanisms in Biological and Social Systems." The meetings were financially supported by the Macy Foundation, a body dedicated to interdisciplinarity as a means to progress in the sciences and hence society. The shared aim of participants was "to see what biological, psychological and sociological problems might be approached with the theoretical tools which, during the war, have created thinking and purposeful machines." De facto, this included a pragmatic semiotics, since such machines required "artifacts, words etc., which operate as signs in society, and which, for their maker, serve as equivalents for circular paths wherein activity can reverberate."[34]

Questions of cybernetic circuitry as they pertained to neurophysiology have been well rehearsed, notably by Heims, but little has been done to discover how cybernetic semiotics was, in this period, embedded in whole-organism, behavioral and ecological studies.[35] Yet shortly before the Second World War, the Rockefeller Foundation identified biology as the branch of natural science that offered the greatest hope of fulfilling their goals of social control and the "science of man": "Can we unravel the tangled problem of the endocrine glands and develop a therapy for the whole hideous range of mental and physical disorders which result from glandular disturbances? Can we develop so sound and extensive a genetics that we can hope to breed in the future superior men? . . . In short, can we rationalize human behavior and create a new science of man?"[36]

The program was developed by Warren Weaver, who, though a mathematician and physicist, chose to devote the majority of the funds at his disposal to the biological sciences—80 percent of the $30 million disbursed between 1933 and 1953.[37] Weaver commented: "Before we can be wise about so complex a subject as the behavior of a man, we obviously have to gain a tremendous amount of information and insight about living organisms in general, necessarily starting with the simpler forms of life. Experimental biology . . . furnishes the basis necessary for progress in solving the sequence of problems which begins with the strictly biological and moves through the mental to the social."[38] Most of this money, of course, went to support the burgeoning science of genetics. Karl von Frisch's receipt of funds for his work on social insects was a highly visible exception to this generalization, and his success in answering to Rockefeller ambitions may have inspired Wilson in the claims that he would shortly make about his own science.

Notwithstanding Schneirla's skepticism about what humans could learn from insects, the Macy Foundation was keen to have him and von Frisch in from the start of the cybernetic discussions. Schneirla's credentials were a 1935 textbook, *Comparative Psychology*, and specific papers on ant learning and their social organization. He had one or two other publications with obvious cross-species interest for the cyberneticians, including one on German psychological warfare. The cyberneticians had not, however, appreciated Schneirla's critical stance on interspecific analogy and psychological methodology, and grouped him along with the apiarist as intellectual and pragmatic comrades. Though von Frisch proved personally difficult to book for the meetings, Macy conference participants nevertheless made sure to foreground his topic, social insects, from the outset, especially in relation to communication and its role in systemic control.

By the second conference, the subject matter had explicitly turned to insect-informed sociology, with Schneirla's talk on ants providing fuel for discussion. The paper related to research on insect orientation that he had been conducting contemporaneously with, though independently of, von Frisch's most important experiments on bees. It was widely accepted that ants find their way back to the nest by laying a trail behind them, Hansel and Gretel style, from the anal "Nasanov gland." But how did ants find their way outward to the point of foraging? Specifically, how did the route there become more direct over successive journeys? To find out, Schneirla set his ants to learning specially modified rat mazes. Schneirla, as ever unwilling to draw cross-phyletic comparisons, considered the ants' ability to learn mazes a distinctive instance of learning, since they did it without the aid of a cerebral cortex.[39] It was a trial-and-error kind of learning without the benefit of goal-directedness commonly attributed to higher mammals.

Some cyberneticians drew from Schneirla's experiments the lesson that the ants' "frightfully rigid social organization" was related to their "frightfully poor communication."⁴⁰ This was not the point that Schneirla had been trying to make at all; nevertheless, the predominant framework of interpretation for such material centered on communication in ants and humans alike.

As the Macy feedback conferences went on, questions about language and communication came increasingly to the fore. Mulling over Schneirla's ants, Northrop wrote to Wiener: "You [and Arturo Rosenblueth] have made it clear that social organization and its character is a function of communication."⁴¹ In general, he concluded, he too was coming to see the significance of communication to the whole issue of feedback and control. Northrop was right about the direction of Wiener's thoughts. Wiener noted: "The analysis of society is still lagging behind. Define society as a group of individuals (or organisms) among whom there is a mutual communication (and understanding) and who undertake tasks in common. The first half would apply . . . to electrons as well as to man, the second half would narrow down the definition to animals capable of (or characterized by) purposive behavior (task!). It seems to include bees and beavers."⁴²

The problem then was to define the nature of the individuals, and the nature of their interrelation. This Wiener took to be the same as the question of "mutual communications and understandings." "Most social interrelations," he went on, "are linguistic, or derived from speech, such as traffic lights or the flags shown on the high seas." He therefore urged that language as a means of communication should form the substance of future discussion. The first day of the very next conference, the fifth, was given over to language; Charles Morris and Roman Jakobson were among the participants.⁴³ But very soon the cyberneticians wanted to return to the topic, and devoted the whole of the seventh and the eighth meetings to it.

Early suggestions for guests at the seventh conference (1950) included Max Delbrück; von Neumann was looking for an excuse to get a geneticist as a regular attendee. Delbrück's refusal (he found the one meeting he did attend "vacuous" and "inane")⁴⁴ came with an alternative proposal: "I suggest C. D. Darlington as a guest, if language is going to be the principal topic. He is a far better interdisciplinary geneticist than I, and specifically interested in language." But G. E. Hutchinson, a philosophically inclined ecologist and regular conferee, had plans to draw material from outside molecular biology and nearer to his own specialism. "Have you considered the possibility of getting Karl von Frisch, who will be around at the time, to begin the session with an account of his extraordinary work on bees and their communication techniques[?]" he wrote in reply to a solicitation for suggestions of guests. "If this seems a good idea, it might be arranged through . . . Donald Griffin," who was then in the Zoology Department at Cornell

University.⁴⁵ As so often with good ideas expressed in committees, responsibility for execution was laid on the shoulders of its originator. McCulloch eventually wrote to Hutchinson, "we will all be grateful if you will come primed [for the seventh meeting] on the language of ants and the bees."⁴⁶ Schneirla had given his apologies for missing the meeting; Frank Fremont-Smith remarked, "I am really sorry about this as I think he could give a good deal to our discussion."⁴⁷

Next time around, perhaps wishing to make space for Schneirla, McCulloch again put forward as a topic for discussion the language of bees. The organizers of the eighth conference sought Schneirla's advice on whom should be invited as expert in this area. Schneirla, suffering from chronic health problems, was unwilling to put himself forward to speak on a related topic; as things turned out, he was not even able to attend the meeting. The tentative program plan had von Frisch speaking on bees, followed by Schneirla's doctoral student Daniel Lehrman speaking on vertebrates. But von Frisch was in Germany, and Lehrman's participation was scotched by Schneirla on the grounds that he was too busy with his dissertation and a heavy new teaching load.⁴⁸ Instead, Schneirla's young colleague Herbert Birch, who worked on problem solving in chimps, was engaged to cover all areas of animal language, and especially encouraged to include invertebrates in his paper. This was to be preceded by a paper by I. A. Richards, invited to participate on the strength of *The Meaning of Meaning*, even though it was by then more than twenty years old.⁴⁹

One imagines that Birch left the New York hotel hosting the conference mightily frustrated and annoyed. Although he had gone to heap scorn on careless analogies between bees, humans, and sundry other organisms and to urge caution in the attribution of language to insects, his audience treated him simply as a convenient textbook on von Frisch. Though Birch was trying to encourage criticism of von Frisch's methods and the interpretation of his observations, his audience only wanted more details of the apiarist's results. An exchange representative of the unproductive (from Birch's viewpoint) discussion ran as follows:

BIRCH: Does this [observation] mean that we now have to impute certain high levels of psychological function like abstraction . . . to the bee itself? I would submit that we do not.

KLÜVER [dryly, one imagines]: We would have more information on the mechanisms involved in imparting information in the dark hive.

The discussants were talking right past each other: Birch was trying to bring the discussion back to a skeptical analysis of the bees' psyche, while Klüver simply wanted a description that *worked* on a systemic level. In fact, many participants were evidently already very familiar with von Frisch's work; von Förster had gone so far as to do calculations of his own based on it. As a final insult to the zoological exactitude insisted on by Schneirla's circle, they all referred to bees as "he," into which denomination even Birch slipped toward the end of the discussion.

Even if the cyberneticians had accepted Birch's points—that bees do not have "intelligence," and that their dance cannot therefore be regarded as purposive—there was still room for discussion. After all, Schneirla too was always keen to point out that ants are comparatively dumb blunderers. He had highlighted this point as early as the second conference, demonstrating it by the fact that they never acquired any generalized understanding of the principles of mazes. Schneirla, however, left space for a discussion of the mystery of their powerful concerted action. His own solution to this was trophallaxis. Trophallaxis provided the chemical and tactile cohesion for the group; the smells and touches of "biosocial intercourse" provided the "organic push" for each phase of behavior and exerted a powerful attraction holding the group together. For example, Schneirla suggested, quiescent pupae secrete something that makes workers tend them quietly without moving. When about to hatch, their encased wrigglings impel workers to handle them roughly, snatching them from one another and so effecting eclosion.

Though some conference attendees were disappointed that Schneirla would not allow purpose or content to ants' representations of the world, they were nonetheless intrigued by his account of the emergence of social complexity and organization from simple forces acting on an individual level. It inspired Claude Shannon to go away and build a maze-solving machine using exactly the same methods attributed by Schneirla to his ants, namely, trial-and-error, remembering, and "forgetting" (that is, nonemployment) of the previous solution should the maze be altered.[50] He too produced a feedback system that gave the appearance of "hunting for the goal."

Shannon's machine, presented at the eighth meeting, not only mimicked Schneirla's successful ants, it also mimicked his ill-fated ones. Like the suicidal ant-mill, Shannon's machine could, under certain conditions, go around in a vicious circle. This generated considerable interest. Right away, the neurophysiologist Ralph Gerard piped up, "[It's] a neurosis." "Yes," agreed Shannon.

The interesting epistemological feature of the situation was that the machine had no way to know that it was going around in a circle, or, as Leonard Savage put it, "It doesn't have any way to recognize that it is 'psycho.'" It was possible to introduce a counting device that changed the strategy after a certain number of circles. But, epistemologically, this still did not amount to an awareness of futility. And another problem was that the machine had no "incentive" to learn the quickest route to a goal. Savage commented: "That really is a very important problem in any kind of real human learning. If you can already peel a potato, why should you take the trouble to find a better way to peel it? Perhaps you are already peeling it correctly. How do you know?"

An epistemological problem was emerging: that goal-directedness was no more a justifiable description for human behavior than it was for ants or robots. Von Förster nudged the problem toward its skeptical extreme:

VON FÖRSTER: What if there is no goal?
SHANNON: . . . the path is repeated again and again. The machine just continues looking . . .
FRANK: It is all too human.
BROSIN: George Orwell should have seen this.

It is immensely frustrating for the historian that Schneirla was not at this meeting. What would have been his response to these human analogies? Would he have persuaded the cyberneticians not to talk about ants if they wanted to make human analogies? Perhaps overawed by the high-powered and robust discussion of the group, Schneirla's replacement Herbert Birch did not contribute anything to this conversation, even though it immediately followed his own presentation, and the neurosis/*Eciton* analogies were surely clear to many members of the group (not least to that precocious ant enthusiast, Wiener himself).

Schneirla's emphasis on the limitation of ant behavior merely served to highlight the other participants' insistence on a sanguine cybernetic interpretation and an emphasis on insect communication rather than, as Schneirla insisted, their specific physiological and psychological capacities. They were determined to interpret ants in their own way, an act encouraged by von Frisch's greater willingness to talk about communication in a sense desirable for humans. Thus the cyberneticians' science created a strong nexus between the study of ants and the study of communication as a universally generalizable phenomenon. In fact, it is not clear from transcripts, letters, and related papers that the cyberneticians even perceived a tension between their interpretation of Schneirla's experiments and his own. They effortlessly assimilated his ants into their own agenda.

The Return of "Vitalism": Representation and Purpose

Notwithstanding their invidious assimilation of Schneirla's viewpoint, the cyberneticians were in another sense entirely true to Schneirla's tradition, stretching back to the neo-Lamarckism of W. M. Wheeler.[51] Indeed, some of what Norbert Wiener wrote sounded as though it could have come straight from Wheeler's pen. "We are not stuff that abides, but patterns that perpetuate themselves."[52] "[The organism is] neither a thing nor a concept, but a continual flux or process."[53] Wiener in fact was greatly taken by that philosopher so influential on the early twentieth-century vitalists, Henri Bergson. He spotted an odd quirk of history in the rise of information engineering, namely, that it represented the triumph of a kind of science—vitalism—that had been considered discredited: "Thus the modern automaton exists in the same sort of Bergsonian time as the living organism; and hence there is no reason in Bergson's considerations why the essential mode of functioning of the living organism should not be the same as that of the automaton of this type. Vitalism has won to the extent that even mechanisms correspond to the time-structure of vitalism."[54]

Wiener felt compelled to qualify this by pointing out that vitalism had not won in the sense of antimaterialism trouncing materialism: "but . . . this victory is a complete defeat, for from every point of view which has the slightest relation to morality or religion, the new mechanics is fully as materialistic as the old. Whether we should call the new point of view materialistic is largely a question of words."[55] However, this mystical version of materialism did not accurately characterize Wheeler and other workers on insect psychology and sociology.[56] Rather, *vitalism* was one of those labels pinned on the losers by winners in a scientific debate; the absence of a better label is a consequence of the vitalists' failure to organize themselves in an effective disciplinary fashion, to define their outlook in their own terms, and to think up better titles for themselves and their detractors. Wiener's hasty rejection of vitalism qua antimaterialism (if such it was; he had likely read Wheeler when writing his ant paper and realized that his "vitalism" was not nonrational) merely highlights the historical loading of the term and invites a reevaluation of what vitalism meant for its proponents, as opposed to its detractors.

Tracing the history of vitalism also opens up a new twentieth-century genealogy of biology that reaches from Jacob von Uexküll to Thomas Sebeok, who christened it zoosemiotics. This story, running parallel to the thread connecting Wheeler and Wilson via the cyberneticians, shows the historical affinities of myrmecology with semiotics and cybernetics as they were developed in the mid-twentieth century. The comparative genealogy enables a better retrospective account of myrmecology as well as showing the new uses of ants. Rather than characterizing the lines as vitalist, one might capture some of their essence by describing them as "representationalist" and "purposivist"; in emphasizing these aspects of natural phenomena vitalists treated energetics as irrelevant rather than wrong.[57] Both cybernetic myrmecology and zoosemiotics (along with Ogden and Richards, and mid-century American scientists in general) had a shared inheritance in the pragmatic philosophy of Charles Sanders Peirce.[58] All, in their various ways, embraced Charles Morris's program of treating semiotics as "an organon ... of all the sciences."[59] But theorizing representation and communication was a matter of theoretical and practical concern, and the subject, though universally acknowledged as important, was constructed in a variety of ways. The alternate route by which Sebeok arrived at zoosemiotics, a would-be discipline that Wilson counted a synonym to sociobiology in 1975, highlights the zoologists' and information scientists' mutual fascination in the cold war period, and the emphasis on representation and purpose.

The semiotician Thomas Sebeok traced the zoological aspect of his linguistics back to Jacob von Uexküll (1864–1944). Uexküll was a Kantian both in his philosophy of science and in his metaphysically skeptical descriptions of the animal's world; what linked the two was the question of representation. He compared science to scaffolding propped against the edifice of nature: one could never know the building in itself, only whether or not the scaffolding stayed up, thus demonstrating that it (scientific theory) conformed to its shape (nature in itself). Similarly the organism itself constructed a "world-as-sensed," with nothing but an evolutionary track record thus far to indicate its accuracy. The key process of nature, therefore, was the use of signs: those things apparently "read" from nature and used to construct a worldview that, in turn, guided organismic behavior. "The investigator finds [in the organism] nothing but a guiding apparatus, which serves to connect the two "fronts" of the body—the one, the receptor, turned towards the world-as-sensed, and the other, the effector, towards a world of action."[60]

This description seems extraordinarily prescient of Wiener's and Shannon's models more than twenty years later. A non-Whig way to analyze the similarity again points up the significance of Ogden and Richards in the period between times; the translation of Uexküll's book from German, undertaken at Ogden's behest, reflected much of the metaphysical skepticism he and Richards were then inscribing into their own theory of meaning: "During the last few years advances of biology, and the psychological investigation of memory and heredity have placed the 'meaning' of signs in general beyond doubt and it is here shown that thought and language are to be treated in the same manner."[61]

Sebeok recalled reading von Uexküll's book shortly after leaving Hungary, and indeed traced all the biological aspect of his work back to this incident. The year was 1936 and Sebeok was, according to his own account, an undergraduate at

Magdalene College, Cambridge (though the college has no record of official affiliation). At this time Richards was just coming to the end of his fellowship, though in practice most of his time was spent elsewhere. Sebeok was reading *The Meaning of Meaning*, and browsed the list of other International Library titles on the flyleaf.[62] *Theoretical Biology* caught his attention and he ordered it. When it arrived, he scarcely understood a word, but somehow it got under his skin. Rereading the book in its original German in the 1960s, he blamed his initial incomprehension on the "wretched" translation achieved under Ogden's "eccentric auspices."[63] Ogden, he claimed, had over-egged the biological side of the book and underplayed the semiotic. What Uexküll had in fact done, claimed Sebeok, was nothing less than establish for the first time the existence of "semiosis in life processes in their entirety . . . we now call [this] *Biosemiotics*."

Sebeok shared a Budapest background with John von Neumann, though having been born in 1920 he was almost a full generation younger than the mathematician. Sebeok moved to the United States in 1937 and studied literary criticism, anthropology, and linguistics at the University of Chicago before obtaining his PhD at Princeton. He joined the Indiana University linguistics faculty in 1943, eventually becoming professor of anthropology and chairman of the university's Research Center for Language and Semiotic Studies. In addition to his human linguistic and cultural interests, animals were a constant feature of Sebeok's academic landscape. He was a founding member of the Animal Behavior Society (founded in 1953; its journal was *Animal Behavior*) and served as a member of its executive committee. Given his principal concern with linguistics, Sebeok's focus on animal behavior unsurprisingly concerned their communication, a topic on which he edited a book in the late 1960s.[64]

Sebeok was on a search for "fundamental laws which govern human behavior."[65] Though he believed that linguistics was the best exemplar of this approach thus far, the key distinction for the research program he advocated was a search for language universals in formal rather than substantive form. The universal basis for diverse languages lay in "neurophysiological and . . . more broadly biological characteristics of man":[66]

> It is possible, therefore, to describe language as well as living systems from a unified cybernetic standpoint. While this is perhaps no more than a useful analogy at present, hopefully providing insight if not yet new information, a mutual appreciation of genetics, animal communication studies, and linguistics may lead to a full understanding of the dynamics of semiosis, and this may, in the last analysis, turn out to be no less than the definition of life.[67]

Despite using the language of cybernetics, Sebeok did not cite Wiener or Shannon in his 1968 collection. In fact, the book's dedication to Claude Lévi-Strauss indicates a greater sympathy with Wheeler in sociological mode, Julian Huxley, and Marcel Mauss. Language universals, traditionally sought in substantive form, were now sought in formal form. *Le langage* had emerged once in evolution but *les langues* had differentiated multiple times through the contingencies of historical linguistics.[68] Here there is a problem of terminology: though Sebeok explicitly aligned himself with Lévi-Strauss's structuralism, nevertheless his approach was, when considered in the context of biology, a much more functionalist approach than that pursued by physiologists. Just as Ogden and Richards were impressed by Pavlov's demonstration of the contingency of representation, so Sebeok was concerned more with whether communication did the job, rather than the form it took. Sebeok's approach might, on the face of it, appear to entail a discussion of the animal psyche, as opposed to the purely behaviorist approach pursued by Loeb and Skinner. It appears to summon up questions such as: How does the animal represent? What is the representational world inhabited by the animal? But the point of his semiotics was that it enabled one to remain skeptical about such questions. The purpose of representation—evolutionary success—was what ultimately drove the phenomenon. Coming from a humanities tradition, Sebeok was able to deal with the function of signs without becoming bogged down in questions of realism.

Forel, on the other hand, had always been interested in the metaphysics of representation. Early in his career, he explained that his belief in the conservation of energy principle underpinned his monism. This was a different way of stating his belief that mind and brain were one and the same; thoughts could not happen without the energy of nerves firing. The activity of cerebral cortex was a "complex of energy."[69] But in 1889 (at least with hindsight) Forel rejected the theory of physical anastomoses and a narrow understanding of energy-conserved neurology. "Braid and Charcot had overlooked the psychic intercalation of the permanent engrams of the subconscious in the human brain and had ascribed their results to . . . reflexes."[70] Having liberated himself from physical constraints, he was able to believe in the power of hypnotism to reconnect the mind, a process conceived as functional and not energy-conservative. Forel provided a social cure and not a physicomedical one, a possibility that was for him guaranteed by the insects' noncerebral powers of learning.[71]

Various writers have commented on the similar path taken by Sigmund Freud: his derivation of the unconscious as a hydraulic system of nervous energy from his earlier work on neurons, and his ultimate development of psychical as opposed to physical energy, which did not follow the same strict rules of conservation. While the ego was wired into the outside world via the nervous system, the id was an internal boiling pot of pent-up energies.[72] Looking beyond the brain, Freud hoped that the economy driving evolution would prove to be psychical, contemplating a Lamarckian explanation that might place human telepathy midway between the ants' mental unity and the consulting-room phenomenon of transference:

> It is a familiar fact that we do not know how the common purpose comes about in the great insect communities: possibly it is done by means of a direct psychical transference... One is led to a suspicion that this is the original, archaic method of communication between individuals and that in the course of phylogenetic evolution it has been replaced by the better method of giving information with the help of signals which are picked up by the sense organs. But the older method might have persisted in the background and still be able to put itself into effect under certain conditions—for instance, in passionately excited mobs.[73]

Freud would have been pleased that Wheeler ultimately came, via economics, to see sociality as a psychical phenomenon. The French engineer Charles Janet had stressed that the dealated queen, by digesting her wing muscles, yielded energetic resources sufficient to remain with her first brood. This energy conservation was the crucial first step permitting the evolution of sociality, because she could remain with the eggs, rather than going out to seek sustenance for herself. Wheeler, however, posited a movement toward eusociality, resourced by the trophallactic exchange of food in sublimated social relations of mother-daughter bonds. This feeding relationship, crucial to the whole process, was finally redefined by him as an unhealthy psychoanalytic *Trieb*, a parasitic interrelation that piggy-backed off the mother's physical and emotional pleasure. The energy-conservative transformation of Janet's mother wasp became, in the American's hands, a Nietzschean dynamic transformation.[74]

Indeed, Wiener's 1909 analysis of ants' nests treated them as a serial development toward perfection that well fitted Wheeler's neo-Lamarckian approach. He continued explicitly and favorably to highlight the non-energy-conservative aspect of Freudianism, which he likened to the physics of Gibbs. Both the psychoanalyst and the physicist, he claimed, provided a way to tell a story about the contingent directionality of history, and showed how disorder could clog up in a system, whether psychic or physical.[75] Wiener did not like the word *libido* inasmuch as it suggested energy; psychoanalysis was about information, and the techniques of psychoanalysis were thus entirely consistent with cybernetics.[76]

One suspects this had a personal element for Wiener, who was forever burdened with the task of trying to unpick the damage his parents had wrought on him in their personal experiment to create a genius. In writing his autobiographies, early examples of a psycho-confessional style that would one day become commonplace, he was trying to make sense of his parents and their effects on him in a culture that recommended Freudian analysis as the means to do so.

The cyberneticians were obsessed with questions of purpose, or rather treated the purpose of systems as the raison d'être of their design.[77] By 1947 Northrop was beginning to see the irrelevance of Newtonian physics to the modern age, where communication was all. He wrote to Wiener that he was coming to appreciate that symbols have an effect out of proportion to the energy involved in their production or propagation.[78] As J. B. S. Haldane put it a few years later, "The signal [communicated by X to Y] usually involves little expenditure of energy by X, and has a large positive or negative effect on Y's energy expenditure."[79] A message, whether internal or external, was energetic, of course, but it did not necessarily spark a response of the same energetic order. Communication did not break the law of the conservation of energy, but the law was an utter irrelevance when it came to explaining the phenomenon. One had to add purpose in order to understand such systems.

T. C. Schneirla, by contrast, was dogged in his rejection of purpose as a means to understand animal behavior. He concluded that an organic "push" rather than a goal-directed "pull" was responsible for the ants' learning abilities. This "drive," when aroused, would impel the subject to continue moving until it instigated a new phase of behavior (which, to the untrained eye, would appear to be the goal of the first behavior). This explained both apparently adaptive (maze running) and maladaptive behavior (suicide mill formation). Schneirla's spokesman Birch protested despairingly at his Macy conference: "We are . . . discussing certain phenomenal similarities that . . . appear in highly different forms of behavior. We are concentrating not on behavioral processes as such, but upon the end result of any of a number of different kinds of behavior processes . . . if we start from a broad, general, end-result category, we must be especially careful to seek out and to enunciate the dissimilarities and the discontinuities of process which may underlie the described behaviors."[80]

In vain Birch produced outlandish examples of "communication" that, in his opinion, demonstrated the ridiculousness of viewing it as a purposive process. Starfish soup "communicated" its predatory presence to scallops, causing their withdrawal. Male mosquitoes could be persuaded to mate with tuning forks if one simply picked one that "communicated" at the right pitch. How could Oliver

Twist, reaching out for more, and an amoeba extending its pseudopodium to engulf a particle possibly be regarded as engaging in the same purposive behavior? But the other conferees were unmoved. Bigelow doubted that there was a distinction between Oliver Twist and the amoeba. Just as an amoeba could be fooled into reaching for an inorganic particle, so a phonograph could elicit an inappropriate communicational response from a human.

The essential distinction was that Schneirla and Birch were interested in mechanism, and could countenance no talk of purpose if that could not be understood as part of a causal process from the bottom up. The cyberneticians, on the other hand, were precisely interested in function construed purposively. Shannon's machine, based on Schneirla's maze-learning ants, kept everyone happy because it was a feedback system that gave the *appearance* of "hunting for the goal." But Schneirla was stubbornly skeptical. Ants' behavior, he insisted, "resembles the action of row of dominoes more than it does the communication of information from man to man. The difference in the two kinds of 'communication' requires two entirely different conceptual schemes and preferably two different words."[81] So long as the cyberneticians had the appearance of purpose, their own purposes were fulfilled. They were, after all, interested in building machines, in imposing their own purposes from above. Schneirla's skepticism highlights the purposive angle of the cyberneticians, both in analytic terms (if something performs a useful function for the organism, you might as well call it purposeful), and in pragmatic terms (their ultimate aim was building physical and social machines to carry out their purposes, of which the machines themselves had no concept). You could say that just as Darwin's sexual selection suggested that handsomeness was in the eye of the evolved beholder, so for the cyberneticians, "clever is as clever does."

Oddly enough, the cybernetic treatment of (human + instrument) as a functional, purposive whole echoed the approach not of Darwin but of that "unscientific" neo-Lamarckian, Samuel Butler. Gregory Bateson, like so many cyberneticians working on devices to aid the blind and deaf, asserted that a blind man's cane should properly be regarded as integral to his body: a true part of him. His postulate echoed Butler's half-joking remark: "If it is wet we are furnished with an organ commonly called an umbrella."[82] Perhaps Butler should have been recognized as the cyberneticians' truer Darwin.

At any rate, the vitalism debate had been won in one sense by biologists, but with quasi-purposeful organization as its entelechy, not theist or deist design. In the process, biologists gave their perspective over to appropriation by the engineers, so that it was no longer recognized as biology. The cyberneticians' insistence on the systemic "pull" of purpose indicates their science's genealogical debt to myrmecology, once accused of vitalism for its use of entelechy, a supposedly mystical appeal to top-down organizational power in the life and evolution of ants. The Macy conferees, thanks to their funding, were not so much interested in the meaning of meaning as they were in the purpose of purpose. Ants could be, and were, used to represent both these interests.

Mindless Models

By the 1960s, representation could be constructed in such a way that it might or might not entail mentalism (or consciousness, or intelligence in its colloquial sense). The old two-kingdom model had emerged in new form: highly intelligent mammals might be trained to perform militarily useful tasks, while insects provided a mindless model for the development of mechanical systems to do the same. The cyberneticians regarded representation for practical purposes as a quasi-purposive phenomenon, analogized by the power of objective representation which they, unlike Ogden, granted to humans. Collectively they asked a pragmatic question, thanks in large part to the applied interests of their funding: "How *well* does the representation function?" An early and long-lived paradigm for this approach was "representation" within the self for the purposes of navigation or orientation, for example a dolphin's representation of its environment through sonar. Freed from Ogden's influence, Richards too pursued the possibility that language could be used reliably for representation; this was the aim of his excursions into the teaching of English, and his attempt to characterize communication as a purposive system of "feedforward."

In combating the representational approach, Schneirla acted almost as though he were engaged in a nineteenth-century debate; he did not see how the cyberneticians could pursue these issues of representation without ascribing consciousness or intelligence to animals, and counted this a mortal sin of science. If he allowed that this was a respectable approach given pragmatic ends he would not have approved of it being pursued under the banner of science. The best Schneirla could manage, as far as the cyberneticians were concerned, was to admit that push could, on occasion, give the appearance of representation or purpose.

The most appropriate historical distinctions in the immediate postwar period are not between mentalists and behaviorists, however, but between what one might term totalizers and particularizers. Although some members of the Macy conferences appeared keen to elaborate an all-encompassing theory of communication, this cluster is best regarded as totalizing in a practical sense, in that they (or their sponsors) wished to impose optimal ergonomic control on society, just as T. H. White had fearfully half-prophesied. Schneirla was wrong in thinking that all the cyberneticians were mentalists, but this was not the crux of his critique, which actually derived from his defiantly particularist approach. He simply did not allow that there could be a general cross-phyletic account of representation, purpose or communication. He did not comment significantly on humans' objective powers of representation; presumably he granted these, but they formed no part of the subject matter of comparative psychology.

Although Schneirla was wrong to tar everyone with the mentalist brush, there were some in immediate postwar period who cared about the content of messages, not just their function or deep evolutionary structure. Granting transferable objective representation to animals (that is, not just the internal representational processes involved in orientation) was a key part of von Frisch's and later Wilson's work. Each had a professional investment in elucidating the internal semantics of their particular varieties of animal communication. The extraordinary thing about von Frisch is that his writing does seem quite naive vis-à-vis questions of animal mind, both from a comparative psychological and from a linguistic viewpoint. Intelligence and conscious purpose always seem to be lurking just over the page in his writing, and Wilson too received early criticism in this regard.[83] The cyberneticians, it must be remembered, lionized von Frisch, revealing perhaps—besides an interest in an apparently purposive natural system—a direct applied interest in such infamous cold war possibilities as trained bomb-carrying dolphins and the like. Such possibilities did require the potential of tinkering with actual code of representation.

On the other hand, relativists among the Macy participants countered the totalizing trend of cybernetic biology. Anthropologists such as Margaret Mead and Gregory Bateson were influential in this regard, questioning the very values of objectivity that some sought to transfer from the human realm to the animal. Julian Huxley also suggested that language be regarded as an example of animal ritual rather than substantive communication.[84] By the 1970s Sebeok had taken these critiques and used them to refine the categorization of communication—which aspects of it were amenable to which forms of analysis.[85]

Did Wilson and von Frisch personally have a totalizing theory of communication, aside from their context of patronage? Von Frisch had so only in a very speculative "grain of sand" sense; in his autobiography he gave an impassioned account of the value of studying one species, which can "challenge us with all, or nearly all, the mysteries of life."[86] Wilson, in his dealings with theoretical biologists, allied himself with an idealist trend of totalization with its roots in von

Uexküll. The 1960s idealist strain of communication as a biological "theory of everything" also characterized the early work of Thomas Sebeok. Wilson arguably also had a totalizing theory inasmuch as he had ambition to establish his science as the most significant in nonmolecular biology, as we shall see in the following chapter.

CHAPTER NINE

From Pheromones to Sociobiology

The plaintive songs of the impoverished American Deep South were more than an ocean away from the worthy strains of Monsieur Emery's teetotal choristers of Lausanne. Yet despite their vastly different origins, Edward Wilson shared with Auguste Forel a lonely childhood. In Wilson's case this was due to the divorce of his parents and a consequent shunting around from place to place, without the opportunity to settle down and make friends.[1] And like Forel, he turned to insects as his consolation. For a man who would later be accused of crude genetic determinism, he has always been remarkably frank about his family's poor heredity.[2] Wilson's autobiography reveals a curious obsession with physical condition, due in large part to the military institutions that loomed over his youth. At times he echoes the masculine rhetoric of entomology advanced by Beebe and Peattie; at others he deprecatingly suggests myrmecology as a suitable alternative for those, like himself, unfit for the higher military calling.[3]

Wilson is an intriguing mixture of a man. He is renowned by many for his abiding Southern courtesy and humility; people with whom he comes into contact in the course of his numerous lectures, visits, and other projects frequently remark on his refreshing lack of bumptiousness. Yet he has also displayed a remarkable political talent throughout his career, collaborating with the right men at the right times to catch and define the latest trend in science and culture. A self-confessed nonmathematician, he has arguably done more than any other biologist of the twentieth century except R. A. Fisher to promote a mathematical understanding of nature, thanks to his more numerate (and less memorable) associates. This knack for politics is one of the reasons Wilson has earned such outspoken enemies along the way. He has also cultivated a good relationship with the media since the very earliest days of his career. This coziness—incidentally producing many of the best scientific photographic portraits ever made—has also raised hackles among his peers.[4]

Taxonomy and After

Wilson studied at Alabama University just after the war, earning his bachelor's degree in only three years. Ernst Mayr's *Systematics and the Origin of Species* and Theodosius Grigorievich Dobzhansky's *Genetics and the Origin of Species* were the twin gospels of the new synthesis biology that formed the basis of his education. As he moved on to his master's, still at Alabama, Wilson became involved in taxonomic research on the imported fire ant. In 1950 Wilson moved to the University of Tennessee to start his PhD with Arthur C. Cole, a key member of the national network of entomologists (which he had entered via the applied route) and a specialist in ant classification. Wilson quickly became bored with the university, apparently finding it an unstimulating environment, and plotted a transfer to Harvard, home of Wheeler's legacy. In 1951 he arrived, somewhat awed, though he himself was immediately labeled a rising star by colleagues, as remarkable for his earnest demeanor as for his sharp mind. In 1953, a visitor to Harvard noted: "Wilson looks to be about 15 years old. They call him 'the growing boy.' He is thin as a beanpole . . . and a celibate by choice as far as I could judge . . . [Frank] Carpenter regards him as a very bright youngster and has arranged a junior fellowship at Harvard for him which will keep him there for the next three years."[5]

One of Wilson's chief backers—indeed, one of his sponsors for the transfer to Cambridge—was William L. Brown, a young man himself, who was as passionate about ants as Wilson. Brown was an uncompromising scientist who actively sought out controversy and debate, making no allowances for seniority or inexperience. To the junior Wilson, the lack of condescension was welcome; to Brown's superiors, the lack of deference was unforgivable. Still awed by his Harvard colleagues, Wilson had set his sights on a project to describe the ants of his home state, Alabama. Brown encouraged him to think bigger—why not describe the ants of the whole nation? In so doing, he set Wilson on a collision course that would cause a disciplinary furor and change the course of his career for ever.

Brown encouraged Wilson to work on *Pheidole*, a genus that spread through very nearly all the contiguous states and thus represented a flag planted in North American taxonomy as a whole.[6] Unfortunately for Wilson, this brought him into direct competition with a number of important myrmecologists. His most dangerous enemy was William S. Creighton, a former student of Wheeler's who had eclipsed his master in terms of taxonomy. Creighton's allies included Cole, Wilson's former teacher, and another ant specialist, Robert Gregg.[7] Although Wheeler's stature was not seriously questioned by any of the next generation of myrmecologists, his taxonomic work was not considered to be up to the standard of his other research. Specifically, Wheeler did not like working up keys for the identification of ants, making it difficult for his successors to repeat his identifications.[8] He also clung for much of his career to an obscure pentanomial system, for which few others had sympathy. Thus Creighton was at that time the country's most senior figure, living or dead, in classification. Creighton's massive *Ants of North America* had only just been published (in 1950) by Harvard's Museum of Comparative Zoology (MCZ), and now Wilson, together with Brown, was trying to tear it to pieces. The essence of their 1953 criticism concerned the perennial taxonomic question of "lumping" versus "splitting." Creighton was a splitter: where local varieties of species existed, he designated them as formal subspecies. Wilson and Brown preferred to designate the varieties informally by geographical description rather than a subspecific name. One major reason for this was that one supposedly definitive "subspecific" trait might be shared by two or more "subspecies" which themselves differed by some other characteristic. By keeping the designation informal, Wilson and Brown could concentrate biological research on the traits themselves, wherever they varied.[9]

To understand why this stance caused so much offence, it is necessary to delve into both the cultural meaning of taxonomy and the personal politics involved. One of the best expressions of the nature of taxonomy in the period is given by Wilson himself:

> It is a craft and a body of knowledge that builds in the head of a biologist only through years of monkish labor. The taxonomist . . . knows that without the expert knowledge accumulated through his brand of specialized study, much of biological research would soon come to a halt. Only a specialist expert enough to recognize the species chosen for study . . . can unlock all that is already known about it in the literature . . . If a biologist does not have the name of the species, he is lost . . . No CD-ROM, no encyclopedia can replace the taxonomic expert.[10]

Taxonomy was the *sine qua non* of biology; in order to do any meaningful research on an organism, the biologist had first to seek accurate information on its classification. This knowledge was not structured in any manner that could be accessed by even the otherwise well-trained biologist; it linked "anatomy, physiology, behavior, biogeography, and evolutionary history,"[11] not to mention the organizational idiosyncrasies of a hundred collections around the globe. In short, it was largely tacit. And of all the animal specializations, entomology was arguably the most arcane when it came to taxonomy, containing approximately three-quarters of known species. Within entomology, ants posed additional problems because they were represented in collections by a single holotype (the definitive species specimen), not one for each caste. Were all ants discovered in a nest of the same species, or were some slaves? And was a fresh specimen a new caste of a known species or a complete novelty?

On top of professional challenges to Creighton's expertise, Brown's and Wilson's work grated for personal and institutional reasons. Perhaps in a postpublication low after *Ants of North America,* Creighton was starting to feel redundant in his work. Discovering that another two of his species had turned out to be synonyms, he wrote to Wilson's former tutor Art Cole, "To be perfectly frank, the discovery . . . has more or less dampened my enthusiasm for the thing. In past years I [have] had occasion to sink so many forms that I am beginning to fear that I may subsequently be called Old Synonym."[12] Creighton was partly sympathetic to Wilson's and Brown's 1953 publication. He liked the fact that it "upset some applecarts" among those who thought that geographical subspecies were easy to recognize (thus questioning the value of Creighton's own work). However, he suspected their reasons for doing so were entirely personal, and moreover thought that their "suggestion for using informal geographical designations instead of formal trinomials [stank]."[13] At this point, Creighton was inclined to blame the older man, Brown, as the bad influence on Wilson, whom he cast as an ingénue but nevertheless the intellectual "spark-plug" in the Brown-Wilson combine.

Creighton thought that Ernst Mayr, head of the MCZ, was deliberately not disciplining Brown but rather letting him have just enough rope to hang himself. But Brown hung on, and in 1954 the row picked up more momentum. Wilson offended again by joining Brown in another reclassification that in Creighton's mind seemed to "draw largely [upon] or . . . repeat" what Creighton himself had done before.[14] A series of very bad-tempered letters ensued between Creighton, Brown, Wilson, and Gregg, Creighton's ally in the affair. Creighton seems to have been offended as much as anything by the youth of his antagonists. Creighton had corresponded with Brown since the latter was a high school student in 1938 and had helped and encouraged him in his study of ants. Indeed, Brown was so taken by the help extended to him that he named his first son, born in 1952, Creighton Brown. Now Creighton complained bitterly, "to think I gave the pup his start."[15] But Creighton was also riled by the Harvard attitude in general; the "Happy Harvard Team," as his friend Bob Gregg dubbed them, were seemingly unstoppable in their hubris. Creighton railed against the scientific "monopolies" of the "pipsqueaks." What most rankled was that Brown and Wilson wrote as though they were representatives of Harvard.[16] Creighton composed what he thought was a gently chiding letter to Wilson, warning him not to be infected by

"Harvarditis." In response, Wilson disingenuously protested that he was "just an Ala[bama] pea-picker."[17]

Wilson's autobiography condemns the weak leadership of Harvard biology at this time, or in his words, "incompetent management."[18] Creighton accused Philip Darlington, curator of entomology at the MCZ, of suffering psychosomatic (if not downright fake) illness to stay out of the ruckus.[19] In November of 1954, Frank Carpenter, chairman of the department of biology and Wilson's doctoral supervisor, was drawn in, and demanded to see the offensive letters. The solution, when it came, was quietly politic. At the end of 1954 Wilson completed his PhD and immediately received funding, arranged by Darlington, to be packed off about as far from the debate as was geographically possible. He spent a year in the South Pacific, winding up in New Guinea.[20]

During Wilson's absence, the reasons for managerial weakness became apparent to the outside world. There was a tussle over the status of various Harvard personnel: members of the biology department who taught, and members of affiliated institutes, such as the MCZ, who did not teach undergraduates. Carpenter, as curator of fossil insects at the MCZ and chairman of the department, found himself in the uncomfortable position of having a foot in both camps—and a double stake also, therefore, in Brown.[21] By the middle of 1955, the tensions were becoming rather strong. No wonder there was doubt over who should take responsibility. These same issues would linger for the next two decades, shaping Wilson's career and science.

While Wilson was away, Carpenter's attitude to the row also seemed to be changing. Apparently he had originally blamed Brown for the spat. Though this information comes from the highly biased Creighton, there is good reason to believe it. Neither Wilson nor Brown was a natural Boston Brahmin, but Brown's mien was far more problematic. It was alleged that his clothes were dirty, that he drank to excess, and he that expressed himself with unrestrained forthrightness. Now, however, Carpenter was coming to see guilt also on the part of Wilson.[22]

Although Carpenter was apparently fed up with Brown's behavior by late 1955 or early the following year, Brown remained at Harvard until 1958.[23] During this period Brown actually had to take a good deal of responsibility, covering for Darlington, who continued to suffer ill health and who, upon his recovery, decamped to Australia for some eighteen months.[24] In 1958, however, Brown was "relieved of his position at Harvard," taking up a post at Cornell two years later. (This job was vacated by Howard Evans, who returned to Harvard, joining Wilson.)[25]

Upon his return to Harvard from the South Pacific, Wilson began to distance himself professionally from Brown.[26] Wilson had picked a fight and had not succeeded in winning approval for either his taxonomy or his fieldwork. Creighton, who at that time remained the doyen of the field, reported with relish that when he accused Wilson of lacking "a proper field acquaintance with the ants of the western United States," Wilson had feebly protested that he had spent three months in 1952 "touring the west from Montana to Arizona."[27] So far as Creighton was concerned, his own taxonomic reputation, grounded as it was in fieldwork expertise, was unchallenged. Through the episode Wilson had learned valuable lessons about professional politics, and the importance of choosing one's collaborators wisely. He would never make the same mistakes again.[28] "If I were called on to make a prediction it would be that Wilson will ease himself out of taxonomy and into Theoretical biology, for which he has a great aptitude," guessed Creighton.[29]

The Agenda for Ant Behavior: Research after Macy

During the taxonomy debacle and Wilson's exile, ant models for communication and organization had been developing further. In fact, "sociobiology" had been discussed for some nine years before Wilson returned to Harvard. Its first mention in a widely disseminated context came in 1948, when C. F. Hockett suggested it as a useful disciplinary term in *American Scientist*.[30] Together with J. P. Scott, T. C. Schneirla helped transform a conference on animal societies into a series of published papers in 1950. They too described their science as sociobiological, meaning "an interdisciplinary science [lying] between the fields of biology (particularly ecology and physiology) and psychology and sociology." "In summary," they suggested, sketching out their ambitions, "this series of papers may serve as an adequate general introduction to the study of sociobiology and animal behavior under field conditions."[31] Scott and Schneirla's shared project soon grew into a Committee for the Study of Animal Societies under Natural Conditions and, in 1953, the Animal Behavior Society.[32]

In February 1953 Schneirla organized a two-day Biological Orientation Conference for the Office of Naval Research (ONR). Initially he had been thinking of a symposium at the Entomological Society of America or American Association for the Advancement of Science meetings, but the offer of military funding swung matters for him.[33] (After his ONR-funded army project, and hosting the New York stage of von Frisch's visit, Schneirla had approached the Macy Foundation to see if they might be a source of general financial support, but without success; he was keenly aware of the importance of taking whatever funding was offer.) Participation was restricted to North Americans and Canadians on security grounds.[34]

Twenty biologists took part, and five psychologists; there were additionally two observers from the Rockefeller Foundation. The conference aims were to look at orientation through the animal series, including humans, and to assess valuable lines of approach for the future.[35] These ambitions met with mixed success; there were problems with identifying paradigmatic questions, methods, and even terminologies. There was some tension between those promoting a physiological approach and those interested in communication *sui generis* (such as Walter Rosenblith, an electronic engineer from MIT).[36]

Moreover, there was a tension between the ONR's needs and the ambitions of researchers. Weighing up the outcomes of the meeting, the ONR's Galler asked, "Can we use . . . organisms as biological systems to study for producing concepts, or ideas for the Navy's task in target detection and identification? Does this general area lend itself to the Navy's objectives? If so, what must the Navy do to exploit this area?"[37] Unconsciously echoing the discourse of the conference itself, Galler summarized, "Are we (the Navy) going in the right direction?" Robert Galambos, chief neurophysiologist at the Army Medical Service Graduate School was, however, disappointed that participants had not had time to get down to philosophical brass tacks, such as the innateness versus learning question. He complained to Schneirla that he had "a general sense that the thing was being run for the convenience of the ONR and its recording staff." Unsurprisingly, Schneirla, as its organizer, defended the validity of the meeting but admitted that there had been some haste and "pressure based on [the] recording procedure."[38] For all its faults, the meeting was at least an attempt to address new areas of research, and there was enthusiasm both on the part of the ONR and the scientific participants for a second conference, but argument over whether it should be the ONR or the scientists who determined the agenda.

Meanwhile the ONR took on its most ambitious insect experiment to date, which involved transporting bees in planes across the Atlantic to determine whether their sense of time was due to the sun's position (in which case they would adjust to Eastern Standard Time) or was innate (in which case their diurnal pattern of activities would continue according to German time).

Traveling the opposite way, Schneirla turned his attention to Europe in order to advance the social insects. In Europe, he found a style of research that matched his own exacting standards. Pierre-Paul Grassé, a Parisian entomologist, experimented with nest construction using various-sized groups of termites.[39] He found that a certain minimum number of termites (about fifty) was necessary for the task to be conducted normally. He also found that the stimuli for building behavior were self-reinforcing; for example, a certain density of earth crumbs in a particular place would trigger reconstruction. This positive feedback was named *stigmergy* by Grassé. It was neither reductive, since it required a certain critical mass to occur, nor holistic, since it consisted of individual workers' responses. It was the perfect complement to the approach Schneirla had been developing.

Working together, Schneirla and Grassé established the International Union for the Study of Social Insects (IUSSI). Grassé was the major driving force; Schneirla set up the North American branch of the organization with Emerson, and sought assistance from the Rockefeller Foundation for the publication of their journal, *Insectes Sociaux*. Warren Weaver refused on the foundation's behalf, citing policy, but the NSF was prevailed upon to stump up cash for the first few issues instead.

The cyberneticians continued in their entomological quest, too. In 1954 the Macy Foundation began another series of conferences, this time on group processes, at Cornell. The conferences were a natural biological successor to the cybernetics series, addressing similar questions but this time through purely zoological, psychological, and psychotherapeutic exemplars. Frank Fremont-Smith provided a bridge between the two series, having attended both, as did Schneirla and his AMNH colleague, Margaret Mead. Wilson's colleague Ernst Mayr was also present. Schneirla attended the first conference, along with his ethological enemy, Konrad Lorenz, whom the former's student, Daniel Lehrman, had recently attacked in the *Quarterly Review of Biology*.[40] If the tone of the published transcript is to be believed, their exchanges at the Ithaca hotel were remarkably civil and constructive.

From a historical disciplinary perspective, the participants' discussions are of particular interest. Margaret Mead described how she had used the term ethology in her own work before it was exclusively identified with zoology; Tinbergen claimed that what the Europeans called ethology was what the Americans called psychology. Schneirla traced the use of the term ethology back to a paper of Wheeler's in 1903; Lorenz insisted that it had not been properly used until 1910, when it was deployed by Oskar Heinroth.[41] Many present were happy to discuss such supposedly disgraced vitalists as Driesch; Lorenz was also particularly keen to highlight the influence of von Uexküll and even Bergson on ethology. Almost everyone expressed a feeling of special debt to Whitman. In their introductory autobiographies, a remarkable number of participants also professed an early interest in the social insects.

Human parallels suffused the whole conference; participants apparently found it even easier to make these connections than had their cybernetic predecessors. Discussions of imprinting, for example, slipped smoothly into issues of neurosis and fetishism. The challenge of understanding sublinguistic communication in animals was connected to the understanding of disturbed humans. Panic stampedes were compared across species, as were the behavioral effects of administering LSD. The psychotherapist Jerome D. Frank even commented with some alarm on the apparent credibility conferred on dianetics by some of their conclusions regarding imprinting.[42]

Communication as a phenomenon worthy of cybernetic study had become a reflexive issue, as it was for Ogden: Fremont-Smith emphasized that the Macy Foundation's principal aim was to remove all obstructions to communication between scientists and hence to smooth the advance of science itself, one of whose chief topics was communication.[43] Mead called on the conferees to reflect on the emotional effects of watching films of wobbly baby goats, pointing out that even they, an objective audience, were not immune from unintended sentimental effects of communication. Perhaps the irony of von Frisch's difficulty in communicating his bees' language some five years previously had not gone unremarked: a concern with the communication of science remained as an undercurrent to more explicit interests. All the Macy conferences were recorded, and transcripts were produced; the journal *Behavioral Science* was inspired to do the same, producing a joint commentary on the papers it contained. The value of von Frisch's films was often remarked (though not the apparently significant financial value to the maker himself). When B. F. Skinner and Wilson met to discuss their lives' work in 1987, they commented favorably on the value of recordings, including the one they were making at the time.[44]

Another indication of the continuing technocratic interest in zoological sociology was *Behavioral Science*, a journal started in 1956, shortly after the Macy conferences on cybernetics had concluded, and including many of their participants on its editorial board. The journal was the organ of the Mental Health Research Institute, established the year before at the University of Michigan. (This was itself an offshoot of the University of Chicago interdisciplinary group known as the Committee on Behavioral Science, many of whose members remained at Chicago but contributed to the ongoing shared project that was reified by the journal.) Both groups aimed to solve some of the mysteries of human relations: "social inequality, industrial strife, marital disharmony, juvenile delinquency . . . war," but most especially mental illness, which in some senses acted as an umbrella term for all of the preceding. Given the variety of explanatory levels pursued by science, from the molecular to the sociological, they sought a "general theory [to] deal with . . . *systems*."[45] The first edition of the journal contained a round-table discussion on a typically cybernetic topic: optimization of purposive behavior using information, and the conditions under which it was most desirable to operate with or without a decision-making brain. Reflecting the importance of invertebrate research to the field, the discussants debated whether the same considerations applied to humans and mollusks.

Schneirla himself chased grants from the National Institute of Mental Health (NIMH) without much success in the late 1950s and early 1960s. Meanwhile the Rockefeller Foundation took an interest in the mental health implications of some of Schneirla's mammalian work. "Thanks a lot for the progress report on the kittens," wrote Robert S. Morison, head of Biological and Medical Research. "It sounds like just about the best-controlled study ever made on the underprivileged child."[46] The implausibility of this sentence would have greatly offended Schneirla's scruples, and his continued obligation to appease such interspecific analogies in order to acquire funding was an ongoing painful feature of his career. (One early example of these compromises came in 1948, when his colleague and boss Lester Aronson had to convince the National Research Council that invertebrate instinct was also applicable to vertebrates.)[47]

Schneirla finally acquired NIMH funding in 1966, and the following year he noted arguments to advance the "human relevance of three research projects." The first of these concerned maternal-young relations in acts, about which Morison had already written. The second concerned the development of orientation in young kittens, which Schneirla related to orientation in blind humans.[48] Schneirla suggested that a valid question would be, "What advantages does a blind-born human have that a kitten lacks, with reference to establishing orientation on a non-visual basis?" But the behavioral patterns of ants were the most intriguing of all: "The leading question of most people, judging from many postlecture discussions, has been the list of striking similarities between human behavior at its supposed worst and many features (e.g. 'predatory war') attributed to army ants. One favorite question concerns the similarities between maladaptive circular milling in army ants and 'subway behavior' or mob behavior in people." Schneirla attempted to have his cake and eat it, offering such comparisons as the only way likely to cultivate funding opportunities, but unable to resist concluding with his customary skepticism: "such discussions . . . usually bring out the weakness of the favorite analogies used by many people in comparing human and insect behavior."[49]

By 1956 the Group Processes meetings were addressing "persuasion," an explicitly pragmatic cold war topic, but Schneirla declined the invitation to attend. Instead he was busy organizing two insect symposia sponsored by the IUSSI for the annual meeting of the AAAS. The theme, reflecting Macy interests, was to be insect communication. Given the traditional connection between communication and orientation, the topic also reflected the ONR's concerns. Schneirla talked Creighton into acting as a discussant on a panel that was to include both Wilson and his one-time mentor, Art Cole. The systematists, though respectful of Schneirla's work, were unsure what to make of the theme. Creighton, for example, did not know how to frame an introduction for Cole in that context: "If it were a matter of fieldwork I could talk about your virtues until the cows come home. But since you have been hooked into discussing insect communication I am somewhat baffled."[50] On a social note, Schneirla carefully arranged the attendant cocktail party so that it was for principals and discussants only, not myrmecologists per se.[51] Thus he excluded Wilson—perhaps to keep him out of Creighton's orbit, or perhaps out of antagonistic feelings of his own. Schneirla was happy to use Wilson to contribute to his project but shared the personal reservations of his colleagues about the trustworthiness and clubbability of the younger man.

Despite the social difficulties and the doubts of some participants, the panel went ahead. Vincent Dethier (of Johns Hopkins University) attempted to prove that flies could be made to dance in a manner somewhat similar to bees, oriented to light and gravity but without stimulating any foraging behavior in other flies. He also observed that under conditions of crowding, the flies would regurgitate food for one another. Dethier's experiments were interesting in that they attempted to prove prototypical social behavior in a nonsocial organism related very distantly to social insects. Schneirla's take on the paper was to use it as a means to a new, Morganesque interpretation of von Frisch's work: it suggested that the bees were not communicating information by means of abstract signs but were simply responding in behavioristic fashion to stimuli.[52]

Similarly, in his own paper, Schneirla argued that ants provided very limited comparators for human communication, since their "communication" was behavioristic rather than based around symbol or code. Trophallaxis was, in Schneirla's opinion, the most important of the apparently communicative behaviors in ants. In this way his paper linked with both Wilson's and Cole's. Wilson described how he had emulated another scientist's experiments with bees, this time with ants. He fed them radioactive isotopes and observed (using a Geiger counter) how the rate of spread through the colony was rapid through eusocial species but slow among *Pogonomyrmex*, which did not appear to engage in mutual regurgitation. His study echoed strongly Wheeler's classic experiments on trophallaxis using blue dye.[53] Cole (still at Tennessee) gave a rather standard account of the senses possessed by insects and their associated physiological apparatus. Even Creighton thought it "conservative." More significantly for the panel, however, Cole suggested that the concept of trophallaxis might be expanded beyond food exchange to cover also stimuli. Essentially, this reshaped trophallaxis (already developed by Wheeler in 1928 as a functional feature of social life) as a communicational phenomenon. Cole had clearly been talking with Schneirla about the topic; the following year Schneirla would publish and argue the same in full.[54]

If one considers these papers as an organized ensemble, a clear theme emerges: Schneirla's aim was to use the papers to suggest that ants' "communication" was pretty much behavioristic, and that trophallaxis, the most important of apparently communicative behaviors in ants, could be studied comparatively in different formic subfamilies, thus providing the key to myrmecology in general. This reading is reinforced by the title Schneirla imposed on Wilson for his paper: "Insect Communication by the Medium of Food Distribution."[55] Creighton was not convinced that other subfamilies displayed the same consistency of behavior as the dorylines studied by Schneirla, but he was certainly willing to participate in the debate that Schneirla was trying to kick-start at the event through this careful conjunction of papers.

The Ants Call Wilson: Communication and Its Potential

Wilson, I suggest, sensed in all this a possible new direction for his work.[56] Theoretically, the focus on communication was intriguing. Wilson also knew that he certainly had a broader myrmecological knowledge than Schneirla and would be better able to make comparative statements about the Formicidae. And perhaps a little of his old taste for controversy was at work too. Just as he had argued in high school that the very strength of biologists' reaction to Lysenko suggested that his theories must be plausible, so now the skepticism of Creighton et al. implied that there was something in Schneirla's ideas, too.[57] But rather than seizing on them uncritically, Wilson shaped his approach on two distinctive theoretical bases. In so doing he created a social myrmecology that was far more amenable to the cyberneticians than Schneirla's had ever been, thanks in part to the encouragement of Caryl Haskins.

Caryl Haskins (1908–2001) is a powerful and surprisingly little-discussed figure in the history of science. He came of a cybernetic heritage; his engineer father was involved in the development of the self-guiding torpedo, though he died when Caryl was very young. Caryl himself had an impeccable academic pedigree, with degrees from Yale and Harvard in the early 1930s. After Yale, Haskins engaged in radiation physics research at GEC. Receiving a substantial inheritance from a wealthy aunt, he proceeded to establish his own laboratories. During the Second World War he was heavily involved in the scientific effort, especially at MIT and through the National Defense Committee. After the war, Haskins returned to his own laboratories; he was also a member of RAND Corporation's board and editor of *Scientific American*. From the mid-1950s Haskins was president of the Carnegie Institute, a post he took over from Vannevar Bush and held for fifteen years. Haskins' tenure was marked by a commitment to the private funding of science research, with the aim of keeping scientists free to research whatever they, not the state, wanted. A paradox thus lay at the center of Haskins' life. On the one hand he was a classic cybernetician, with a strong interest in the theory of systems and participation in technocratic institutions. Yet thanks to his own financial circumstances he was able to maintain the belief that scientists could study without becoming involved in the sociopolitical systems that framed their research.

Throughout all of this, Haskins' first interest was the supposedly amateur (and hence supposedly politically disinterested) field of myrmecology. He went on collecting trips whenever he could and always kept artificial colonies in his laboratories. He and Schneirla planned a book in honor of Wheeler at the time of his death, though this never came to fruition.[58] *Of Ants and Men* was Haskins' first book (1939), based largely on an intimate knowledge of Wheeler's observations and theories. In particular, he advanced Wheeler's technique of looking at the behavior of extant "primitive" ants as an insight into the precursors of socially advanced varieties. He also laid out communication as the vital question to solve for the future, as well as the issue of how ants recognize each other.[59]

The sequel, *Of Societies and Men* (1951) was also argued almost entirely using ant data. In attempting to answer the questions of communication, the influence of Schneirla, whose army ant observations had flourished in the interim, came clearly to the fore. Following Schneirla, Haskins did not load communication in insects with informational or abstract content. Rather his account of the phenomenon was much more about stimulation and reinforcement of behavior—a functional glue (like trophallaxis) that held the nest together. There was even a little echo of Ogden and Richards in Haskins' distinction between two sorts of language, the contentful and the emotive. Haskins further proposed that there were three kinds of society: associative, familial, and integrative. Social insects were of the third variety, producing entirely specialized individuals that were hardly individuals at all. Human society tended to be a combination of the familial and the associative; however, there was also a "human culture society," a metaphysically separate or parallel kind of society that was integrative to a greater or lesser extent. This too was held together by a functional glue of "trophallaxis," whether in physical forms like breast-feeding or on the psychic plane in modes such as gossip, barter, and speech.[60] The two levels of human society, biological (associative/familial) and cultural (integrative), were, according to Haskins, held in a precarious balance, sometimes meshing rather too well, as in the case of totalitarianism, and sometimes conflicting. Democracy was the delicate device to hold the various aspects in a desirable tension or combination.

Where Schneirla was cautious in drawing analogies, Haskins had the touch of the bold popularizer. His interspecific comparisons pleased the cyberneticians and echoed, even anticipated their concerns in forums such as the Circular Causal and Group Processes conferences and the journal *Behavioral Science*. Haskins was interested in the interrelation of computing and behavioral work. "I particularly want to hear about the computer processing of the behavioral assay, which sounds most interesting. At the lab we have been thinking hard about the computer field too—not to mention the business of self-organising machines!"[61]

Haskins the well-connected bureaucrat was a valuable source of advice and advocacy to resource-stricken myrmecologists. Schneirla and his contemporary Neal Weber, both of whom struggled in their careers, benefited from his help. Haskins was equally happy to advance the prospects of the young Ed Wilson and his junior mentor Brown, who both had many more academic advantages. It was Haskins who recommended Wilson, Brown, and Weber as participants in Schneirla's mooted ESA/AAAS 1953 symposium.[62] Haskins was a natural diplomat, and apparently enjoyed advancing the efforts of whomever had interesting things to say about his beloved ants. He maintained good relations with many myrmecologists and was one of the few not polarized by the Brown and Wilson versus Creighton and Gregg feud. Alfred Emerson, who found Weber antagonistic, liked and respected Haskins even though he disagreed with his interpretations of ant behavior.

Schneirla, by contrast, did not cultivate a warm relationship with Wilson (though he did not go to the extremes of antipathy of Creighton). Before the taxonomy furor, Schneirla was apparently happy to have Wilson on the editorial committee for *Insectes Sociaux*.[63] However, Wilson and Brown raised Schneirla's hackles in their interactions concerning the 1956 symposium. Schneirla was a formal man whose junior colleagues never dreamed of calling him "Ted,"[64] and the confidence and familiarity of the Harvard pair, who did just that, rubbed him the wrong way. By 1957, Schneirla seemed somewhat set against Wilson; together with Michener, he agreed to overlook Wilson and ask his enemy Bob Gregg and the little-known LaBerge (Iowa State College) to depute for them at the forthcoming IUSSI International Congress in Paris.[65]

The personal dynamics among these three men, Wilson, Haskins, and Schneirla, shaped the science for which each became known. In both his populism and his cybernetic perspective, Haskins had a natural affinity with Wilson, whom he was coming to know in the early 1950s. Haskins joined Wilson's 1954–1955 expedition to the South Pacific. Together they searched for a missing link predicted by Wheeler between primitive, solitary ants and modern, social ants; their lack of success in this mission did not prevent Haskins from reporting enthusiastically on the trip to Weber and Schneirla.[66] Haskins had reinterpreted and popularized Schneirla, but he was only an amateur (if gifted and well-connected) myrmecologist; with Haskins' encouragement, Wilson could potentially give the cybernetic popularization academic credibility.

Pheromones: The Sweet Smells of Success

Creighton's guess about Wilson's future direction in theoretical biology was spot on. In the 1950s, Wilson showed interest in theoretical questions of allometry and mathematization of all kinds. One senses that he was trying out different theoretical approaches, hoping to find something big, something revolutionary—and more successful than taxonomic splitting—with which to make his name. Even when he had settled on pheromones as the key, the identity of actual pheromones lagged behind the theorization in the context of functional communication.[67] Wilson wasn't the only one for whom theory ran ahead of experimental proof; Sebeok's work on animal communication was also quite consciously a model in search of physical confirmation.

One attempt at theoretical synthesis came through allometry, the study of the growth of a part measured in relation to the whole.[68] For Wilson, the significance of allometry lay in its potential to explain the evolution of various castes by gradual accentuation and diminution of species-generalized physiological features. By relating this to the relative numbers of each caste in the nest, Wilson claimed to have a new theory of "adaptive demography." Another and rather interesting attempt came in collaboration with the biological mathematician William H. Bossert. Together they considered whether evolutionary pressure itself might be considered as a kind of communication with the organism, specifically whether isolation mechanisms might be considered as such.[69] Wilson's famous island experiments (for which Mayr did and Haskins could have claimed precedence)[70] were obviously connected to this early work, though the Uexküllian elements of Bossert's thinking were by then out of the picture.

Ernst Mayr apparently took exception early on to Wilson's faith in his ability to create "new" theories simply by describing old ideas in numerical terms (and, moreover, terms for which Wilson had to rely on more mathematically inclined collaborators). William Creighton was, of course, even harsher, accusing Wilson of "covering [his] ignorance with an umbrella of mathematics."[71] Eventually, Wilson redirected his theoretical efforts toward communication and information, biological phenomena that centered on pheromones. Fire ants, in which he "discovered" pheromonal communication, were his organisms of metamorphosis.

At the remarkable age of nineteen, Wilson was briefly employed by the state of Alabama—then suffering from an irruption of fire ants—as a professional entomologist.[72] Wilson had come to the attention of the Department of Conservation via a series of articles on the subject in the local press, in which he had been quoted. He took time out of his undergraduate degree to study the ants in the field, together with a colleague. Wilson's bold conclusions regarding the ants' ecology, taxonomy, origin, and behavior by turns impressed, amused, and outraged various experts in entomology. He claimed that the ants were a serious threat to Alabama's agricultural economy, that a new red form of the ant had very recently arisen through mutation, and that this aggressive new form was better adapted to life in North America than its premutated, darker relative.[73]

Wilson continued to work—now with Brown—on fire ants in 1956–1957. It was a natural area for anyone interested in ants, since the U.S. Department of Agriculture was at that time very liberal in its funding for related projects.[74] By 1958 Wilson had recanted his novel mutation theory, accepting that immigration was the most likely origin of the ants.[75] His colleagues had given the idea short shrift; the Georgia entomologist Murray S. Blum said it "could have been fabricated equally well by Lewis Carroll."[76] Wilson's hopes for his research's theoretical significance nevertheless remained pinned on phylogeny, inspired no doubt by his elder colleague Ernst Mayr and his treatment of species and race as biological, not formal units. In particular, the hunt for "missing links" had been a theme of Wilson's 1954–1955 expedition to the South Pacific. In a state of high excitement, he and Caryl Haskins searched (fruitlessly) in Western Australia for a missing link between primitive and eusocial ants, an extant species known as *Nothomyrmecia macrops* that had not been spotted since it was first discovered in the 1930s. Wilson's biggest phylogenetic success came later on the trip: the discovery in then Ceylon of another missing link, this time between two subfamilies, the Myrmicinae and Dolichoderinae.[77] As late as his 1958 article on fire ants for *Scientific American*, Wilson made no mention of communication or trails; as in his recent work the focus was on the ants as an example of evolutionary change in action. Nevertheless, the seeds had been sown; 1956 was also the year that Wilson participated in Schneirla's symposia on insect communication.

In the course of his work with Bill Brown on fire ants, Wilson had discovered that he could use the ants' venom to lay an artificial trail that workers would then follow. As ever, it did not take long for Wilson's exploits to make the papers. An Associated Press release announced this, Wilson's latest discovery, in February 1959; he had found that "the fire ant lays a trail of venom that other fire ants follow." From a practical point of view, this opened up the possibility that the substance could be isolated and laid as a trail to lead the ants into insecticide traps. Wilson, however, was groping his way toward a more theoretical set of ramifications. One can read in some newspaper articles from these months a tension between the "fire ants soon to be eradicated!" story that the papers wanted to tell and the academic angle that Wilson wished to get across. The *Worcester Gazette* concluded its March 16 story with a point clearly undigested by its author: "The substance, which seems to be chemically allied to or 'even identical with' the toxic substance of the venom, apparently serves to orientate ants along the right path in addition to acting as a 'releaser' in which the venom, for example, provides the trail."[78]

The most interesting feature of the fire ant was now defined by Wilson as its pheromonal trail. From the point of view of control and eradication, the discoveries about odor trails were of limited use, but by 1959 Wilson had long departed Alabama. At just the time in Harvard that he was moving away from classification to theoretical biology, the fire ants—setting aside their contested taxonomy—promised exciting and pertinent experimental avenues. Odor trails were a communicational feature of ant life; they were a matter of information. And information, as Wilson astutely surmised from recently prominent cybernetic and military research, was where the future of biology lay. A new word provided the necessary rallying point. In 1959 the term pheromone was adopted, capturing, as new terminology so often does, a whole hoped-for research program along with the entity itself.

By 1963, when Wilson wrote another article for *Scientific American*, he had formulated matters clearly, shedding all traces of the mundane practical concerns that had sparked the original research. The piece was confident and forward-looking. In it, Wilson mapped out future research, placing himself at its frontier. He claimed that previous research into animal communication had been hampered by an overemphasis on humanoid senses of sight and hearing. "It is becoming increasingly clear, however," he wrote, "that chemical systems provide the dominant means of communication in animal species, perhaps even in most."[79] Wilson distinguished two types of chemical effect: releasers (which elicit a behavioral response in the recipient) and primers (eliciting physiological changes in the first instance). Of these, he devoted more space to the former, largely because it was more dramatic and instantaneous in effect, making for easier and more striking experiments. For instance, a live ant smeared with juice from a dead nestmate would be carried off, wriggling, by its co-workers and dumped on the nest's refuse pile. Trials like this also invoked something much nearer to human communication than the slower effects of primers.

Wilson has not yet made public enough material to explore his thinking in the crucial years between 1958 and 1963. One notebook reveals that in 1961 Wilson was reading up on cybernetics and making notes on the formulaic relation between the number of choices made in selecting a message, the frequency limit of the signal, and the duration of the determining interval. Wilson himself has emphasized the importance of Stuart Altmann for his scientific development.[80] By the late 1960s, Wilson was having cyberneticians and information theorists read his manuscripts, and their language comes through very clearly in *The Insect Societies*. Wilson discusses the amount of information, measured in bits, that can be conveyed,[81] and the optimization of molecule size, where distinctiveness (which increases with molecule size) trades off against volatility (inversely proportional to size). "Vocabulary," "code," "syntax," and intensity versus frequency of signal are all cybernetic issues covered by Wilson in relation to pheromones in this 1971 book. Above all, while Wilson disciplined himself to understand difficult mathematics and theory, he was obeying his urge to synthesize: to create a biological "theory of everything." Following his notes, drafts, and publications, one can see him trying all combinations. At one point he tried missing links; at another, the evolutionary process was redefined by fire ants; for a brief moment it even seemed that primers could be a way to draw allometry into a paradigm of information.

Schneirla's critique of Wilson's professional review of the state of entomology, published in 1963, neatly captures the cybernetic departure that Wilson had made from his own work, filtered through Haskins. In composing the review, Wilson had delivered a typical, perhaps unintentional, snub: he explained that he had not covered Schneirla's army ants as Schneirla's own reviews of the area were "too recent and too good to bear duplication." Schneirla was not happy with the postdating of a reference to his work on reciprocal tactile and chemical stimulation. Wilson had named a 1957 paper; Schneirla wanted reference to a 1952 publication instead. The timing was crucial, because the earlier paper predated Wilson's much-hyped work on fire ants. Schneirla also took issue with some of Wilson's language, which he implied was somewhat anthropomorphic. Rather than referring to an "alarm" chemical, Schneirla suggested Wilson use the term "excitant-disturbance substance." Schneirla's most interesting critique, however, pinned down the difference between the atomistic and holist approach: "It is really a *theory* that: 'most social behavior in ants is mediated by chemical releasers.' [Besides,] isn't this really a theory of certain components of social behavior in the individual rather than of social behavior qua collective, organized behavior?"[82] But Wilson's atomized view of society was only to become further entrenched. He asserted in 1971: "The remarkable qualities of social life [in insect societies] are mass phenomena that emerge from the integration of much simpler individual patterns by means of communication. If communication itself is first treated as a discrete phenomenon, the entire subject becomes much more readily analyzed."[83]

Steven Heims has characterized the epistemological framework and cognitive style of the Macy cyberneticians along Wilson's lines, as "reductionist, atomistic, positivistic, pragmatic, conservative, mechanistic, and empiricist."[84] He goes on to explain the significance of this approach for the areas of human life into which the conferees hoped to transplant their science:

> Atomism in the social sciences manifested itself in the tendency to reduce social and political issues to individual psychology, usually to Freudian psychoanalysis of the individual. Underlying the focus on individual behavior and psychology is the premise that the understanding of societies can be built up from the understanding of individuals, just as in physics the knowledge of atoms forms the basis for understanding macroscopic matter.[85]

This description is both convincing and helpful. It contrasts rather neatly with the myrmeco-sociology of the Wheeler circle shortly before the war. Wheeler's science was sociological, not psychological; group-based, not atomistic; functional, not purposive; homeostatic, not ergonomic; representational, not energetic. What it *did* have in common with cybernetic science was the promise, whether realistic or not, of controllability. A grant proposal prepared by Wilson in 1968 made such a claim, with what seems like extraordinary confidence: "Our ultimate goal is the *in vitro* reconstruction of social behavior by means of integrative mechanisms experimentally demonstrated and the proof of that explanation by the artificial induction of the complete repertory of social responses on the part of isolated members of insect colonies."[86]

Nevertheless, it takes a little thought to clarify the differences between Schneirla and Wilson in this regard. They do not quite fit a simple atomist/holist dichotomy. After all, Schneirla's ideal was to reduce as much behavior as possible to an approach-withdrawal type analysis, an apparently reductionist method. Pushed for a definition of group behavior, Schneirla responded by quoting a student of his: "'Group behavior is what happens when you get more than one individual together.'—I offer that definition for discussion. I think it is not simple."[87] Perhaps the clearest way to understand the approach that Schneirla represented is to look at his colleague Grassé. Grassé was interested in the signals that make individual ants (or, in his case, termites) perform their various roles, but with the understanding that these signals are themselves a function of the social medium.

Wilson's atomistic approach was reflected in his enthusiastic adoption of Hamilton's haplodiploid genetic explanation for the natural selection of worker ants' apparently altruistic behavior.[88] Wheeler (as argued in chapter 3) made trophallaxis the functional phenomenon accounting for the origin and maintenance of the superorganism. His approach resonated with sociologists of the 1930s. But in opting for a resolutely individualistic account of genetic selection in the 1960s and 1970s, Wilson rejected accounts that focused on the emergent whole of the colony.

Thus Wilson did not like the idea of using trophallaxis as the basis for the comparisons between different types of ant, for he eschewed Schneirla's holist basis for choosing it. Whereas Schneirla wanted to avoid representational discourse in his work, Wilson actively chose to construct his science in a context of representational communication. Where Schneirla had been diplomatically skeptical about von Frisch's bees' ability to talk, Wilson gave whole-hearted endorsement: they really were communicating abstract information. "What is truly different about the waggle dance . . . is that it is a truly symbolical message that guides a complex response after the message has been given."[89] It is also possible, of course, that there was something of the young Turk approach in action, and that Wilson simply wanted to stake out new behavioral territory for himself. At any rate, Wilson was anxious to minimize the importance of trophallaxis. In doing so, he needed to discredit Wheeler's extension of the concept in the late 1920s. In a paper of 1967, "The Superorganism Concept and Beyond," published in *The Insect Societies,* and again in *Sociobiology,* he claimed that after its coinage in 1918, the term trophallaxis was applied with increasing laxity, soon coming to explain everything and nothing in insect society.[90] Although he had no particular axe to grind for Schneirla, William Creighton was predictably waspish in his abuse of Wilson's paper. He wrote to Bob Gregg:

> It is sad to realize that you and I were influenced by two such ninnies as Wheeler and Emerson. Ed hasn't a good word for either man. Or, for that matter, for most of the other 27 workers whose views he cites. Having disposed of the superorganism concept and later with trophallaxis Ed plops for Hamilton's haplodiploid theory of altruism as the proper explanation for social hymenoptera and actually refers to the "altruism" of larval wasps in feeding the males. He evidently got the raspberry when he proposed this to the Royal Entomological Society in 1966—and no wonder. It appears we have another Maeterlinck with us. . . . I was trained by men who thought that there was no greater sin than anthropomorphism. Why doesn't Ed join a Jesuit College and get into his element?[91]

As a result of his rejection of trophallaxis, Wilson has been dogged throughout his career by the specter of the superorganism, another of Wheeler's ideas and one tightly bound up with trophallaxis. Unfortunately for Wilson, the superorganism has been one of those ideas that has made its way into general culture, inspiring what popular enthusiasm there is for myrmecology as well as being reconstructed to suit a number of other fashionable topics, notably neuroscience and computing.[92] Not only this, but the superorganism and trophallaxis in particular have also remained key themes for his valued collaborator, Bert Hölldobler, whose transfer to Harvard in 1969 he arranged. Aware of Wilson's commitment to pheromones to the exclusion of trophallaxis and the superorganism, Bob Gregg commented at that time, "it is a surprise to hear what Ed Wilson is planning for Hölldobler, and it will be interesting to watch, especially if Hölldobler doesn't go along with Ed on trophallaxis. This would warm Ted Schneirla's heart I'm sure. Actually, I don't see why trophallaxis and pheromones can not each play an important role in the explanation of ant behavior and that of other social insects as well."[93] At the time of writing, Wilson has finally confronted his intellectual ghosts, and has a manuscript in preparation on the superorganism with Hölldobler.[94]

Wilson's pheromone work, in the fertile context of animal communication and cybernetics, found immediate favor. *Encyclopaedia Britannica*'s 1967 revision of its entry on ants took account of Wilson's new work, even under the unsympathetic direction of William Creighton. He recommended that the earlier article should be updated to cover pheromones. Unable to hold back his anti-Wilson bias, he wrote, grudgingly: "The subject is such a spectacular one that a great many of the younger workers are scrambling to get on the band-wagon and it is certain to attract much general interest."[95] This was a major revision, and to make space for it Creighton suggested deleting "ants as nuisances," which would "not be much of a loss," since professional opinion on the matter was so subjective. So, it was out with the fire ants; Wilson's new work had literally eclipsed the old.

The Building of "Semiocracy"

In addition to the push from taxonomy and the pull from theoretical, informational biology, departmental politics at Harvard played a dramatic part in Wilson's science, forcing him to define it rather decisively and assertively. The background history was of the usual kind: the politics of space and an ongoing feeling that funding was insufficient to requirements.[96] The Harvard Biological Laboratories had been built under the direction of G. H. Parker in 1930 and funded by Rockefeller Foundation International Education Board. Things were rather cramped from the outset, with particular problems regarding the housing of live animals. Another section was intended to house a new department of applied biology, containing Bussey staff, who were to be moved out of Jamaica Plain. During President Conant's first year of tenure (1933–1934), the biological sciences at Harvard underwent massive reorganization, with consolidation of separate departments. During the war the then still empty north wing, intended for applied biology, was taken over by a government-sponsored radio research lab operated by Harvard and MIT, conducting top-secret research on radar. After the war they vacated the building, but funding from the Office of Scientific Research and Development (OSRD) and the ONR continued. The newly formed National Institutes of Health and National Science Foundation provided extensive support for pre- and post-doctoral fellowships, and the number of scientists grew massively, with postdoctoral researchers outnumbering graduate students in many laboratories.

Shortly after the war, molecular biology was on the rise, creating further pressure on laboratory space and increasing animosity in competition for funding.[97] Building and renovation to make space for molecular practitioners began in 1960, at which time Wilson's caucus formalized itself as the Committee on Macrobiology. A press release from Harvard University on 30 March 1964 confirmed the centrality of information theory not just to molecular biology but to all areas of biology. "Edward O. Wilson, who studies how chemicals carry messages from ant to ant, will become Professor of Zoology; and Matthew S. Meselson, who studies how DNA, the chemical of heredity, carries messages from generation to generation, will become Professor of Biology."[98] Wilson's astute co-option of information theory not only enabled him to develop his own theories, but also provided a motivation (and perhaps model) to attempt a reconstruction of his department.

But the pressure was on for Wilson. Meselson had a new lab built for himself in the early 1960s, as did two other professors in molecular biology. In 1967 a breakaway Department of Biochemistry and Molecular Biology (BMB) was formed, physically split across the Biology and Chemistry buildings. Of the remaining biologists, about forty faculty members studied whole organisms; twenty, cellular and subcellular phenomena; and there was talk of a further division. Frank Carpenter, meanwhile, was concerned about the issue of space and the disproportionate amount being allocated to cell biologists. He was also unhappy that some courses were taught entirely by MCZ staff, while some members of the department proper had hardly any students.

It was at this time that Howard Evans, Alexander Agassiz Professor of Zoology at the MCZ, decided to resurrect the reputation of Wheeler, co-writing a biography with his wife, Mary Alice Evans. The Evanses' subtitle for the book, "Biologist," was significant. In defining Wheeler as a biologist—not a myrmecologist, not a natural historian—they made the case for the integration of various kinds of biology, and the recognition of entomology within it, such as Wilson lobbied for in real life.

In July of 1968 Wilson had wind of possible split and wrote to the Dean, Franklin Ford, proposing that the department should remain united, but that its professorships should mainly be devoted to cell biology and physiology—new professors in evolutionary biology being hired and resourced by associated institutions such as the MCZ. These would, de facto, take over the role of what had now become known as the Center for Environmental and Behavioral Biology.[99] Why was Wilson opposed to the split? One can posit a number of plausible reasons. Wilson may have feared that the cellular faction would merge with the previously fissioned BMB, forming a superdepartment on molecular lines. He may have felt that it was better to be a minor member of a powerful group than to have impotent autonomy. Certainly, he seems genuinely to have believed that biology needed both organismic and cellular elements in order to retain intellectual integrity. Keeping the two together would also maintain the widest possible territory for his ambitious "theories of everything." More conspiratorially, he may have feared that hormone research (a burgeoning field, construed at Harvard as molecular physiology) would subsume his nascent work on pheromones. And perhaps he simply misunderstood the funding situation and thought that he could attract massively increased support for the MCZ, thus building organismic/evolutionary biology into a viable alternate power base within a single department.

Wilson evidently had inside information, because in August 1968, members of Cellular and Developmental Biology (CDB) indeed wrote to Dean Ford proposing splitting away, and suggesting a parallel department of Evolutionary and Population Biology.[100] Wilson sent copies of his July letter to other colleagues after the August proposal had been made public. In the covering letter, his main concern was that the new department of CDB would converge with BMB, resulting in a huge and powerful nonorganismic department of biology.[101]

Ernst Mayr, meanwhile, was extremely negative about Wilson's proposals, calling them a "bull-in-the-china-shop approach." There simply weren't enough Agassiz professors to cover all the required areas, he retorted, and besides, it was contrary to the terms of their bequest to have them teach:

> What you are then proposing, is to give 15 quota posts to cellular-physiological specialists (most of whom unquestionably are and will be molecular) and to have four professorships not paid for by the Faculty of Arts and Sciences, for the entire vast area of animal behavior, ecology, population dynamics, population genetics, population cytology and ... environmental physiology and other aspects of organismic biology. Frankly speaking, I had thought that you would have a greater interest in evolutionary and environmental biology that to sponsor such a lopsided proposal.[102]

In September Ford convened an ad hoc committee on the future of biology, to include outsiders as well as insiders. Wilson continued to flog his cherished plan: "The MCZ, in my opinion, is the key to the whole thing."[103] He suggested Richard Lewontin and Alfred Crompton, the director designate of the MCZ, as members of the committee. Wilson wrote again to Ford the following month. The problem, he intimated, stemmed largely from Meselson and Jim Watson of the BMB being present at biology committee of professors meetings. Getting them off it would be a great start in diffusing tensions. The second tension, according to Wilson, was that appointments in evolutionary biology were largely autonomous, and not open to scrutiny by others. In response, he proposed giving the cellular men a say in MCZ and similar appointments.

In January 1969, a compromise was proposed by John Torrey and circulated to the members of the committee of professors. He proposed having two separate graduate committees in CDB and "Evolutionary, Behavioral and Environmental" (EBE) biology.[104] This answered the problem of involving those who were not regarded as teaching enough, while leaving open the option of a total split if the arrangement worked well. By February, Carpenter was circulating "John Torrey's case against the formation of two departments," with an "introductory statement" by E. O. Wilson. The extraordinary confidence and optimism of Wilson's claims, when considered in their departmental context, go a long way toward explaining his ongoing disciplinary posturing:

> We believe that during the next ten to twenty years ecologists will learn how to predict the effect of environmental change on populations, evolutionists will develop the capability of forecasting short-term evolutionary change, and behavioral biologists will explain whole patterns of behavior as the product of organized cell activity. These coming developments, so vital to the future well-being of man, will be the intellectual counterpart of the molecular biology revolution that has extended over the past twenty-five years. We feel that in order to place Harvard in the forefront of the new environmental, behavioral, and evolutionary biology, it is necessary for those of us who represent these subjects to organize our efforts and to devise a plan for the future.[105]

These breathtaking predictions bear a remarkable parallel to Warren Weaver's ambitions for the future of biology:

> The welfare of mankind depends in a vital way on man's understanding of himself and his physical environment. Science has made magnificent progress in the analysis and control of inanimate forces, but it has not made equal advances in the more delicate, more difficult, and more important problem of the analysis and control of animate forces.[106]

Weaver was von Frisch's chief patron, and that patronage was mediated through the scientific support of Donald Griffin. It seems plausible, therefore, that Wilson, working alongside Griffin at Harvard, recognized in von Frisch a model for interesting research that could be sold to influential funders. Wilson, with his outstanding communicational skills, also shared with his German colleague a media-friendly approach, another factor that must have persuaded him to emulate his rhetoric of appeal. In fact, Wilson's objections to a departmental split may also have been a matter of alignment with Griffin. Some five years earlier, Ernst Mayr had wanted to incorporate the MCZ and other affiliated associations into the department, changing the balance of power in favor of old-fashioned zoology. Griffin had objected on the grounds that this was an inappropriate "purification" of biology.[107]

By September 1969, the lobby for separate committees within a single department was gathering force, as was the call for more appointments in areas such as ecology, genetics of higher organisms, and animal behavior.[108] The evolutionary biologists were concerned not only about their numeric representation in comparison to the CDB group, but also about their level of credibility. The OEB set feared that any additional professorships gained at the MCZ would be at a lower level of credibility than if they had been approved by the departmental route, thus imperiling the overall credibility of organismic biology at Harvard.[109] Wilson and Paul Levine had an acrimonious discussion in the hall around this time; Wilson understood Levine to be accusing him, along with Carpenter and Mayr, of blocking appointments in molecular biology. Moreover, Levine seemed to intimate that standards exercised by certain evolutionists were not as high as the other biologists', lowering the tone overall and putting people off coming to Harvard. Levine intimated that Crompton, the next director of the MCZ (taking over from Mayr, and coming from the Peabody Museum at Yale) was just such an appointment, likely not to be admitted to the departmental committee of professors.[110]

Eventually, two separate committees were formed in 1969: Organismic and Evolutionary Biology (OEB) and Cellular and Developmental Biology (CDB). That was also the year that Hölldobler came to Harvard from University of Frankfurt, to work as Wilson's research associate (he stayed and was appointed professor of biology in Invertebrate Behavior in 1973). Hölldobler had been a student of von Frisch's protégé Martin Lindauer at Würzburg, and represented an appointment in the growth area of cybernetic communication.[111]

As the 1970s got underway, Wilson was beginning to see things turn more his way, perhaps due to his increasing age and consequent stature, and the retirement of the old guard, such as Frank Carpenter. Early in 1970, Wilson was the source for a story in *The Harvard Crimson* that a split in the Biology Department was imminent.[112] The two committees were drifting ever further apart, and by 1973 the biological laboratories were under great spatial pressure. The MCZ built a new laboratory wing in 1973, where Wilson moved with the newly promoted Hölldobler. Their fourth floor was a splendid space, with a bridge to the insect collections on the fourth floor of the old building. But even this was not *his* space; the MCZ was very strict about retaining control over it.[113]

Meanwhile, Wilson attempted to develop his own particular idea of biology by collaborating with selected colleagues in other fields. In 1973 Wilson was co-teaching a course on sociobiology with the primate anthropologist Irven DeVore. The newly appointed Robert Trivers also shared with Wilson and DeVore the desire for a synthetic evolutionary story, a biological "theory of everything." Trivers was a history graduate, and it was his children's stories on animal behavior that had brought him into contact with DeVore and inspired him to return to graduate school and work on animal behavior. Trivers wanted to see a general account produced: an interspecific history of psychology (including the question of altruism) based on evolution.[114] Wilson busied himself trying to set up a Committee on Behavioral Biology with responsibility for an interdepartmental program in behavioral biology, drawing in the departments of biology, anthropology, psychology, and sociology. He suggested himself as provisional chair.[115]

"It seems that what Harvard is trying to do is to set up an Institute of Insect Psychology." This, Creighton's assessment, was justified insofar as it applied to Wilson's ambitions, even if the plan was not explicit. Less plausibly, he sensed that Wilson that was "uneasy" with this state of affairs, suspecting that Hölldobler was being imported to "run" the planned institute.[116] Hölldobler was known to differ from Wilson on some key issues: "I find it interesting that Hölldobler does not go along with Ed in thinking that any ant activity can be explained on the basis of pheromones. Ed figured that he had buried trophallaxis but now Hölldobler may resuscitate it."[117] Given the history of their relationship—Hölldobler was marginally junior to Wilson and owed his initial appointment to him—his different perspective was challenging but not threatening to Wilson.

Nevertheless, Wilson was feeling the need to establish himself as a formic specialist. In the period between publishing *The Insect Societies* and *Sociobiology*, Wilson attempted to mend bridges with the myrmecologists, or, as Creighton put it, he made "an attempt to set more firmly in the saddle."[118] In 1972 he set himself up as one of them by giving the Harvard Founder's Memorial Lecture on W. M. Wheeler. At the same time he was building up "ant archives"—a "rogues' gallery" of portraits and documents relating to earlier myrmecologists—which Creighton surmised was "only part of a larger scheme to set up an ANT Institute at Harvard."[119] Wilson's love of ants is undeniable, but one suspects that he understood the need to establish himself as a master of one area, ready to move out to his big synthetic theory. The back cover of *Sociobiology* needed to say "acknowledged world expert in ants shows how they are relevant to the whole of life on earth," or words to that effect. Synthetic theories from nonspecialists are not so convincing; an apprenticeship in the specific must first be demonstrated.

In summary, Wilson's general modus operandi has not been intimidation but the control of the self-same substance peddled by his ants: information.[120] Through the machinations of Harvard institutional politics, Wilson could always be seen placing himself at the center of communication, offering to act as provisional secretary for ad hoc committees, talking to political players, and sending a constant stream of letters back and forth. He has also been adept at operating the media, frequently appearing in the *Harvard Crimson* as a means to control various institutional issues—not to mention national and international media when it came to the sociobiology furor.

Wilson himself occasionally appears to recognize the analogical connections between the content of his research on communication and his own position, mediated by the self-same process. He rewrote Wheeler's "Termitodoxa" idea (chapter 4) in 1979 along these lines. Ostensibly the lecture highlighted the illusion of "culture" qua paragenetic phenomenon. But of course, Wilson ended up taking pot-shots at his favorite targets. In one crossed-out aside, he considered saying, "An organization called Science for the Termites has protested. The International Committee Against Caste Equality [sic] has called for the burning of books and the eating of the authors."[121] And certainly, when Wilson's theory on insect communication was first published it was instantly perceived by the wider world as a managerial technique. *Sociobiology* and other works of what is now known as evolutionary psychology (such as Robert Ardrey's *The Territorial Imperative* or Desmond Morris's *The Naked Ape*) were reviewed in the new range of business psychology magazines. *Time* ran a "Behavior" section in 1975 in which a piece on scientific ethics derived from Wilson's *Sociobiology* appeared alongside a review of *Power! How to Get It, How to Use It* and *Winning Through Intimidation*.[122] Wilson has, perhaps unknowingly, taken the outworking of cybernetics as the science of management into his political machinations at Harvard's Department of Biology.

Communication in insects had been a major theme in entomology (especially social entomology) for some dozen years before Wilson shaped his program of sociobiology around it. Indeed, "sociobiology" itself had been advanced as a valuable description of interdisciplinary science by the now largely forgotten myrmecologist T. C. Schneirla. However, Schneirla took a scrupulously Morganesque approach to ants, denying them any powers of abstraction or intentionality. He also objected to the politically technocratic context in which this cybernetic view of insects was naturalized. In so doing at Macy conferences and on military- and Rockefeller-funded platforms, Schneirla inadvertently sharpened the cybernetic focus on insects. Meanwhile, Caryl Haskins raised the profile of Schneirla's work, which otherwise might have slipped between disciplinary stools (not to mention falling afoul of cold war politics). Haskins, a powerful patron of science, encouraged the formation of broad human analogies while maintaining a rhetoric of scientific disinterestedness. Wilson stepped forward to fill the spot suggested by Haskins' musings.

Wilson's science, in both content and form, could well be described by the neologism "semiocracy": rule by signs, or rule by control of communication. Even the semiotician Sebeok paid him homage: "your book is certain to become a classic, *the* cornerstone of evolutionary sociobiology."[123] The example of Sebeok's historical counterpoint to Wilson's story is, in fact, instructive. His zoosemiotics was very similar to Wilson's sociobiology, and yet the word is now entirely forgotten. His approach was all wrong, and highlights what was so right, or at any rate so successful, about Wilson's. Wilson's was not top-down control but self-generated; it was the hidden hand of society; it fitted military concerns; it fitted media presentation; it fitted the ambitions of the military-industrial-academic complex. Moreover, Wilson's construction of sociobiology as the science of communicating individuals complemented, and was advanced by, his strategies for institutional management at Harvard and in the world beyond.

Conclusion

Sick of hearing King Solomon's exhortation quoted in every banal piece of writing about ants, the myrmecologist George Wheeler penned an article in 1957 feelingly entitled, "Don't Go to the Ant."[1] With similar skepticism, Max Beerbohm noted, "the ant has a lesson to teach us all, and it is not good."[2] Ezra Pound, however, was more cagey about the didactic moral value of the Formicidae:

> The ant's a centaur in his dragon world.
> Pull down thy vanity, it is not man
> Made courage, or made order, or made grace,
> Pull down thy vanity, I say pull down.
> Learn of the green world what can be thy place
> In scaled invention or true artistry.[3]

These lines, taken from the most famous of Pound's *Pisan Cantos*, encapsulate the most contentious interpretative problem of the poem sequence.[4] Did Pound renounce fascism during his imprisonment, or did he write the series as a defiant celebration of its values? Proponents of the former theory interpret the repeated call to "pull down thy vanity" as an indication that Pound had indeed perceived the boastful emptiness of the totalitarian fantasy and had repented of it. Those who wish to castigate Pound wonder whether he was not in fact addressing his captors at the time of writing. Either way, Pound's choice of the ant as man's humbler allows his meaning to remain ambiguous, for ants themselves were an object of keen scrutiny during the twentieth century, equally provoking fascination and contempt in the minds of their observers.

In no sense can it be claimed that myrmecology showed a trend toward increasing objectivity over the course of the century. Along the way, there were numerous subtle but powerful shifts in the choice of paradigmatic organism, caste and behavior. These choices of scientific study underwrote the moral lesson that respective researchers read into their ants; correlatively, differing cultural perspectives shaped choices of scientific study.

Mutual feeding was obviously central to both Wheeler and Forel. To Forel it represented the wonderful extreme to which ants took their sharing responsibilities; to Wheeler, the tawdry instincts that underpinned the greater economic forces of growth and equilibrium. Accordingly, there were various important differences between mutual feeding as it was construed by Wheeler and Forel. After reading Roubaud, Wheeler's focus shifted from adult-adult reciprocal feeding to the exchange between adults and larvae. Forel, although recognizing that both phenomena existed, focused primarily on the exchange between adult workers as the normal act. Forel did not discuss the fact that, after bringing up her first brood, the queen ant was dependent on food regurgitated by her progeny.[5] This was apparently a point on which Forel, a believer in the importance of hygienic maternity, did not wish to dwell. Nor, despite publishing after Wheeler's 1918 paper on the subject, did he address the fact that the ant larvae exuded a fatty substance through their chitinous skin that was assiduously licked up by their nurses. Forel's discussion of trophallaxis in Wheeler's sense was confined to two species, making it sound fairly unusual; it was also given as only the eighth in a list of eleven worker behaviors.[6]

The different foci of the two researchers also extended to their choice of paradigmatic caste. For Forel, talk of the "queen" was invidious; the workers were the "true queens of the formicaries," the "hard-working communists" that in fact ruled the nest.[7] Exchanges involving larvae or the queen were of less interest to him; his writing on mutual feeding between workers (and on the honey-ants, the logical evolutionary extension of mutual feeding) was given considerably more emphasis than trophallaxis in Wheeler's sense. Wheeler, on the other hand, made trophallaxis the centerpiece of his entire analysis of ant feeding.

The different choices of paradigmatic reciprocal feeding arrangements arose partly because Wheeler's principal interest was in the origin of sociality while Forel was more concerned with its maintenance. For contingent historical reasons, it happened that maternal-offspring relations were more likely to hold the key for Wheeler than worker-worker ones. One reason for difference about which we may be certain was that each man was searching for a different human angle on the phenomenon of mutual feeding. Forel was looking for a high-minded moral while Wheeler was happy to find something that appealed to his sense of humor. Forel's workers evolved through a Lamarckian perfection of cooperative behavior; Wheeler's workers emerged from a parasitic relationship within family life: mothers parasitized their offspring and in so doing unintentionally created society.⁸ The twisted, sublimated psychology of maternal exchanges explained the behavior of socialized individuals and provided a case study of what the unsocialized, creative individual (such as the scientist) should avoid. Forel, having compared the ants' addiction to symphilic exudates with alcoholism, could not bear to explain the raison d'être of his eugenic utopia by the same mechanism. His workers were models of civic-minded socialists.

The long-running themes of instinct and intelligence also framed choices of ant behavior for observation and description by myrmecologists. Their starting materials were the many stories from the nineteenth century regarding the apparent intelligence of ants, able to thwart the most fiendish systems for protecting human food stores, particularly in the colonial setting. A particularly famous narrative concerned a nest whose route home from foraging was one day bisected by the construction of a railroad. After a number of ants had been crushed by passing trains, their colleagues allegedly figured out that in order to avoid further deaths en route, they would need to burrow under the track—which they then proceeded to do.⁹ Identical stories, it should be noted, were quoted by both evolutionists and antievolutionists, to underline the wonder of animal intelligence and instinct respectively.

Forel reversed the expected order of instinct and intelligence in the construction of his mnemic theory of inherited memory. The intelligence of Forel's ants was not the Thomist rationality that Wasmann was so eager to identify and condemn in the psychiatrist's work; rather, it was simply behavioral plasticity. Instinct, in his schema, was an economic and convenient encapsulation of all the successful flexible behaviors that the species had essayed in its phylogenetic past. Instinct now became more problematic than ever before: it was revealed as both a phenomenon with a more complex history and identity than previously imagined and also, potentially, as an evolutionary threat for humans, since in the changing world the ability to remain flexible was all-important. The Chinese, with their "stultified" idiographic language and Mandarin governmental system, were commonly considered the exemplar of humanity that had allowed its evolution to freeze up like instinct, and had hence come to a full stop.

Forel's attempts at demonstrating the ants' capacity to learn were mainly to do with inducing tolerance between different nests or species, as befitted his internationalist politics. Others, attempting to disprove Fabre's insistence on insects' rigidity of instinct, created a narrative about the evolution of maternal provisioning behaviors, and hence sociality. Wheeler, taking Forel's explanation (or something very like it) as read, returned to the "Chinese problem" (or, for him, the "Soviet problem"), and reframed the issue as one of creative individualism versus socialized conformity.

T. C. Schneirla returned to the intelligence of ants, setting them mazes to learn just like rats. His point, underlined by observation of the suicidal milling Ecitons, was that ants were "intelligent," but in a very limited sense. Yes, they could learn, but that process could be modeled as a simple approach-withdrawal mechanism susceptible to fatal systemic errors. It was not intelligence in anything like the sense of rationality. Schneirla's lifelong battle against anthropomorphism was perhaps best summarized by his student Herbert Birch, who asserted at a Macy conference that if animals "communicated" alarm chemicals, then so did starfish soup—a *reductio ad absurdum* that did nothing to convince the cyberneticians, desperate to hear the possibility of nonhuman abstract representation confirmed. Others took animals' apparent cleverness as a well-made rebuke to human hubris. Huxley and Bateson considered the possibility that "intelligent" human behavior was more ritualized than commonly conceived; Sebeok took the feats of Clever Hans as a timely reminder that more human communication was unwitting than his fellow species like to imagine.[10]

For the cyberneticians, von Frisch's bees fulfilled perfectly the phenomena that they wanted to see in animals: rational navigation, goal-seeking behavior, and, most important, abstract communication. They were intelligent animals almost in the sense of "cleverness" that early twentieth-century scientists—the anti-nature-fakers—mocked. Of these zoological cyberneticians, Donald Griffin was an important player, and a good deal more deserves to be written on him. He was patron of von Frisch in America and a colleague of Wilson's at Harvard, a keen advocate for the appreciation of animal mentalism. Following in von Frisch's footsteps, and encouraged by Griffin, Wilson's ants were also clever, communicating their needs by pheromonal language.

Thus the question of intelligence elided, as it so often had for at least the past century, with the question of language; the two were still, it seemed, coterminous.

Forel's and Fielde's early interest in language lived on for a time in the Basic English movement and in the often conflicting efforts of Ogden and Richards. Pound too, as it happened, was an admirer of Basic English; his paean to the language was quoted in "Basic English (In Basic)," a 1939 pamphlet issued by the Orthological Institute.[11] He judged Basic "a magnificent system for measuring extant works ... as a training exercise for young poets, as a means for the diffusion of ideas," concluding that "the advantages of Basic should be obvious to any man of intelligence." Like the Pisan Cantos, these words capture the ideological ambivalence of myrmecology, this time in the context of language. The masses would find it convenient to read and speak, while the academically superior would understand, study, and improve its basis. In the career of Richards one traces a similar line from the elitist practical criticism of the interwar period, with all its social overtones, to the covert plutocracy of the cold war.

The early myrmecologists' concerns for world peace and internationalism were better reflected in Ogden's approach than that of his younger colleague. Both Fielde and Forel had framed the subject in terms of international harmony, and indeed, these were the terms in which the Rockefeller Foundation was keen to support the work of von Frisch after the Second World War. In the cold war period, anxieties about the interrelatedness of mutual incomprehension and aggression persisted. Vietnam survivor Joe Haldeman predicated his *Forever War*, a science fiction allegory of his own experience in that conflict, on the futility of fighting an enemy with whom one could not speak. Haldeman's hero Mandella ponders, "What might have happened if we had sat down and tried to communicate?"[12] At the novel's end, he is vindicated. Arriving in the future at the end of the war (after some 1,143 years of fighting), Mandella is informed that miscommunication with the Tauran enemy was indeed to blame. The Taurans had been unable to communicate with humans, he discovers, because "they had no concept of the individual; they had been natural clones for millions of years." Once humans, in the millennial interim, had evolved to the same antlike condition (complete with a sterile or "homosex" worker caste), they were able to talk with the enemy. The first question mutually exchanged was, "Why did you start this thing?" to which the answer was, "Me?"[13]

But communication was no straightforward issue during the cold war, that era of misinformation, double agents, and double bluff. Language had, until then, been held to be naturally translatable. Even the most fiendish code did not pathologize the nature of language itself. Now, this assumption came under attack on two fronts: by extreme McCarthyite suspicion, and, from academics on the opposite political wing, by the specter of relativism.[14]

Or *was* language previously held to be reliable? Forel had, from a clinical perspective, always taken a strong interest in lying and deceit, and in particular its ability to be expressed by a person who, on some level, believed he was telling the truth. Forel was torn between his natural trusting attitude and the evolutionary implications of his engramic theory—namely, that deceit (such as the ants' fatal taste for the exudate of parasitic symphiles) could also become hardwired into behavior. Ogden and Richards divided language into two sorts: the lucidly communicative, and the emotive, which could be used by the right sort of people as poetry or by the wrong sort as propaganda. It was the danger from the emotive sort of language that caused them most concern, since deception could be practiced on an emotional level without intellectual judgment ever becoming engaged. Wheeler too, toward the end of his life, increasingly saw deceit as a central phenomenon of nature. Much impressed by Freud, perhaps in the wake of his breakdown, he came to the conclusion that most apparent virtues, notably sociality, were rooted in self-deception. Wheeler also followed Nietzsche in explaining this ubiquitous phenomenon according to the phylogenetically ancient impulse to power within the organism. Around the time that Lorenz was starting out on his aggression and dominance theory, Wheeler was writing something rather more interesting (if less experimentally grounded) on the subject. What made it more interesting than Lorenz's conceptualization of aggression was that it was based on "bluff"; dominance was not achieved by acts of aggression but by *convincing* competitors of dominance—a matter of communication—whether or not that could be backed up by physical acts. Thus Wheeler saw deceitful communication at the heart of social order right through the animal kingdom, from ants to parsons, from monkeys to lawyers.[15]

NIH beneficiaries and Macy conferees addressed similar issues, mainly within a psychiatric discourse. Communication sane and insane, rational and irrational, were matters of principal concern. Margaret Mead made the point that discussants attuned to the insidiousness of neurosis were already reaching: that "normal English" was anything but.[16] The zoological strand of the Macy conferences highlighted the possibility of bluff in ritualistic communication. G. E. Hutchinson described the activity of certain Empid fly; normally the male wrapped an insect offering in silk and presented it to the female as part of a mating ritual. The purely ritualistic aspect of this behavior was demonstrated by the fact that the female did not actually appear to eat it. In some species, the male wrapped up nothing in silk and presented it, and Hutchinson claimed this was a version of the "Armstrong effect," the ritualization or displacement of important behavior. Bigelow, however, claimed right away "that is a jilting trick."[17]

Notwithstanding Bigelow's suspiciousness, many other Macy participants and cyberneticians were more trusting in their outlook. The Manicheans von Neumann, Bigelow, and Bateson were balanced by the Augustinians, including Wiener himself.[18] Wiener recognized a whole slew of counterinformational strategies in modern society: secrecy, message jamming, and bluff. But unlike Wheeler and von Neumann, who saw games of bluff as part of life's very fabric,

Wiener saw them as undesirable and potentially avoidable.[19] It was an article of faith with Wiener that, unlike humans, nature does not bluff. (Proof of this postulate, he admitted, was impossible, since it amounted to the inductive proof of induction.) The zoologist Helen Spurway and her husband, J. B. S. Haldane, hypothesized the possibility that there was such a thing as lying in nature, but that it was very much the exception rather than the rule. It appeared to apply only to interspecific communication (deception of predators, for example); intraspecific communication was considered as a rule to be "advantageous for the species."[20]

E. O. Wilson has been a surprisingly wide-eyed Augustinian, sharing Wiener's view of nature, not von Neumann's. There is no sign in *The Insect Societies* (1971) of dissimulation in the ants' use of language. Even in *On Human Nature* (1978), where Wilson fearlessly naturalized such undesirable features of human culture as violence, there was little sense of nature—human or otherwise—as innately deceitful. Overall, the book was slightly darker in tone than its two predecessors, but there was only one mention of deceit, in the context of the self-deceit and anxiety that results from human uncertainty about which level of altruism (familial or "soft," that is, stranger-directed) one should practice in a given situation.[21] Wilson has never really taken on the depths of the selfish gene theory and its worrying implications for reliability. As he remarked, "As an entomologist who has spent most of his career on hands and knees studying the highly organized and super-efficient colonies of ants I am often asked the following question: what can human beings learn from these insects? And the answer I give, with my fingers steepled, my mouth pursed, and my eyes squinted, is—*Nothing*. Not a thing and it would be a major Aesopian error to believe otherwise."

Wilson's analogical reticence in this 1998 oration to his old fraternity appears admirable, entirely in line with the objectivist principles of modern science. Yet as he continued, even he could not resist doing just what he condemned, albeit placing the words in someone else's mouth:

> However, we can learn certain broad principles by comparing ants with people. For example, as President Lowell said in 1930 when conferring an honorary degree on W. M. Wheeler, my predecessor as curator in entomology at Harvard, (and I paraphrase him here), "Professor Wheeler has shown that social insects, like human beings, can create civilizations without the use of reason."[22]

Wilson ended up at an extraordinary statement, and a most unpalatable one at that. His words conjure up the incredible belief in the power of analogy that has marked the entire century under study. Whether one took the ant's ways as a positive or a negative injunction, it was apparently impossible for myrmecologists not to consider them and become wise.

Forel's overtness in drawing human-ant analogies made it easy for his critics to accuse him of naiveté. At the time Carlo Emery felt free to disagree; even Forel's somewhat weak and overawed brother-in-law Edouard Bugnion found it possible to express his unease at the socialist lesson allegedly provided by the ants. Subsequently, Forel has been routinely described as an eccentric.

Wheeler was more complex, more subtle. His participation in the nature-fake furor at the outset of his myrmecological career set out his methodological commitment to avoid anthropomorphism. His ongoing struggle not to be labeled a "mere" natural historian meant that the conspicuous eschewal of anthropomorphism (and its linked values of sentiment and anecdotalism) marked his entire life. But he couldn't help himself. "The Termitodoxa" was right out of the Swiftian tradition, and the trajectory of its moral meaning derived additional significance from its context of delivery. The social governance exerted by the biologist caste of the termites was, suggested Wheeler, the role that should be taken up by members of the American Society of Naturalists to whom he spoke.

Wheeler always withdrew at the last moment from following through on his "jokes" with concrete recommendations. Putting his thoughts in the mouth of termite was a literary strategy for keeping the author at one remove from the narrator's opinions. In real life, Wheeler was infamous for unleashing his humor to full caustic effect. A frequently repeated anecdote concerns an unsolicited visitor from the California Academy of Sciences who severely tried Wheeler's patience in his time-consuming attempt to scrounge specimens. During the course of their tediously long conversation, the visitor mentioned that he had been given a new bug by a Mr. Nast, but hesitated to name it *nasti*. Sensing a way to be rid of his unwanted guest, Wheeler dryly riposted, "why don't you call it *nastianus?*"[23]

Another target for Wheeler's humorous analogizing was, predictably, the female of the species. In many insects, the female "appetition" necessitated that she should be brought a food offering, or a fake food offering, lest she turn her unfortunate suitor into lunch. Wheeler explained the instinct in human terms:

> That Virgil's *varium et mutabile semper femina* is not strictly true and that the female of such a highly endowed mammal as man has a similar persistent instinct is only too apparent. Perhaps the cave women had nothing to do with the cave men until the latter brought them steaks of the aurochs or the mammoth. But we need not go so far back in history to find analogies. There are females in our midst whose coyness has been overcome by a lobster and champagne supper, or the present of a dia-

mond ring, a motor-car, or a bank account ... Some, however, have been known to succumb to such easily procure trifles as a bunch of violets or a lock of hair.[24]

This paper and the "Termitodoxa" were part of an anthology of similar essays, *Foibles of Insects and Men*. The overall effect of the collection is that of the self-gainsaying claim: "just kidding!" In fact, Wheeler counted *Foibles* as one of his five most important publications, suggesting that the collective morals of the text *were* of significance to his biological mission.[25]

Wheeler's overriding collective moral was, needless to spell out, a eugenic one. A eugenic plan was central to the whole Termitodoxa scheme: "by confining reproduction to a special caste, by feeding it and the young of other castes on a peculiarly vitaminous diet and by promptly and deftly eliminating all abnormalities, we have been able to secure a physically and mentally perfect race."[26] Wheeler served on the advisory council of the Eugenics Society of the United States of America (dedicated to preventing the immigration of "inferior" races) and considered the task "a very great honor."[27] Other members of the committee included Wheeler's close friend (and former entomological colleague) David Fairchild, and his close friend of latter years, the evolutionist, philosopher, and sometime entomologist George H. Parker.[28]

Wheeler's trophallactic view of ants, by naturalizing an economic vision of nature, created a covert social agenda for humans. It contrasted with the overtly political analogies drawn by Forel, where lessons from the ant world could be directly adopted by humans. Rather than attempt to justify his analogies, which in any case were not explicit, Wheeler's strategy was implicitly to justify his authority to analogize. His introduction to *Foibles* recapitulated in brief the techniques of elite natural history outlined in chapter 6. Curiosity, he began, was often considered a foible of entomologists, written off as eccentricity. But curiosity, Wheeler continued, was in fact a virtue, and the greatest foible of all was the failure to face up to the objective lessons of nature: "We are coming to see that there are two very different types of thinking, one characteristic of a small portion of mankind, the scientists, when functioning as such, and one which is the only type indulged in by most humans and also by the scientists when they are off duty."[29] Once that authority was granted, the emotional, irrational "autistic thinking" of his critics was established, and any subsequent dissent could be written off as invalid by definition.

Wilson's use of analogy has been subtler still. In fact it might not even be appropriate to describe him as having a "strategy" for the use and justification of analogy in an intentional sense. Rather, Wilson's analogy has gained credibility from being embedded in a system beyond his control—perhaps, even, his awareness. His communicative ants were entrenched in a bigger program to normalize communication: for mental health, productivity, national security, and military effectiveness—a program of technocratic control, incidentally, dreamed of by Wheeler and his sociological and literary colleagues. Whereas Wheeler labored hard to distinguish himself, qua scientist, from the antlike Paretian mass, cyberneticians either convinced themselves that they were creating machines, not people, in this mold, or else saw it as a necessary mirroring of the communist foe.[30] Ants, models singly of autonavigation and jointly of nonintelligent purposiveness, were perfect to represent this system. Schneirla, it should be remarked in passing, rejected the ideology of the technocratic agenda and was about the only myrmecologist in this study who set out to avoid analogizing in his work with genuine scruple, though he could not prevent it from being interpreted within a cybernetic framework (not least because of the military origin of his funding).

Cybernetics, for those who would accept it, also solved the old anthropomorphism problem of natural historians. "Two legs better" had been the cliché of Victorian evolution-deniers struggling to come to terms with Darwinism; T. H. White protested that Julian Huxley et al., in retaining the language barrier between humans and animals, were still doing the work of the pre-evolutionists. In fact, they had done so to avoid the charge of anthropomorphism, but worrying about this charge had actually had the opposite effect to that intended by materialists, for biologists balked at treating humans like animals when it came to language. It was not until both humans and animals could be described as machines (in the discourse of cybernetic biology) that the Gordian knot was finally cut: Wilson's ants were the perfect linguistic cyborg.

If Forel could travel through time, he would be delighted to read Wilson's remark that "socialism really works under some circumstances. Karl Marx just had the wrong species."[31] What he would not realize would be that, thanks to the changing cultural framing of analogy in the interim, the meaning of Wilson's words had swung 180 degrees from their apparent agreement with his own socialist dream. Wheeler's ants modeled the machine, which then was used by Wilson to model the ants. The analogy had been black-boxed. Richards, perhaps the most obvious person to participate in both stages of this process, not surprisingly found himself in a bit of a cognitive tangle. Ignoring T. S. Eliot's warnings about the "technological momentum" of mass culture, he wrote a bleakly comic play for "tele-screen" set in the futile and thinly disguised "Institute for Advancing Studies." "A Leak in the Universe" features, literally, a black box: a metaphor for how representation consumes itself. As the scenes unfold, a series of academics fail to make sense of the box, just as the prologue forewarns:

> This play's about itself, and so are you—
> Stalking yourselves and studying what you lost . . .

The myrmecologists were stalking themselves all along.

Introduction

1. Mary Shelley, *Frankenstein, or, the Modern Prometheus* (Oxford: Oxford University Press, 1994), 77.
2. Moses Harris, "Glanville's Fritillary," in *The Aurelian or Natural History of English Insects namely Moths and Butterflies* (London, 1766).
3. Sir William Petty (1677), cited in Arthur O. Lovejoy, *The Great Chain of Being: A Study in the History of an Idea* (Cambridge, MA: Harvard University Press, [1936] 1964), 190.
4. On early insects, see Jonathan Sheehan, "The Mind and Metaphysics of Early Modern Ants" (manuscript, 1999), and Jean-Marc Drouin, "L'Image des sociétés d'insectes en France à l'epoque de la Révolution," *Revue de Synthèse* 4 (1992): 333–45.
5. See, for example, Sally G. Kohlstedt, Michael M. Sokal, and Bruce V. Lewenstein, eds., *The Establishment of Science in America: 150 Years of the American Association for the Advancement of Science* (New Brunswick, NJ: Rutgers University Press, 1999).
6. See the essay on sources for publications on U.S. economic entomology.
7. W. Conner Sorensen, *Brethren of the Net: American Entomology, 1840–1880* (Tuscaloosa: University of Alabama Press, 1995), 92–106; Herbert H. Ross, *A Textbook of Entomology* (New York: John Wiley, 1948), 17; E. O. Essig, *History of Entomology* (New York: Hafner Publishing, [1931] 1965), 811–952; Paulo Palladino, *Entomology, Ecology and Agriculture: The Making of Scientific Careers in North America, 1885–1985* (Amsterdam: Harwood Academic Publishers, 1996), 47–73; P. W. Reingart, *From Arsenic to DDT: A History of Entomology in Western Canada* (Toronto: University of Toronto Press, 1980).
8. Michael Worboys, "Tropical Diseases," in *Companion Encyclopedia to the History of Medicine*, ed. W. F. Bynum and Roy Porter (2 vols., London: Routledge, 1993), vol. 1, 512–36.
9. For biographies, see Arnold Mallis, *American Entomologists* (New Brunswick, NJ: Rutgers University Press, 1971); Essig, *History of Entomology*; Sorensen, *Brethren of the Net*.
10. Mallis, *American Entomologists*, 119–73. Paulo Palladino gives a skeptical account of the land colleges' foundation, presenting it as the government's attempt to strengthen the alliance between the eastern and western states during the Civil War. Palladino, *Entomology, Ecology and Agriculture*, 24–26.
11. Charles E. Rosenberg, *No Other Gods: On Science and American Social Thought* (rev.

and expanded ed., Baltimore: Johns Hopkins University Press, 1997), 135–72. On the dogma of pure research, see Roger L. Geiger, *To Advance Knowledge: The Growth of American Research Universities, 1900–1940* (New York: Oxford University Press, 1986).

12. L. O. Howard, *A History of Applied Entomology (Somewhat Anecdotal)*. Smithsonian Miscellaneous Collections 84 (Washington, DC: Smithsonian Institute, 1930), 105–9.

13. Historians disagree about the relative significance of the federal government and provincial entomologists in the professionalization. See Rosenberg, *No Other Gods*, 135–84; Philip J. Pauly, "Modernist Practice in American Biology," in *Modernist Impulses in the Human Sciences 1870–1930*, ed. Dorothy Ross (Baltimore: Johns Hopkins University Press, 1994), 272–89; Conner Sorensen, "The Rise of Government Sponsored Applied Entomology, 1848–1870," *Agricultural History* 62 (1988): 98–115; Margaret W. Rossiter, "The Organization of the Agricultural Sciences," in *The Organization of Knowledge in Modern America, 1860–1920*, ed. Alexandra Oleson and John Voss (Baltimore: Johns Hopkins University Press, 1979), 211–48; L. O. Howard, *Fighting the Insects: The Story of an Entomologist. Telling the Life and Experiences of the Writer* (New York: Macmillan, 1933), 65–66.

14. Mallis, *American Entomologists*; Essig, *History of Entomology*; Sorensen, *Brethren of the Net*, 86–89. Another possible though apparently less common career path was to progress from practical farming to state appointments in agriculture. Thus, B. M. Lelong (1858–1901) started as an orchard manager and, through his interest in pest control, was appointed to the California State Board of Horticulture in 1886 (Essig, *History of Entomology*).

15. See the essay on sources for publications relating to these strategies.

16. J. F. M. Clark, "Eleanor Ormerod (1828–1901) as an Economic Entomologist: 'Pioneer of Purity Even More than Paris Green,'" *British Journal for the History of Science* 25 (1992): 431–52.

17. Maxwell Lefroy, editorial, *Annals of Applied Biology* 1 (1914): 1–4.

18. Arthur Shipley, *An Infinite Torment of Flies* (Cambridge: privately printed at the University Press), 1905.

19. K. Jordan, "President's Address," *Proceedings of the Entomological Society of London* 4 (1929): 128.

20. Frederick Augustus Dixey, "President's Address," *Transactions of the Entomological Society of London* 58 (1909): lxxxvii–cxli.

21. Council of the Entomological Society of London, letter to the Prime Minister, *Transactions of the Entomological Society of London* 66 (1917): civ–cv.

22. "Harold Maxwell Lefroy," *Entomologist's Monthly Magazine* 61 (1925): 259–60.

23. Karl Escherich (1871–1951) was the most important figure in German hygienic forest entomology. See Sarah Jansen, "Chemical-Warfare Techniques for Insect Control: Insect 'Pests' in Germany before and after World War I," *Endeavor* 24 (2000): 28–33; idem, "*Shädlinge*" *Geschichte eines wissenschaftlichen und politischen Konstrukts, 1840–1920* (Frankfurt: Campus, 2001).

24. Arthur Shipley and David Sharp were two important insect zoologists making such claims in the early twentieth century. Charlotte Sleigh, "Empire of the Ants: H. G. Wells and Tropical Entomology," *Science as Culture* 10 (2001): 33–71.

25. Jansen, "Chemical-Warfare Techniques for Insect Control"; idem, "Ameisenhügel, Irrenhaus und Bordell: Insektenkunde und Degenerationsdiskurs bei August Forel (1848–

1931), Entomologe, Psychiater und Sexualreformer," in *Kontamination*, ed. N. Haas, R. Nägele, and H.-J. Rheinberger (Eggingen: Edition Isele, 2001), 141–84; *idem, "Shädlinge" Geschichte eines wissenschaftlichen und politischen Konstrukts*.

26. Clare Lloyd, *The Travelling Naturalists* (London: Croom Helm, 1985).

27. Wallace's theories about evolution and mimicry became the centerpiece of Edward Poulton's dominance of British entomology at the turn of the twentieth century. Recent studies on Wallace include Peter Raby, *Alfred Russel Wallace: A Life* (Princeton, NJ: Princeton University Press, 2002); Jane R. Camerini, ed., *The Alfred Russel Wallace Reader: A Selection of Writings from the Field* (Baltimore: Johns Hopkins University Press, 2002); and Martin Fichman, *An Elusive Victorian: The Evolution of Alfred Russel Wallace* (Chicago: University of Chicago Press, 2004).

28. For example, Colonel William Sykes (1790–1872) viewed ants in India, while the American medical missionary T. S. Savage (1804–1880) did the same in Africa.

29. N. Annandale, "The Retirement of Lieutenant-Colonel Alcock," *Records of the Indian Museum* (1908): 1–3.

30. E. E. Green, "President's Address," *Transactions of the Entomological Society of London* 72 (1923): cxiii–cxxvii.

31. J. E. Collin, "President's Address," *Proceedings of the Entomological Society of London* 2 (1927): 102–13. The address commemorated Lieutenant-Colonel Yerbury (1888–1927).

32. Emile Hegh, *Comment nos planteurs et nos colons peuvent-ils se proteger contre les moustiqes qui transmettent des maladies?* Etudes de Biologie Agricole 4 (London: L'Imprimerie Belge, 1918); *idem, Les termites* (Brussels: Imprimerie Industrielle et Financière, 1922).

33. Other colonial figures cited in academic works of entomology include R. C. Wroughton in Bombay and George Arnold in Rhodesia.

34. P. A. Buxton, "Physical Factors Controlling Harvesting in an Ant," *Transactions of the Entomological Society of London* 72 (1923): 538–43.

35. J. F. M. Clark, "'The Ants Were Duly Visited': Making Sense of John Lubbock, Scientific Naturalism and the Senses of Social Insects," *British Journal for the History of Science* 30 (1927): 151–76; John Lubbock, *Ants, Bees and Wasps: A Record of Observations on the Habits of the Social Hymenoptera* (London: Kegan Paul, Trench, Trubner, [1882] 1929).

36. H. Maxwell Lefroy and F. Howlett, *Indian Insect Life: A Manual of the Insects of the Plains (Tropical India)* (Calcutta: Thacker, Spink; London: W. Thacker, 1909), ix. The Belgian Maurice Maeterlinck also linked the alien psychology of insects with their threat to human beings: "The insect does not belong to our world. The other animals, even the plants . . . do not seem wholly strangers to us . . . They surprise us: even make us marvel, but they fail to overthrow our basic concepts. The insect, on the other hand, brings with him something that does not seem to belong to the customs, the morale, the psychology of our globe . . . There is, without doubt, amazement . . . incomprehension . . . instinctive and profound inquietude inspired by these creatures so incomparably better armed, better equipped than ourselves, these compressions of energy and activity which are our most mysterious rivals in these latter hours, and perhaps our successors." In J. H. Fabre, *The Life of the Spider* (London: Hodder and Stoughton, 1912), vii–viii.

37. On ornithology, see Helen Macdonald, "'What Makes You a Scientist Is the Way

You Look at Things': Ornithology and the Observer 1930–1955," *Studies in History and Philosophy of Biological and Biomedical Sciences* 33 (2002): 53–77. On televisual natural history, see Gregg Mitman, *Reel Nature: America's Romance with Wildlife on Film* (Cambridge, MA: Harvard University Press, 1999).

38. Though the term *myrmecologist* was not coined until around 1906, for convenience I use it to designate students of ants throughout the time span 1874–1975.

39. As will become clear, Forel's, Wheeler's, and Wilson's respective arenas of claimed expertise were not consistent through the century, so it is not appropriate to talk about how representative each man was in his respective era of myrmecology. Insofar as myrmecology was not an autonomous discipline, it was more simply the case that each one managed to set the agenda for a majority of ant students for a time.

40. N. Katherine Hayles, "Designs on the Body: Cybernetics, Norbert Wiener, and the Play of Metaphor," *History of the Human Sciences* 3 (1990): 212–28; 213.

41. See the essay on sources for publications on early insects and their signification.

42. Richard W. Burkhardt, "Le comportement animal et la biologie française, 1920–1950," in *Les sciences biologiques et médicales en France, 1920–1950*, ed. Claude Debru, Jean Gayon, and Jean-François Picard (Paris: CNRS Editions, 1994), 99–111.

Part I • Psychological Ants

1. Auguste Forel, *Les fourmis de la Suisse* (Basel: H. Georg, 1874), i. Forel's claim to priority in this regard is entirely defensible.

2. Auguste Forel, *Out of My Life and Work* (London: George Allen & Unwin, 1937), 85.

One • Evolutionary Myrmecology and the Natural History of the Human Mind

1. On Forel's biography, see Auguste Forel, *Out of My Life and Work* (London: George Allen & Unwin, 1937), and Hans H. Walser, ed., *August Forel: Briefe, Correspondance: 1864–1927* (Berne: Hans Huber), 11–44.

2. Forel, with extraordinary frankness, referred to this in his preface to his *Les fourmis de la Suisse* (Basel: H. Georg, 1874), i, even though his mother was at that time alive (if not entirely well). He describes all this in more detail in *Out of My Life and Work*, chaps. 1 and 2.

3. Michel Sartori and Daniel Cherix, "Histoire de l'étude des insectes sociaux en Suisse à travers l'œuvre d'Auguste Forel," *Bulletin de la Société Entomologique de France* 88 (1983): 66–74; 67.

4. Forel, *Out of My Life and Work*, 32–33.

5. Sartori and Cherix, "Histoire de l'étude des insectes sociaux," 66–74.

6. On the history of the canton of Vaud, see Paul Maillefer, *Histoire du Canton de Vaud dès les origines* (Lausanne: Payot, 1903), and La Société Vaudoise d'Histoire et d'Archéologie: *Cent cinquante ans d'histoire Vaudoise 1803–1953* (Lausanne: Payot, 1953). A social history of life in Forel's rural context is provided by David Birmingham, *Switzerland: A Village History* (London: Macmillan, 2000).

7. Forel participated in one example of this. Perhaps stung by his failure in the med-

ical examination, he weighed in on the debate over the lien fédéral of 1873 that sought to override local Vaudois regulation of doctors. He considered three possible systems of regulation and concluded that even total liberty in medical practice, with its threat of charlatanism among an undiscriminating public, was less dangerous than a federal law. Forel papers.

8. Maillefer, *Histoire du Canton de Vaud*, and Société Vaudoise, *Cent cinquante ans d'histoire Vaudoise*, 513.

9. Forel, *Out of My Life and Work*, 34–35.

10. For extensive genealogical information, see Extrait de *Receuil de généalogies Vaudoises*, n.d. [1909?], Forel papers.

11. Forel, *Out of My Life and Work*, 264.

12. Forel, *Out of My Life and Work*, 88.

13. Forel, *Out of My Life and Work*, 34.

14. Auguste Forel, *The Social World of the Ants Compared with that of Man* (2 vols., London: G. P. Putnam's Sons, [1921–22] 1928), vol. 2, 336.

15. Forel, *Les fourmis de la Suisse*, iii. A single colony could therefore possess several nests but would still count as a single entity.

16. For all these observations, see Forel, *Les fourmis de la Suisse*, 242–87.

17. Adele Field, "Artificial Mixed Nests of Ants," *Biological Bulletin* 5 (1903): 320–25; idem, "A Case of Feud between Ants of the Same Species Living in Different Communities," *Biological Bulletin* 5 (1903): 326–29; idem, "Power of Recognition in Ants," *Biological Bulletin* 7 (1904): 227–50; idem, "The Progressive Odor of Ants," *Biological Bulletin* 10 (1905): 1–16; idem, "The Sense of Smell in Ants," *Annals of the New York Academy of Science* 16 (1905): 394. The formicary represented to Fielde a model socialist community, much as it did to Forel. One journalist recounted how an inquiry about parliamentary law led her to Fielde. Arriving at Fielde's house, the journalist almost laughed to discover that she had turned up to hear a disquisition on ants. Yet within a few days of this surprising start, she reported that as a result of what she had heard she was "deep in the study of the adaptation of rules and methods for the proper government of corporate bodies and the easiest ways of systematizing the work of the organization, the framing of constitutions, etc." Helen Norton Stevens, *Memorial Biography of Adele M. Fielde, Humanitarian* (Seattle: Pigott Printing Concern, 1918), 270–71.

18. Forel, *Out of My Life and Work*, 340.

19. Forel's distinction is described in William M. Wheeler, *Ants: Their Structure, Development and Behavior* (New York: Columbia University Press, 1910), 198.

20. Auguste Forel, "Aux étudiants Macédoniens persécutés," 1927. Forel papers.

21. Forel, "Aux étudiants Macédoniens persécutés."

22. Title of a pamphlet written by Forel in 1914. Forel papers.

23. Forel, *Out of My Life and Work*, 193.

24. Wilfred Trotter, *Instincts of the Herd in Peace and War* (London: Ernest Benn), 1916.

25. Auguste Forel, "Avant-projet d'une constitution de la société des nations," 1914(?). Forel papers. Interestingly, the bilingual character of Switzerland shaped Forel's work; most of the neurological literature that he read and wrote was in German, while a great deal of observational entomological literature was in French. Auguste wrote several letters to his

fiancée, Emma, in French (his preferred language) before realizing she could scarcely understand them. Later in life, Forel added the international language of Esperanto to his other hobby-horses.

26. Forel's combination of racial eugenics and species egalitarianism is perhaps best captured in the following pronouncement: "We overestimate the powers of the educated negro and the trained dog and underestimate the powers of the illiterate individual and the wild animal." Auguste Forel, *Ants and Some Other Insects* (London: Kegan Paul, Trench, Trubner, 1904), 13. Forel's internationalist eugenics bears a strong resemblance to that of his friend from Munich days and fellow-Utopianist, Alfred Ploetz. See Paul Weindling, *Health, Race and German Politics Between National Unification and Nazism, 1870–1945* (Cambridge: Cambridge University Press, 1989), 70–80.

27. See letter from Forel to unidentified recipient, 29 July 1912, in Walser, ed., *August Forel*, 426–27; Auguste Forel, *Crimes et anomalies mentales constitutionelles* (Geneva: Kundig, 1902); and idem, *La question exuelle exposée aux adultes cultivés* (3rd ed., Paris: Steinheil, 1911), 447. See also Sarah Jansen, "Ameisenhügel, Irrenhaus und Bordell: Insektenkunde und Degenerationsdiskurs bei August Forel (1848–1931), Entomologe, Psychiater und Sexualreformer," in *Kontamination*, ed. N. Haas, R. Nägele, and H.-J. Rheinberger (Eggingen: Edition Isele, 2001), 141–84.

28. See, for example, Emile Zola, *La bête humaine* (Paris: Fasquelle, 1984), chap. 11. Latterly, fauvists and primitivists reclaimed this kind of instinct in a more positive sense, as an antidote and vivifying alternative to the stifling effects of learning and "civilization."

29. Forel, *Out of My Life and Work*, 332.

30. Forel, *Out of My Life and Work*, 249–50.

31. Forel, *Out of My Life and Work*, 162. Forel claimed to have anticipated Ramon y Cajal in this discovery. Whether or not he did so is less interesting in this context than the question of why he created this alternative model.

32. Auguste Forel, *Hypnotism, or Suggestion and Psychotherapy: A Study of the Psychological, Psycho-Physiological and Therapeutic Aspects of Hypnotism* (London: Rebman, [1889] 1906), 4–5; idem, *Social World of the Ants*, vol. i, 182–85. Forel acknowledged his terminological debt to Ricard Semon. Forel found himself unable to account satisfactorily for the phylogenetic (inheritable) incorporation of engrams. He described some of Kammerer's experiments, but without particular certainty. (See Forel, *Social World of the Ants*, vol. i, 16–19.) Elsewhere he suggested that "engraphs might serve as an explanation for De Vries' mutations" (Forel, *Hypnotism*, 5) In the end, Forel always returned to Ewald Hering's aphorism: "Instinct is the memory of the species" (Forel, *Social World of the Ants*, vol. i, xliii). This was the best answer he seemed to find.

33. Edouard Bugnion, *The Origin of Instinct: A Study of the War between the Ants and the Termites* (London: Kegan Paul, Trench and Trubner, 1927).

34. Bugnion, *The Origin of Instinct*, 29.

35. Bugnion, *The Origin of Instinct*, 43.

36. Forel, *Les fourmis de la Suisse*, 440–48; idem, *Social World of the Ants*, vol. i, 1–12, 125–46, and vol. ii, 298–301. Notably, Forel found that ponerines did not display mutual feeding, his paradigmatic behavior of social life among the ants. See also Eugène Bouvier, *Le communisme chez les insectes* (Paris: Flammarion, 1926), 118–19.

37. Auguste Forel, *The Senses of Insects* (London: Methuen, [1878–1906] 1908), 1.

38. Forel, *The Senses of Insects*. See also Harry Eltringham, *The Senses of Insects* (London: Methuen, 1933).

39. Forel, *Out of My Life and Work*, 190.

40. See Ricard Semon, *The Mneme* (London: Ruskin; New York: Macmillan, [1904] 1921). Wheeler had taken notes on Eugenio Rignano's *Upon the Inheritance of Acquired Characteristics* (Chicago: University of Chicago Press, 1911), in which he explained teleology as due to "mnemic peculiarities." Notes taken by Wheeler, n.d. Wheeler papers.

41. On German monism, see Weindling, *Health, Race and German Politics*, 131–33.

42. Forel papers. In this respect the Germans were much like the Vaudois, as Forel saw them: "Of course the Vaudois have many good qualities; they are, on the whole, very good-natured, tolerant, intelligent, and peaceable; but they are absolutely lacking in backbone. Consequently they are easy to rule, and this is a temptation to ambitious and masterful people." Forel, *Out of My Life and Work*, 88.

43. Forel, *Out of My Life and Work*, 303–4.

44. Like Forel, Ploetz would reject Pan-Germanism. See Weindling, *Health, Race and German Politics*, 70–80.

45. Weindling, *Health, Race and German Politics*, 98.

46. Weindling, *Health, Race and German Politics*, 93, 99.

47. Carlo Emery, "Les insectes sociaux et la société humaine," *Revue d'Economie Politique* 15 (1901): 26–40; 26. See also idem, *La vita delle formiche* (Turin: Bocca, 1915).

48. Luke 17: 5–6.

49. Personal communication from Daniel Cherix, derived from Kütter, Forel's protégé, who died in 1990.

50. Letter to Abdul Baha Abbas, 28 December 1920. Forel papers.

51. Auguste Forel, *Sexual Ethics* (London: New Age Press, c.1907), 31.

52. In *Sexual Ethics*, Forel hastened to assure his readers that this was not the same thing as property, but elsewhere he wrote, "if [prominent persons] have come to wealth and importance this is usually by no means due to external good fortune and good education, but much more to a double portion of the good qualities of the ovum and spermatozoon from which they sprang." Auguste Forel, *Hygiene of Nerves and Mind in Health and Disease* (London: John Murray, [1903] 1907), 278. Forel recommended tolerance of sexual "deviants" (mainly homosexuals), so long as they did not pass their "perversions" on to another generation. Forel, *Sexual Ethics*, 55. See also Weindling, *Health, Race and German Politics*, 104–5.

53. Forel papers.

Two • A (Non-)Disciplinary Context for Evolutionary Myrmecology

1. Eugène Bouvier, *The Psychic Life of Insects* (London: T. Fisher Unwin, [1918] 1922), xi–xv.

2. Mary A. Evans and Howard E. Evans, *William Morton Wheeler, Biologist* (Cambridge, MA: Harvard University Press, 1970), 211–12.

3. Although Emery's studies at the Naples Stazione were mostly systematic in nature,

his zoogeographical work did reflect a desire to construct a "natural" system of classification, and in this limited sense his project reflected Forel's aims regarding behavioral analysis.

4. Auguste Forel, "Ants and Some Other Insects: An Inquiry into the Psychic Powers of These Animals with an Appendix on the Peculiarities of Their Olfactory Sense," *Monist* 14 (1903–4): 33–66 and 177–93. Reprinted in the Religion of Science Library (Chicago: Open Court, 1904).

5. T. C. Schneirla's disciple Daniel S. Lehrman reflected on what he saw as the careless and ambiguous use of the closely related terms "innate" and "inherited" in 1970. He pointed out that they were used to refer to both the geneticist's ability to predict the distribution of a behavior in a population and to a kind of genetic predestination in the animal itself. Daniel S. Lehrman, "Semantic and Conceptual Issues in the Nature-Nurture Problem," in *Development and Evolution of Behavior: Essays in Memory of T. C. Schneirla*, ed. Lester R. Aronson, Ethel Tobach, Daniel S. Lehrman, and Jay S. Rosenblatt (San Francisco: W. H. Freeman, 1970), 17–52. Lehrman's essay is a polemical reflection on the instinct debate that he and Schneirla had with Lorenz in the 1950s.

6. Emile Roubaud, "Recherches biologiques sur les guêpes solitaires et social d'Afrique: la genèse de l'instinct maternel chez les vespides," *Annales des Sciences Naturelles*, ser. 10 (Zoologie), 1 (1916): 1–160; 62.

7. Romanes' focus on the latter of these mechanisms caused at least one reviewer to accuse him of deriving his ideas from Samuel Butler. See Robert J. Richards, *Darwin and the Emergence of Evolutionary Theories of Mind and Behavior* (Chicago: University of Chicago Press, 1987), 350, n. 54. Romanes' hierarchy of mind also paralleled animal evolutionary progress with the psychological development of the human embryo and child. Romanes' principal publications were *Animal Intelligence* (1881), *Mental Evolution in Animals* (1883), and *Mental Evolution in Man* (1888). For an account of Romanes' life and work, see Richards, *Darwin and the Emergence of Evolutionary Theories of Mind and Behavior*, 334–81.

8. The secondary literature on taxonomic exchange has discussed well nineteenth-century examples. See Anne Secord, "Corresponding Interests: Artisans and Gentlemen in Nineteenth-Century Natural History," *British Journal for the History of Science* 27 (1994): 383–408; Jim Endersby, "A Garden Enclosed: Botanical Barter in Sydney, 1818–1839," *British Journal for the History of Science* 33 (2000): 313–34; idem, "'From Having No Herbarium.' Local Knowledge vs. Metropolitan Expertise: Joseph Hooker's Australasian Correspondence with William Colenso and Ronald Gun," *Pacific Science* 55 (2001): 343–58. See also John Elsner and Roger Cardinal, eds., *The Cultures of Collecting* (London: Reaktion Books, 1994); and Sue Ann Prince, ed., *Stuffing Birds, Pressing Plants, Shaping Knowledge* (Philadelphia: American Philosophical Society, 2003).

9. Letter from John Clark to Forel, 22 March 1923, Forel papers.

10. Letters from W. C. Crawley to Forel, 9 February 1922 and 16 February 1922, Forel papers.

11. Letter from Clark to Forel, 18 November 1922, Forel papers.

12. Forel papers. The letter is marked "sent" by Forel, and dated 4 March 1918. Forel also notes that forty-five papers have been sent to Crawley.

13. Letter from Horace Donisthorpe to Forel, 1 December 1921, Forel papers.

14. See, for example, the letter from Carlo Emery to Forel, 13 December 1922, Forel papers. "I envy you your optimism and faith in humanity," he wrote in another on 30 August 1920 (ibid.).

15. Auguste Forel, "I Fourmis mexicaines récoltées par M. le professeur W. M. Wheeler. II A propos de la classification des fourmis," *Annales de la Société Royale d'Entomologie de Belgique* 45 (1901): 123–41.

16. Letter to Forel, 26 February 1908, in *August Forel: Briefe, Correspondance: 1864–1927*, ed. Hans H. Walser (Bern: Hans Huber), 398–99.

17. Letter from Clark to Forel, 18 November 1922, Forel papers.

18. Letters from Auguste Forel to Alexis Forel, undated, Forel papers. Auguste renders "colonists" with a spurious "e"—*colones*—just one letter away from *colonies*, the groups of ants under investigation.

19. Letter from Bédot to Forel, 24 October 1921, Forel papers.

20. Personal communication from Daniel Cherix. The story was told to him by Kütter, Forel's protégé, who died in 1990.

21. Daniel Cherix, "Heinrich Kütter, digne successeur d'Auguste Forel," *Actes des Colloques Insectes Sociaux* 4 (1988): 3–14.

22. For a review of disciplinary context, see the essay on sources.

23. Roger Smith remarks critically on the tendencies to present psychology as a "unified endeavour" in his review of Graham Richards, *Putting Psychology in Its Place: A Critical Historical Overview* (2nd ed., New York: Routledge, 1996). Review in *British Journal for the History of Science* 36 (2003): 374–75. Smith praises the book for avoiding the tendency to generalize in this manner.

24. Kurt Danziger, *Constructing the Subject: Historical Origins of Psychological Research* (Cambridge: Cambridge University Press, 1990).

25. See Albrechte Bethe, "Duerfen wir den Ameisen und Bienen psychische Qualitaeten zuschreiben?" *Archiv fur die Gesamte Physiologie* 70 (1898): 15–100. Wasmann discusses this in Erich Wasmann, "A New Reflex Theory of Ant Life," *Biologisches Centralblatt* 18 (1898): 577–88.

26. Gregg Mitman, *The State of Nature: Ecology, Community, and American Social Thought, 1900–1950* (Chicago: University of Chicago Press, 1992), 24–25. Loeb's and Whitman's academic incompatibility caused Loeb to depart for Berkeley in 1903. Whitman's perspective on natural history positively influenced his students Wallace Craig, C. H. Turner, Samuel Jackson Holmes, and Wheeler.

27. John Lubbock, *Ants, Bees and Wasps: A Record of Observations on the Habits of the Social Hymenoptera* (London: Kegan Paul, Trench, Trubner, [1882] 1929). J. F. M. Clark, "'The Ants Were Duly Visited': Making Sense of John Lubbock, Scientific Naturalism and the Senses of Social Insects," *British Journal for the History of Science* 30 (1927): 151–76.

28. This bears a strong resemblance to Hacking's philosophical contention that scientific palpability equates with reality. Ian Hacking, *Representing and Intervening: Introductory Topics in the Philosophy of Natural Science* (Cambridge: Cambridge University Press, 1983). Cf. William M. Wheeler, "Present Tendencies in Biological Theory," *Scientific Monthly* 28 (1929): 97–109; 101.

29. On "natural" systems of classification, see Harriet Ritvo, *The Platypus and the Mer-*

maid and Other Figments of the Classifying Imagination (Cambridge, MA: Harvard University Press, 1997).

30. In this and the following four paragraphs I draw on Marion Thomas, "Rethinking the History of Ethology: French Animal Behaviour Studies in the Third Republic (1870–1940)" (PhD thesis, University of Manchester, 2003), chap. 2.

31. Thomas, "Rethinking the History of Ethology," 96–97.

32. Unlike the French, who did not greatly discuss German science, Forel was also happy to engage Ernst Haeckel (1834–1919) and his socioevolutionary account of social polyps.

33. Espinas, once he had read Forel's *Les fourmis de la Suisse,* quickly revised his *Des sociétés animales,* refocusing it largely on ants; Perrier and Giard wrote mainly on marine invertebrates and their social properties. See Thomas, "Rethinking the History of Ethology," chap. 3. Richard Burkhardt has noted the predominance of invertebrate studies in *L'Année Psychologique* between 1914 and 1950; see Burkhardt, "Le Comportement Animal," 102.

34. See especially pp. 104–5, where Bouvier recounted the mnemic theory. The translator, L. O. Howard, drew attention to the book's cumulative argument, apologizing for the necessity of the duller initial chapters on tropisms (ix–x).

35. Henri Piéron, *Principles of Experimental Psychology* (London: Kegan Paul, Trench and Trubner; New York: Harcourt, Brace, 1929), 14.

36. Piéron, *Principles of Experimental Psychology,* 112.

37. The most famous of Ferton's publications were eight memoirs in the *Annales de la Société Entomologique de France,* which appeared between 1901 and 1914. A more popularly accessible account appeared in book form shortly after his death as *La vie des abeilles et des guêpes* (Paris: Chiron, 1923).

38. This demonstrated an additional form of memory, much advanced by Piéron: the kinaesthetic or "muscular memory."

39. Etienne Rabaud, *How Animals Find Their Way About: A Study of Distant Orientation and Place-Recognition* (London: Kegan Paul, Trench, Trubner; New York: Harcourt, Brace, 1928), 120–22; 127.

40. William M. Wheeler, "On Instincts," *Journal of Abnormal Psychology* 15 (1921): 295–318; 298. Aquinas' source, Aristotle, had claimed in his *History of Animals* that "just as in man we find knowledge, wisdom and sagacity, so in certain animals there exists some other natural capacity akin to these." See *Parts of Animals,* 641˙17–641˙1.10; *History of Animals,* 588˙1.4, 588˙1.24–25.

41. See, for example, Gilles A. Bazin, *The Natural History of Bees. Containing an Account of the Production, their Œconomy, the Manner of their Making Wax and Honey, and the Best Methods for the Improvement and Preservation of Them* (London, 1744). The book is itself a loose reworking of Réaumur's volume on bees. On Republican bees, see Jean-Marc Drouin, "L'image des sociétés d'insectes en France à l'époque de la Révolution," *Revue de Synthèse* 4 (1992), 333–45.

42. Janet Browne "Darwin in Caricature: A Study in the Popularisation and Dissemination of Evolution," *Proceedings of the American Philosophical Society* 145 (2001): 496–509.

43. Beginning in the 1860s, the Sanskrit scholar Friedrich Max Müller successfully

defended the view that reason (Wheeler's "ratiocination") was the defining characteristic of humanity, and that this faculty was necessarily and sufficiently demonstrated by the capacity for language. "No reason without language, no language without reason," went the common maxim. In the use of language, man displayed the ability to form abstract concepts and to refer to objects in their absence. See Gregory Radick, "Morgan's Canon, Garner's Phonograph, and the Evolutionary Origins of Language and Reason," *British Journal for the History of Science* 33 (2000): 3–23. Müller's opinions were shared by the English biologist and philosopher C. Lloyd Morgan. For more on the definitions of "intelligence" and "reason" in this period see Robert A. Boakes, *From Darwin to Behaviourism: Psychology and the Minds of Animals* (Cambridge: Cambridge University Press, 1984), 23–52.

44. Thomas Belt, *The Naturalist in Nicaragua* (London: John Murray, 1874), 28.

45. Harold Bastin, *Insects: Their Life-Histories and Habits* (London: T.C. & E.C. Jack, 1913), 2.

46. Bouvier, *Psychic Life of Insects*, 357–58.

47. R. W. G. Hingston, *Problems of Instinct and Intelligence* (London: Edward Arnold, 1928), 272–76. Wheeler spoke highly of Hingston's books, indicating that his was not just a popular opinion (Hingston's books were mostly nonspecialist travelogues) but also a professionally respectable belief.

48. Godfrey H. Thomson, *Instinct, Intelligence and Character: An Educational Psychology* (London: George Allen and Unwin, 1924), 14. Thomson was referring to the "vivid" writings of Fabre.

49. Julian Huxley, *Ants* (London: Dennis Dobson, [1930] 1949), 3–4.

50. W. C. Allee, "Cooperation Among Animals," *American Journal of Sociology* 37 (1931): 386–98. On Allee, see Mitman, *The State of Nature*, 79, where this diagram is reproduced.

51. Henri Bergson, *The Two Sources of Morality and Religion* (London: Macmillan, [1932] 1935), 17, quoted in Thomas, "Rethinking the History of Ethology," chap. 4.

52. Letter from Clark to Forel, 30 August 1922, Forel papers.

53. Quoted in Maurice Maeterlinck, *The Life of the White Ant* (London: George Allen, 1927), 187.

54. Jean-Henri Fabre, *Souvenirs entomologiques: Etudes sur l'instinct et les mœurs des insectes* (10 vols., Paris: Delagrave, 1879–1907), vol. iv, 32–49.

55. Fabre, *Souvenirs entomologiques*, vol. i, 171–74.

56. Jean-Henri Fabre, *The Wonders of Instinct: Chapters in the Psychology of Insects* (London: T. Fisher Unwin, 1918), 170–71.

57. Fabre, *The Wonders of Instinct*, 128–47.

58. Jean-Henri Fabre, *Social Life in the Insect World* (London: T. Fisher Unwin, 1912), 124.

59. Fabre, *The Wonders of Instinct*, 36–48.

60. Abigail Lustig, "Ants and the Nature of Nature in Auguste Forel, Erich Wasmann, and William Morton Wheeler," in *The Moral Authority of Nature*, ed. Lorraine Daston and Fernando Vidal (Chicago: University of Chicago Press, 2004), 282–307.

61. Erich Wasmann, *Comparative Studies in the Psychology of Ants and of Higher Animals* (translation of 2nd German edition; St. Louis, MO: B. Herder; London: Sands, 1905), 34.

62. Piéron, for example, was forced to deal with "instinct" in his *Principles of Experimental Psychology*, conceding that it was a term useful for describing a coordinated clutch of reflexes. Thus, letting down milk, licking, and responding to nearby sounds might, together with other reflexes, constitute the "maternal instinct." Piéron, *Principles of Experimental Psychology*, 23–27.

63. See Jacques Loeb, *The Mechanistic Conception of Life* (reprint ed.; Cambridge, MA: Harvard University Press, 1964). Similarly, see Piéron, *Principles of Experimental Psychology*, 23: "The sharp distinctions made among the reactional processes [including those considered as "adapted"], however, are the work of theorists. They do not express facts but speculative conceptions."

64. See Wasmann, *Comparative Studies in the Psychology of Ants and Higher Animals*, 4–5, and chaps. 7 and 8 of *idem, Instinct und Intelligenz im Thierreich: Ein kritischer Beitrag zur modernen Thierpsychologie* (Freiburg-in-the-Breisgau: Stimmen aus Maria Laach, 1897).

65. Erich Wasmann, "Die psychischen Fähigkeiten der Ameisen," *Zoologica* 11 (1899): 132 pp. August Forel, "Nochmals Herr Dr. Bethe und die Insekten-Psychologie," *Biologisches Centralblatt* 23 (1903): 1–3. See also Rudolf Brun, *Raumorientierung der Ameisen und das Orientierungsproblem im Allgemeinen: Eine kritisch-experimentelle Studie; zugleich ein Beitrag zur Theorie der Mneme* (Jena: Gustav Fischer, 1914).

66. Letter from Jennings to Davenport, 5 January 1900. Quoted in Philip Pauly, "The Loeb-Jennings Debate and the Science of Animal Behavior," *Journal for the History of the Behavioral Sciences* 17 (1981): 504–15; 510. Jennings later refuted Loeb's ideas in Herbert S. Jennings, *Behavior of the Lower Organisms* (Bloomington: Indiana University Press, [1906] 1962). Here he described experiments showing that the stimulus presented to an animal was insufficient to elicit the characteristic "tropism" expressed in the situation. Donald Fleming considers that Loeb's theories were thoroughly refuted by Jennings book; see his article "Jacques Loeb" in the *Dictionary of Scientific Biography* as well as his introductions to Loeb, *The Mechanistic Conception of Life* and to Jennings, *Behavior of the Lower Organisms*. See also Jensen's foreword in ibid.

67. Emery commented as follows on Wheeler and Wasmann: "No doubt you have received Wasmann's most recent memoir where he takes issue with Wheeler and his trophallaxis. Wheeler is too simplistic and too inclined towards extreme theories. Wasmann [is] too wily, a picky logician, a sophist and a lawyer." Letter from Emery to Forel, 30 August 1920, Forel papers.

68. Quoted in David Elliston Allen, *The Naturalist in Britain: A Social History* (London: Allen Lane, 1976), 180.

69. Roubaud's friend was Viscount R. du Buysson. Emile Roubaud, "The Natural History of the Solitary Wasps of the Genus Synagris," *Annual Report of the Smithsonian Institution* 33 (1910): 507–25. Translated from *Annales de la Société Entomologique de France* 79 (1910): 1–21. Roubaud thanks the viscount and others again in Emile Roubaud, "Recherches biologiques sur les guêpes solitaires et social d'Afrique: La genèse de l'instinct maternel chez les vespides," *Annales des Sciences Naturelles*, ser. 10 (Zoologie), 1 (1916): 1–160; 3.

70. "Emile Roubaud (1882–1962)." Obituary, *Bulletin de la Société de Pathologie exotique*.

71. Belt, *The Naturalist in Nicaragua*; Lubbock, *Ants, Bees and Wasps*.

72. L. Berland, "Charles Janet (1849–1932)," *Annales Société Entomologique de France* 101 (1932): 157–64.

73. Charles Janet, *Observations sur les guêpes* (Paris: C. Naud, 1903), 81.

74. Charles Janet, "Sur la *Vespa Crabro*. Ponte. Conservation de la chaleur dans le nid," *Comptes rendus de l'Académie des Sciences* 120 (1895): 384. Janet's most celebrated publication was also focused on nests: idem, "Etudes sur les fourmis, les guêpes et les abeilles. Note 9. Sur *Vespa crabro*. Histoire d'un nid depuis son origine," *Mémoires de la Société Zoologique de France* 8 (1895).

75. Berland, "Charles Janet," 157.

76. Janet, *Observations sur les guêpes*, 9.

77. Charles Janet, "Remplacement des muscles vibrateurs du vol par des colonnes d'adipocytes chez les fourmis, après le vol nuptial," *Comptes Rendus de l'Académie des Sciences* 142 (1906): 1095; idem, "Histolyse, sans phagocytose, des muscles vibrateurs du vol, chez les reines des fourmis," *Comptes Rendus de l'Académie des Sciences* 144 (1907): 393; idem, "Histogénèse du tissu adipeux remplaçant les muscles vibrateurs histolysé après le vol nuptial, chez les reines des fourmis," *Comptes Rendus de l'Académie des Sciences* 144 (1907): 1070; idem, "Histolyse des muscles de mise en place des ailes après le vol nuptial," *Comptes Rendus de l'Académie des Sciences* 144 (1907): 1205. Wheeler discusses Janet's research in *Ants*, 183–91.

78. Wheeler, *Ants*, 185–86.

79. Spencer's model of adult consociation is rejected in Wheeler, *Ants*, 100, 116–18.

80. Roubaud, "The Natural History of the Solitary Wasps of the Genus Synagris," 507–25.

81. The French entomologist Paul Marchal had explicitly (and successfully) set out to counter Fabre's claim about the immutability of instinct in "Etude sur l'instinct du *Cerceris ornata*," *Archives de Zoologie Expérimentale et Générale*, ser. 2, 5 (1887): 27–60. Marchal's aim was to strike a blow for Romanes' Darwinian project of explaining mental evolution by reinterpreting the behaviour of Fabre's supposedly inflexible solitary wasp. See also Paul Marchal, "La reproduction et l'evolution des guêpes sociales," *Archives de Zoologie Expérimentale et Générale*, ser. 3, 3 (1896): 1–100.

82. Roubaud, "Recherches biologiques sur les guêpes," 80.

83. William M. Wheeler, "Review of Bouvier, *La vie psychique des insectes*," *Science* 52 (1920): 443–46; 444.

84. Alphaeus S. Packard, *Lamarck: The Founder of Evolution; his Life and Work, with Translations of his Writings on Organic Evolution* (New York: Longmans, Green, 1901). See W. Conner Sorensen, *Brethren of the Net: American Entomology, 1840–1880* (Tuscaloosa: University of Alabama Press, 1995), 197–213, on the evolutionary theories of American entomologists.

85. David L. Krantz and David Allen, "The Rise and Fall of McDougall's Instinct Doctrine," *Journal for the History of the Behavioral Sciences* 3 (1967): 326–38; Hamilton Cravens and John C. Burnham, "Psychology and Evolutionary Naturalism in American Thought, 1890–1940," *American Quarterly* 23 (1971): 635–57; John C. Burnham, "The Medical Uses and Origins of Freud's Instinctual Drive Theory," *Psychoanalytic Quarterly* 43 (1974): 193–217; Boakes, *From Darwin to Behaviourism*. See also Roger Smith, *Fontana History of*

the Human Sciences (London: HarperCollins, 1997), 755–64. Many of these histories give 1930 as the end of the period in which the study of instinct had academic credibility. In this chapter I implicitly argue for a more nuanced account of the fortunes of instinct theory during the twentieth century. Although it ceased to be explicitly invoked as an explanatory mechanism during the 1930s, it was back soon after with the work of Lorenz and Tinbergen on stereotyped patterns of behavior. They worked within a research program that respected all three of the features of instinct cited in this chapter. More important even than this, apparently "nurture" based accounts of behavior and society retained many of the connotations of instinct theory. That is to say, functional accounts of these rely on an innate predictability of the bulk of human behavior—instinct, by any other name. Contemporary critiques of instinct theory included Knight Dunlap, "Are There Any Instincts?" *Journal of Abnormal Psychology* 14 (1919): 307–11; L. L. Bernard, "The Misuse of Instinct in the Social Sciences," *Psychological Review* 28 (1921): 96–119; and L. L. Bernard, *Instinct: A Study in Social Psychology* (New York: Holt, 1924). Wheeler, meanwhile, was among those critiquing human psychologists for their total ignorance of biological and evolutionary analysis; see Wheeler, "On Instincts," 316.

86. Phil Rau and Nellie L. Rau, *Wasp Studies Afield* (Princeton, NJ: Princeton University Press; London: Oxford University Press, 1918), 357–58.

Part II • Sociological Ants

1. L. J. Henderson, T. Barbour, F. M. Carpenter, and Hans Zinsser, "William Morton Wheeler," *Science* 85 (1937): 533–35. Wheeler's friendship with Charles K. Ogden (of which more in chapter 7) was symptomatic of his identity as a "renaissance man of science."

2. Philip Pauly, *Biologists and the Promise of America Life: From Meriwether Lewis to Alfred Kinsey* (Princeton, NJ: Princeton University Press), 2000.

3. W. M. Wheeler, "'Natural History,' 'Œcology' or 'Ethology?'" *Science* 15 (1902): 971–76. See also Keith R. Benson, "The Emergence of Ecology from Natural History," *Endeavour* 24 (2000): 59–62.

Three • From Psychology to Sociology

1. Wheeler, however, had a similar reaction to German students during his 1893–94 visit: "I cannot say I admire them . . . They spend most of the time drinking beer, dueling and running after girls . . . And to see these animals enjoying the most superb health—rosy and hearty in every limb, while we Americans go about like living skeletons—because our ideals are so high!" Quoted in Mary Alice Evans and Howard Ensign Evans, *William Morton Wheeler, Biologist,* (Cambridge, MA: Harvard University Press), 86. By his return in 1907, Wheeler's views—or perhaps the Germans—had mellowed.

2. It was here that Wheeler met his close friend David Fairchild (see chapter 4).

3. See Lynn K. Nyhart, *Biology Takes Form: Animal Morphology and the German Universities, 1800–1900* (Chicago: University of Chicago Press, 1995).

4. Letter from W. C. Creighton to Howard Evans, 14 March 1968, Creighton papers.

Wheeler exchanged civil, even friendly, letters with Loeb later in his life, but his anti-Semitism was deeply felt, and such sentiments can hardly have made for easy collegial relations. Moreover, Wheeler had a methodological axe to grind with Loeb. Loeb himself left Chicago shortly after.

5. Letter from Creighton to Evans, 14 March 1968, Creighton papers.

6. See Evans and Evans, *William Morton Wheeler*, 9.

7. Quoted in Evans and Evans, *William Morton Wheeler*, 157.

8. Letter from Creighton to Evans, 14 March 1968, Creighton papers.

9. W. M. Wheeler, "Ants and Bees as Carriers of Pathogenic Microörganisms," *American Journal of Tropical Disease and Preventative Medicine* 2 (1914): 160–68.

10. T. D. A. Cockerell, quoted in Evans and Evans, *William Morton Wheeler*, 158.

11. W. M. Wheeler, *Ants: Their Structure, Development and Behavior* (New York: Columbia University Press, 1910), 7.

12. W. M. Wheeler, "Jean-Henri Fabre," *Journal of Animal Behavior* 6 (1916): 74–80.

13. Wheeler's audience was the Harvard philosophical Royce Club. The piece was later published in the *Journal of Abnormal Psychology*—hardly the audience to proselytize for the cause of economic entomology. Economic rhetoric often emphasized the alienness of insects, in order to provoke horror and the urge to wipe them out. Wheeler and the pure entomologists also used this feature of the insects in their writing, sometimes for the sake of interest, and also perhaps to challenge the reader with their human parallels.

14. W. M. Wheeler, "The Termitodoxa, or Biology and Society," *Scientific Monthly* 10 (1920): 113–24; 113.

15. Typed MSS, Wheeler papers. The lectures are unfortunately not dated, but judging by their references and phraseology, I would put them somewhere near the end (1926) of Wheeler's economic responsibilities. Wheeler's insistence on the importance of ethology within economic entomology was apparently not unfounded. In North America from the 1920s on, the study of insects' behavior as a means to their control was abandoned in favor of chemical treatment, using little if any knowledge even of their life-cycles. Palladino's description of this process constitutes the bulk of *Entomology, Ecology and Agriculture*. Axel Melander, who trained under Wheeler, used population genetics in 1913 to solve the problem of a scale infestation that was resistant to the insecticide in use. He recommended a faulty spraying so as not to create a wholly resistant strain the following season, but his advice fell on deaf ears. (Ibid., p. 34.)

16. Evans and Evans, *William Morton Wheeler*, 286. His sister Clara Julia Wheeler had been certified insane by the Milwaukee County Court on 4 April 1914. (Notice, Wheeler papers.)

17. W. M. Wheeler, "The Ant-Colony as an Organism," *Journal of Morphology* 22 (1911): 301–25. An example of ahistorical referencing may be found in Kevin Kelly, *Out of Control: The New Biology of Machines* (London: Fourth Estate, 1994); Steven Johnson, *Emergence: The Connected Lives of Ants, Brain, Cities and Software* (London: Allen Lane, 2001).

18. On the subject of genetics, see also W. M. Wheeler, "Ethology and the Mutation Theory," *Science* 21 (1905): 535–40. On behaviorism see *idem*, "A New Word for an Old Thing," review of John B. Watson, *Behaviorism*, *Quarterly Review of Biology* 1 (1926): 439–43. See

also W. M. Wheeler, "Present Tendencies in Biological Theory," *Scientific Monthly* 28 (1929): 97–109. Wheeler's antipathy to genetics is discussed in depth in Evans and Evans, *William Morton Wheeler*, 231–46.

19. Personal remark from Stefan Cover, curator of the Wheeler collection. In the late 1990s the leftovers were still stacked from the shelves to the ceiling all around the (large) hallway.

20. The detailed development of Wheeler's philosophy of science is described in Charlotte Sleigh, "'The Ninth Mortal Sin': The Lamarckism of W. M. Wheeler," in *Darwinian Heresies*, ed. Abigail Lustig, Robert Richards, and Michael Ruse (Cambridge: Cambridge University Press, 2004), 151–72.

21. "Darwin had convinced men of the continuity of human with animal evolution as regards all bodily characters, [and hence] the similar continuity of man's mental evolution with that of the animal world." William McDougall, *Introduction to Social Psychology* (3rd ed., London: Methuen, [1908] 1910), 2–5.

22. "Instinctive actions are displayed in their purest form by animals not very high in the scale of intelligence ... Insect life affords perhaps the most striking examples of purely instinctive action." McDougall, *Introduction to Social Psychology*, 24.

23. W. H. R. Rivers, *Instinct and the Unconscious: A Contribution to a Biological Theory of the Psycho-Neuroses* (2nd ed., Cambridge: Cambridge University Press, [1920] 1922).

24. There were some exceptions to this rule. Ludwig Büchner, *Mind in Animals* (London: Freethought Publishing Co., 1880), showed the evolution of insect societies in response to their increasing intelligence. Termites were monarchists; bees, constitutional monarchists; ants, republicans. The implication was that with their superior intelligence, humans should or perhaps would tend toward the same state.

25. In fact, Rivers in *Instinct and the Unconscious* conceptualized such noble acts as a triumph over the instinct of fear/flight. Maurice Maeterlinck discussed the morality of the self-sacrificing soldier termites in *idem, The Life of the White Ant* (London: George Allen, 1927), 155–8.

26. See chapter 2, note 85.

27. Abigail Lustig, "Ants and the Nature of Nature in Auguste Forel, Erich Wasmann, and William Morton Wheeler," in *The Moral Authority of Nature*, ed. Lorraine Daston and Fernando Vidal (Chicago: University of Chicago Press, 2004), 282–307.

28. This work earned McDougall the criticism of Floyd H. Allport, then based at the University of North Carolina. Allport's textbook *Social Psychology* (Boston: Houghton Mifflin, 1924) made a stand against treating the group as a unit of analysis, and against the assignation of agency on any level other than the individual. Allport's book stands firm in the Anglo-American tradition of thinking about society in terms of its constituent, separate human units; it also marked the beginning of what Farr has called "the individualization of social psychology in North America." Farr likewise states that the "flowering [of individualism] is a characteristically American phenomenon." See Robert M. Farr, *The Roots of Modern Psychology 1872–1954* (Oxford: Blackwell, 1996), 103–18. Graumann wrote on this subject with respect to a specific form of behaviorism in C. F. Graumann, "The Individualization of Social and the Desocialization of the Individual: Floyd H. Allport's Contribution

to Social Psychology," in *Changing Conceptions of Crowd Mind and Behaviour*, ed. C. F. Graumann and S. Moscovici (New York: Springer-Verlag, 1986), 97–116.

29. W. M. Wheeler, "Animal Societies: Biology and Society," *Scientific Monthly* 39 (1934): 289–301; 290. See also Marion Thomas, "Rethinking the History of Ethology: French Animal Behaviour Studies in the Third Republic (1870–1940)" (PhD thesis, University of Manchester, 2003), chap. 3.

30. The first edition of *Des sociétés animales* was in 1877; Espinas revised its contents in the light of Forel's researches the following year. Alfred Espinas, *Des sociétés animales* (2nd ed., Paris: Germer Ballière, [1877] 1878).

31. Espinas, *Des sociétés animales*, 9.

32. See Paul Weindling, "Ernst Haeckel, Darwinismus, and the Secularization of Nature," in *History, Humanity and Evolution: Essays for John C. Greene*, ed. James R. Moore (Cambridge: Cambridge University Press, 1989), 311–27.

33. See Friedrich Engels, *The Origin of the Family, Private Property and the State in the Light of the Researches of Lewis H. Morgan* (London: Lawrence and Wishart, [1884] 1940). Engels covers previous theories of maternal-based societal evolution, and advances his own, in a striking parallel to the queen-dominated evolution of the formicary.

34. Espinas, *Des sociétés animales*, 519–21.

35. Espinas, *Des sociétés animales*, 531–32.

36. Espinas, *Des sociétés animales*, 542.

37. Durkheim too was accused of proposing a mystical group mind to explain the function of society. See Mary Douglas, foreword, in Marcel Mauss, *The Gift: The Form and Reason for Exchange in Archaic Societies* (London: Routledge, [1950] 1990), vii–xviii; xi–xii. In "Emergent Evolution and the Social," Wheeler again states that "the organization [of complex emergent wholes] *is entirely the work of the components themselves* and not [external] "entelechies" (Driesch) . . . or "élan vital" (Bergson)." (Ibid., 437; original emphasis.)

38. In her foreword to *The Gift*, Douglas claims that the continentals then no longer regarded British liberal utilitarianism as the opponent of socialist democracy; this role was also usurped by communism and fascism. Mauss, *The Gift*, vii–xviii.

39. Wheeler, "The Ant-Colony as an Organism," 142.

40. W. M. Wheeler, "Social Life Among the Insects," *Scientific Monthly* 16 (1923): 5–33; 5.

41. W. M. Wheeler, *The Social Insects: Their Origin and Evolution* (London: Kegan Paul, Trench, Trubner, 1928), 5.

42. Evans and Evans, *William Morton Wheeler*, 268.

43. Charles Janet, "Sur la *Vespa Crabro*. Ponte. Conservation de la chaleur dans le nid," *Comptes rendus de l'Académie des Sciences* 120 (1895), quoted in L. Berland, "Charles Janet (1849–1932)," *Annales Société Entomologique de France* 101 (1932): 157–64; 158.

44. Berland, "Charles Janet," 158–59.

45. Roubaud, quoted in W. M. Wheeler, "A Study of Some Young Ant Larvae with a Consideration of the Origin and Meaning of Social Habits among Insects," *Proceedings of the American Philosophical Society* 57 (1918): 293–343; 321

46. Wheeler, *Ants*, 74.

47. Wheeler, *Ants*, 225–26.

48. Auguste Forel, "Mélanges entomologiques, biologiques et autres," *Annales de la Société Royale d'Entomologie de Belgique* 47 (1903): 249–68; Wheeler, *Ants*, 243–45.

49. W. M. Wheeler, *The Social Insects: Their Origin and Evolution* (London: Kegan Paul, Trench, Trubner, 1928), 11–14.

50. On Wheeler and emergence philosophy, see W. M. Wheeler, "Emergent Evolution and the Social," *Science* 64 (1926): 433–40; idem, *Emergent Evolution and the Social* (London: Kegan Paul, Trench and Trubner, 1927); idem, *Emergent Evolution and the Development of Societies* (New York: W. W. Norton, 1928).

51. Published as Wheeler, *The Social Insects*.

52. Wheeler, *The Social Insects*, 244.

53. Thus I disagree with E. O. Wilson, who claims that Wheeler expanded his definition of trophallaxis in 1928. E. O. Wilson, *Sociobiology. The New Synthesis* (Cambridge, MA: Belknap Press of Harvard University Press, 1975), 29–30. Wheeler's biographers claim that Wheeler's latter treatment of trophallaxis was provoked by Wasmann's criticisms (Evans and Evans, *William Morton Wheeler*, 257); it seems to me, however, that his response was little more than a restatement of his former position.

54. Wheeler, "A Study of Some Young Ant Larvae," 315.

55. Wheeler, "A Study of Some Young Ant Larvae," 315; emphasis added.

56. Wheeler, *The Social Insects*, 230–1.

Four • The Brave New World of Myrmecology

1. The Wheelers' daughter, Adaline, apparently destroyed Morton's more personal materials, and the whole of Dora's (none of which was deposited in the institutional safety of a library). The argument presented here does not overstep this frustrating paucity of documentation, rather locating Wheeler and Hoover in a shared culture.

2. W. M. Wheeler, *The Social Insects: Their Origin and Evolution* (London: Kegan Paul, Trench and Trubner, 1928), 226.

3. Wheeler, *The Social Insects*, 229.

4. W. M. Wheeler, "A Study of Some Young Ant Larvae with a Consideration of the Origin and Meaning of Social Habits among Insects," *Proceedings of the American Philosophical Society* 57 (1918): 293–343; 322. Trophallaxis was derived from two Greek words signifying "nourishment" and "interchange," whereas Roubaud's *œcotrophobiosis* owed its origin to the words for "home," "nourishment," and "life."

5. Wheeler, "A Study of Some Young Ant Larvae," 326.

6. Hoover's core values have been described as "efficiency, enterprise, opportunity, individualism, substantial laissez-faire, personal success, material welfare." Richard Hofstadter, *The American Political Tradition and the Men Who Made It* (2nd ed., New York: Vintage, 1989), 372.

7. Quoted in Hofstadter, *The American Political Tradition*, 386.

8. Wheeler, *The Social Insects*, 22–23.

9. Wheeler, *The Social Insects*, 228.

10. Wheeler, *The Social Insects*, 227.

11. Wheeler, *The Social Insects*, 227.

12. Wheeler, *The Social Insects*, 22–23.

13. L. J. Henderson started his Seminar on the Sociology of Pareto at Harvard in 1932, and it continued for seven or eight years. During this time, Talcott Parsons, Pitirim A. Sorokin, and Robert Merton all attended. Henderson also used Pareto as the basis of his Sociology 23 lectures, beginning in 1938—another important source of influence for the succeeding generation of sociologists.

14. Vilfredo Pareto, *Compendium of General Sociology* (Minneapolis: University of Minnesota Press, [1916/1920] 1980).

15. Wheeler, *The Social Insects*, 2; see also W. M. Wheeler, *Foibles of Insects and Men* (New York: Alfred A. Knopf, 1928). It is striking that Wheeler's interpretation of Pareto resembles nothing so much as a pessimistic version of Espinas' society-as-group-mind.

16. Wheeler was a committed eugenist. For more on Wheeler's eugenics, see Charlotte Sleigh, "'The Ninth Mortal Sin':" The Lamarckism of W. M. Wheeler," in *Darwinian Heresies*, ed. Abigail Lustig, Robert Richards, and Michael Ruse (Cambridge: Cambridge University Press, 2004), 151–72.

17. Aldous Huxley, *Brave New World Revisited* (London: Heron Books, [1958] 1968), 266. The story of *Brave New World*, briefly, is as follows. Society is eugenically bred into castes suited for different levels of work. Their obedience and happiness is maintained by subliminal conditioning, the drug "soma," the constant distraction of entertainment, and by compulsory promiscuity, which eliminates the passion of love. The book's hero, Bernard Marx, comes to doubt whether such things are the route to true human fulfillment, a suspicion that is further provoked by his acquaintance with the "Savage," who, having had the misfortune of being born by natural means (a social taboo), has grown up free from the effects of the New World. The Savage is disgusted by his encounter with "civilization," and the book ends with his ritualized death.

18. See Wolf Lepenies, *Between Science and Literature: The Rise of Sociology* (Cambridge: Cambridge University Press, 1988), for an admirable discussion of the scientific and literary approaches to sociology, and E. Gaziano, "Ecological Metaphors as Scientific Boundary Work: Innovation and Authority in Interwar Sociology and Biology," *American Journal of Sociology* 101 (1996): 874–907, for an account of the circulation of metaphors between sociology and biology.

19. W. M. Wheeler, "The Termitodoxa, or Biology and Society," *Scientific Monthly* 10 (1920): 113–24; 115. Even T. C. Schneirla, who generally despised interspecific analogy, described human breast-feeding as trophallaxis. T. C. Schneirla, "Problems in the Biopsychology of Social Organization," *Journal of Abnormal and Social Psychology* 41 (1946): 385–402. Reprinted in *Selected Writings of T. C. Schneirla*, ed. Lester R. Aronson, Ethel Tobach, Jay S. Rosenblatt, and Daniel S. Lehrman (San Francisco: W. H. Freeman, 1972), 417–39; 426.

20. On Aldous Huxley and Pareto, see David Bradshaw, ed., *The Hidden Huxley: Contempt and Compassion for the Masses 1920–36* (London: Faber and Faber, 1994).

21. Julian Huxley, *Ants* (London: Dennis Dobson, [1930] 1949), 39.

22. Huxley had read Wheeler's seminal work *Ants* sometime between its publication in 1910 and writing his own book, *The Individual in the Animal Kingdom* (Cambridge: Cambridge University Press, 1912). In this book, Julian Huxley also played with the idea that

anything could be called an organism so long as it performed the functions associated with organism status.

23. Wheeler, "The Termitodoxa," 115.

24. Aldous Huxley, *Brave New World* (London: Heron Books, [1932] 1968), 58–61, 87–90.

25. Huxley, *Brave New World*, 79–85.

26. Huxley, *Brave New World*, 94; idem, *Brave New World Revisited*, 326.

27. E. G. Hundert, *The Enlightenment's Fable: Bernard Mandeville and the Discovery of Society* (Cambridge: Cambridge University Press, 1994); cf. Wheeler, *The Social Insects*, 306–7.

28. Bronislaw Malinowski, *Argonauts of the Western Pacific: An Account of Native Enterprise and Adventure in the Archipelagos of Melanesian New Guinea* (London: Routledge, [1922] 1978), 167.

29. W. M. Wheeler, "Emergent Evolution and the Social," *Science* 64 (1926): 433–40, note 4.

30. Marcel Mauss, *The Gift: The Form and Reason for Exchange in Archaic Societies* (London: Routledge, [1950] 1990).

31. Herbert Noyes, *Man and the Termite* (London: Peter Davies, 1937), 1.

32. Anson Rabinbach, *The Human Motor: Energy, Fatigue, and the Origins of Modernity* (Berkeley and Los Angeles: University of California Press, 1990), discusses the establishment of the standardized body in this era; the Čapek brothers based their drudging robots (in *R. U. R.*) on ant workers. See also D. Erlich, and T. P. Dunn, eds., *Clockwork Worlds: Mechanized Environments in SF* (Westport, CT: Greenwood Press, 1983), and Rosalind Williams, *Notes on the Underground: An Essay on Technology, Society, and the Imagination* (London: MIT Press, 1990).

33. Wheeler papers. The letter is undated, but the other correspondence that is filed with it suggests a date around 1927.

34. Pitirim Sorokin, *A Long Journey* (New Haven: College and University Press, 1963), 241.

35. Letter from David Fairchild to W. M. Wheeler, 9 October 1925, Wheeler papers.

36. Letter from Fairchild to Wheeler, 26 March 1929, Wheeler papers.

37. Letter from Fairchild to Wheeler, 26 May 1927, Wheeler papers. Capitalization original. The genus *Eciton* were army ants that Fairchild observed in Barro Colorado. They were swarming in a huge mass that hung from a tree.

38. Letter from Fairchild to Wheeler, 20 December 1929, Wheeler papers. Capitalization and underlining original.

39. W. M. Wheeler, "The Organization of Research," *Science* 53 (1921): 53–67; 54.

40. Wheeler, "The Organization of Research," 65–6.

41. In 1927, Wheeler's university set up the Fatigue Laboratory at the Harvard Business School; his friend and colleague Henderson was its first director. Henderson discusses the kind of time and motion studies deplored by Wheeler in his 1936 paper "The Effects of Social Environment," reprinted in L. J. Henderson, *On the Social System*, ed. Bernard Barber (Chicago: University of Chicago Press, 1970), 214–34.

42. Huxley, *Brave New World Revisited*, 265–66.

43. W. M. Wheeler, "On Instincts," *Journal of Abnormal Psychology* 15 (1921): 295–318; 316.

44. Remarks drawn from W. M. Wheeler, "The Dry-Rot of Our Academic Biology," *Science* 57 (1923): 61–71, and idem, "A New Word for an Old Thing," review of John B. Watson, *Behaviorism, Quarterly Review of Biology* 1 (1926): 439–43.

45. Wheeler gives food, reproduction and protection as the three problems to be solved by all organisms, including man, in W. M. Wheeler, "Notes about Ants and Their Resemblance to Man," *National Geographic Magazine* 23 (1912): 731–66.

46. W. M. Wheeler, "Animal Societies: Biology and Society," *Scientific Monthly* 39 (1934): 289–301; 293. The original notes for this essay are found on the back of a reading list from Wheeler's 1932 comparative sociology course, and appear to be the material for his concluding lecture. The lecture is entitled "The Extent to Which a Study of Animal Societies Can Contribute to the General Problems of Social Psychology." Wheeler papers.

47. Wheeler, "Animal Societies." On the basis of the heckling that interrupted this lecture, the *Brooklyn Eagle* ran the headline, "Women Resent Charge They Would Stagnate the World." Lena Philips, the president of the National Council of Women, remarked tartly, "Why, the first curious person in the world was Mother Eve. Adam was contented enough." (Evans and Evans, *William Morton Wheeler,* 268–69.)

48. Wheeler was, of course, by no means alone in his oedipal struggles. See, for example, E. M. Forster, "The Machine Stops" (1909), in *The Machine Stops, and Other Stories,* ed. Rod Mengham (London: André Deutsch, 1997). The story is a techno-entomological nightmare centered on a biological mother and a technological matrix.

49. W. M. Wheeler, "The Queen Ant as a Psychological Study," *Popular Science Monthly* 68 (1906): 291–99; O. E. Plath, "Insect Societies," in *A Handbook of Social Psychology,* ed. Carl Murchison (Worcester, MA: Clark University Press, 1935), 83–141; 127–29; William M. Mann, "Stalking Ants, Savage and Civilized," *National Geographic Magazine* 66 (1934): 171–92; 172–73. In his time, Forel had been dubbed "the sexless worker" by his colleagues in the Munich asylum on account of his preference for ants over women. Forel seems to have been rather pleased by his nickname, noting in his memoirs that the "massive femininity of the women of . . . Munich . . . confirm[ed] me in my habit of chastity." Auguste Forel, *Out of My Life and Work* (London: George Allen & Unwin, 1937), 99.

50. Here, Huxley seems to be reverting to instinct in its nonacquired sense, for in the real world, maternal instinct was usually given as the paradigmatic exemplar of instincts in humans. William McDougall compared the actions of a mother in saving her baby unfavorably to those of a man going to war for his country. In the latter case, the man acted in full knowledge of the situation; the woman was merely acting on her instinct. William McDougall, *Introduction to Social Psychology* (3rd ed., London: Methuen, [1908] 1910), 174, 208. In "Notes about Ants and their Resemblance to Man," Wheeler explains that maternity is in fact the pivotal instinct of the colony, upon which its successful features (care for the brood by workers) and pathological phenomena (toleration of parasites) are based.

51. Wheeler, "The Organization of Research," 62–63.

52. Compare this with W. B. Yeats' infamous Steinach operation. Through his vasectomy, Yeats sought to liberate himself from draining reproductive sexuality and to revitalize himself from within, literally with his own creative juices. Tim Armstrong, *Modernism,*

Technology and the Body: A Cultural Study (Cambridge: Cambridge University Press, 1998), 133–34, 143–50. One of Wheeler's favorite authors on the antisociality/biology question was Félix LeDantec, *L'Egoïsme, base de toute société: Etude des déformations résultant de la vie en commun* (Paris: Alcan, 1911).

Five • The Generic Contexts of Natural History

1. Wheeler was also at Woods Hole. Philip Pauly, "Summer Resort and Scientific Discipline: Woods Hole and the Structure of American Biology, 1882–1925," in *The American Development of Biology*, ed. Ronald Rainger, Keith R. Benson, and Jane Maienschein (Philadelphia: University of Pennsylvania Press, 1988), 121–50.

2. Letter from T. H. Morgan to W. M. Wheeler, 10 February 1922, Wheeler papers.

3. W. M. Wheeler, "What Is Natural History?" *Bulletin of the Boston Society of Natural History* 59 (1931): 3–12; 4. See also Sally Gregory Kohlstedt, "From Learned Society to Public Museum: The Boston Society of Natural History," in *The Organization of Knowledge in Modern America, 1860–1920*, ed. Alexandra Oleson and John Voss (Baltimore: Johns Hopkins University Press, 1979), 386–406. Kohlstedt describes the Boston Society of Natural History's efforts to redefine its identity and purposes as the institutions of science changed around it.

4. W. M. Wheeler, "The Termitodoxa, or Biology and Society," *Scientific Monthly* 10 (1920): 113–24 (given as a talk in 1919); idem, "Animal Societies: Biology and Society," *Scientific Monthly* 39 (1934): 289–301 (given as a talk in 1933). In his review, "A Notable Contribution to Entomology," *Quarterly Review of Biology* 11 (1936): 337–41, Wheeler made a late spirited stand for naturalists in the face of academic biology, admiring Tarlton Rayment, the review's subject, for being in the tradition of Réaumur, Fabre, Forel, and the Peckhams. There is scope for a broader study on the unique U.S. identity of the term "naturalist," combining both general cultural constructions of the term along with its specific disciplinary placement with respect to other scientific disciplines.

5. William Beebe, *The Book of Naturalists* (New York: Knopf, 1944), 87.

6. Beebe, *The Book of Naturalists*, 250.

7. For a discussion of how other biologists have constructed their work in literary terms, see Greg Myers, *Writing Biology: Texts in the Social Construction of Scientific Knowledge* (Madison: University of Wisconsin Press, 1990).

8. Joan Shelley Rubin, *The Making of Middlebrow Culture* (Chapel Hill: University of North Carolina Press, 1992).

9. Michael Denning, *Mechanic Accents: Dime Novels and Working Class Culture in America* (London: Verso, 1987); Kirsten Drother, *English Children and Their Magazines, 1751–1945* (New Haven, CT: Yale University Press, 1988); Antony Griffiths, *Prints and Printmaking: An Introduction to the History and Techniques* (London: British Museum Publications, 1980).

10. W. J. Holland, *The Butterfly Book* (Country Life Press, Garden City, NY: Doubleday, Doran, [1915] 1935).

11. See also Matthew D. Whalen and Mary F. Tobin, "Periodicals and the Popularization of Science in America, 1860–1910," *Journal of American Culture* 3 (1980): 195–203.

12. Of the large publishers, Macmillan tended to deal with nature textbooks and practically oriented books; Scribners and Houghton Mifflin dealt with general nature books and the more theist end of the market; Doubleday, Page covered nature literature quite widely. Ginn and Company and Harper and Brothers tended to concentrate on the more imaginative end of the nature range.

13. M. Louise Greene, "A Nature Creed in the Concrete," *Nature-Study Review* 7 (1911): 187–90.

14. Peter A. Fritzell, *Nature Writing and America: Essays upon a Cultural Type* (Ames: Iowa State University Press, 1990), 3–36.

15. Marina Frasca-Spada and Nicholas Jardine, eds., *Books and the Sciences in History* (Cambridge: Cambridge University Press, 2000).

16. Henry Dircks, *Nature Study: Or the Art of Attaining those Excellencies in Poetry and Eloquence which are Mainly Dependent on the Manifold Influences of Universal Nature* (London: E. Moxon and Son, 1869).

17. Dircks, *Nature Study*, 111.

18. Gillian Beer, *Darwin's Plots: Evolutionary Narrative in Darwin, George Eliot and Nineteenth-Century Fiction* (2nd ed., Cambridge: Cambridge University Press, 2000), 79–103.

19. Clara Barrus, ed., *The Life and Letters of John Burroughs* (2 vols., Boston: Houghton Mifflin, 1925), vol. 1, 247.

20. John Burroughs, *Literary Values and Other Papers* (London: Gay and Bird, 1903), 27–28.

21. Burroughs, *Literary Values*, 32.

22. W. M. Wheeler, "The Dry-Rot of Our Academic Biology," *Science* 57 (1923): 61–71. Reprinted in *idem*, *Foibles of Insects and Men* (New York: Alfred A. Knopf, 1928), 187–204; 198.

23. Quoted in Burroughs, *Literary Values*, 27.

24. Ernest Thompson Seton and Julia M. Seton, *The Gospel of The Redman: An Indian Bible* (London: Psychic Press, [1937] 1970).

25. Karen Jones, "Writing the Wolf," paper given at British Society for the History of Science Annual Conference, University of Leeds, July 2005.

26. The psychologist James Baldwin also subscribed to a similar belief. In a very early piece of work he analyzed the philosophy of Spinoza to show how, correctly interpreted, it entailed an absolute realism and a direct epistemological intuitionism. Throughout his career, he stuck to the principle that only nature as known could be an object of science. However, since Baldwin held to a monism of mind (albeit realist, not idealist in the strict sense), the implication of this principle was that "nature is intelligent and that the laws of thought are the laws of things." As it was for Burroughs, a harmony of mental and physical laws was the essential basis of Baldwin's work, developed as his principle of "organic selection" (showing how evolution works on mind and body as a unit). See Robert J. Richards, *Darwin and the Emergence of Evolutionary Theories of Mind and Behavior* (Chicago: University of Chicago Press, 1987), 451–503. For a discussion of Bergsonian biology, including further references, see Charlotte Sleigh, "Plastic Body, Permanent Body: Czech Representations of Corporeality in the Early Twentieth Century," *Studies in History and Philosophy of Biological and Biomedical Sciences* (forthcoming).

27. Besides the texts discussed here see also the following: Henry Chester Tracy, *American Naturalists* (New York: Dutton, 1930); Norman Foerster, *Nature in American Literature* (New York: Macmillan, 1923); Philip Marshall Hicks, *The Development of the Natural History Essay in American Literature* (Philadelphia: University of Pennsylvania Press, 1924); N. Bryllion Fagin, *William Bartram: Interpreter of the American Landscape* (Baltimore: Johns Hopkins University Press, 1933). S. W. Geiser, *Naturalists of the Frontier* (2nd ed., Dallas, TX: Southern Methodist Press, 1948); Pamela Regis, *Describing Early America: Bartram, Jefferson, Crevecoeur, and the Rhetoric of Natural History* (DeKalb, IL: Northern Illinois University Press, 1992). Beebe's *Book of Naturalists,* though international in scope, includes more Americans than any other nationality, and the history of natural history that it implicitly and explicitly tells culminates in a distinctively American approach.

28. See the essay on sources.

29. Joseph Wood Krutch, ed., *Great American Nature Writing* (New York: Sloane, 1950).

30. See also Marjorie Hope Nicolson, *Mountain Gloom and Mountain Glory: The Development of the Aesthetics of the Infinite* (Seattle: University of Washington Press, 1959).

31. Donald Culross Peattie, *Green Laurels: The Lives and Achievements of the Great Naturalists* (New York: Simon and Schuster, 1936), 212.

32. Peattie, *Green Laurels,* 213–14.

33. John Burroughs was similarly skeptical about the value of European—particularly English—attitudes toward nature:

> In English literature there is the artificial Nature of Pope and his class . . . as dead and hollow as fossil shells. Earlier than that, the quaint and affected Nature of the Elizabethan poets . . . and lastly the transmuted and spiritualized Nature of Wordsworth . . . Thus, from a goddess Nature has changed to a rustic nymph, a cloistered nun, a heroine of romance . . . till she has at last become a priestess of the soul. (Burroughs, *Literary Values,* 203–4.)

Uncle John "reprinted" a letter from a boy in 1904 (apparently actually written by himself) that rehearsed similar themes. The "boy" recounted how a man had come to his school and told them about the superiority of European art—but why, asked the boy, were the "Dagoes" who worked on the railroad so dirty if they were so sophisticated? Uncle John "replied" that some people would call this man a "Miss Nancy." This would be very wrong, commented Uncle John, while managing to imply the very opposite; he was sure that the man truly appreciated art, or at least thought that he did. Uncle John went on to warn of the dangers for a boy who had only girls to play with, or of becoming a "Little Lord Fauntleroy." A really beautiful school was one that had a well-made garden, an achievement that entailed manly activity. Cornell University Press and Comstock Publishing Co. records.

34. William Martin, *Natural History and the American Mind* (New York: Columbia University Press, 1941), 337. Martin advocated a revisitation of taxonomy in the light of modern evolutionary theory, suggesting that more "splitting" was necessary—thus ensuring that biologists would not be able to do without the services of naturalists for the foreseeable future. Compare to Peattie, *Green Laurels,* 346–47.

35. Peattie, *Green Laurels,* 329–30.

36. Peattie was also a classmate of Wheeler's son, Ralph.

37. M. Elsie Gullick's *Betty and the Little Folk* was advertised as a "delightful fairy story of Ant Life." Gullick was a prolific writer of such tales, which were pitched for different age groups. *Mother Nature and her Fairies* was by Hugh Findlay. The signs could, however, be misleading. The Rev. Charles A. Hall's *Open Book of Nature*, for example, looks unpromising; besides its ecclesiastical provenance it boasts a sentimental frontispiece, "A Happy Hunting Ground for the Naturalist," showing improbably well-fed, well-kempt rural children in a pastoral-woodland setting. Yet the book is actually rather technical, with an excellent scientific glossary and stretching yet practical advice on nature photography and the construction of a simple microscope. Charles A. Hall, *The Open Book of Nature: An Introduction to Nature-Study* (2nd ed., London: A. & C. Black, 1911).

38. Philip Pauly counts Bailey (along with Roosevelt) as the last of the late nineteenth-century generation of biologists presenting themselves as "experts." Philip Pauly, *Biologists and the Promise of American Life: From Meriwether Lewis to Alfred Kinsey* (Princeton, NJ: Princeton University Press, 2000). See also Morris Bishop, *A History of Cornell* (Ithaca: Cornell University Press, 1962), and Gould P. Colman, *Education and Agriculture: A History of the New York State College of Agriculture at Cornell University* (Ithaca: Cornell University Press, 1963). On other birthplaces of nature study (notably the teaching colleges of Chicago and Columbia), see Sally Gregory Kohlstedt, "Nature Study in North America and Australasia, 1890–1945: International Connections and Local Implementations," *Historical Records of Australian Science* 11 (1997): 349–454; 443–45, and idem, "Nature, Not Books: Scientists and the Origins of the Nature-Study Movement in the 1990s," *Isis* 96 (2005): 324–52.

39. Foreword to Bulletin 160, Jan. 1899, of Horticultural Division of Cornell University Agricultural Experiment Station. Cornell University Press and Comstock Publishing Co. papers.

40. On the former movement, see Ralph H. Lutts, *The Nature Fakers: Wildlife, Science and Sentiment* (Charlottesville: University Press of Virginia, 1990), passim.

41. Pamela Henson, "The Comstock Research School in Evolutionary Entomology," *Osiris* 8 (1993): 159–77; idem, "The Comstocks of Cornell: A Marriage of Interests," in *Creative Couples in the Sciences*, ed. Helena M. Pycior, Nancy G. Slack, and Pnina G. Abir-Am (New Brunswick, NJ: Rutgers University Press, 1996), 112–25; idem, "'Through Books to Nature': Anna Botsford Comstock and the Nature Study Movement," in *Natural Eloquence: Women Reinscribe Science*, ed. Barbara Gates and Ann Shteir (Madison: University of Wisconsin Press, 1997), 116–43.

42. In the year 1904–5, for example, entomology received $2,400 to fund both the school and the experiment station; the next smallest (horticulture, dairy and chemistry) all received around $7,500, while agriculture, the largest, was in receipt of $12,000. Liberty Hyde Bailey papers.

43. C. F. Hodge, "Nature-Study Work with Insects," *Nature-Study Review* 2 (1906): 265–70.

44. Letter from Anna M Patton, St. Troy, NY, 10 August 1903, Cornell University Press and Comstock Publishing Co. papers.

45. Philip Pauly discusses the significance of high school biology in biologists' profes-

sional aspirations in *Biologists and the Promise of America Life*, 171–93. Interest in the nature study movement was, however, largely confined to elementary schools.

46. See also Lutts, *The Nature Fakers*, 25–30.

47. On the history of the American Nature Study Society, see E. Laurence Palmer, "Fifty Years of Nature Study and the American Nature Study Society," *Nature Magazine* November 1957: 473–80.

48. L. H. Bailey, quoted in 1925 Nature education notes by ANSS; ANSS minutes & records 1925–41, ANSS papers.

49. L. H. Bailey, n.d., quoted in ANSS files, ANSS papers.

50. There was a certain amount of overlap with nature teaching in the United Kingdom. The Perthshire Museum was among the first to transplant the movement to Britain; in collaboration with the Carnegie Museum, Pittsburgh, it began offering prizes to children for essays as early as 1898. Hedger Wallace corresponded closely with the Cornell / New York network. T. G. Rooper entered into an exchange with American educationalists, and, as a government inspector of schools, he went on to promote its use widely in the United Kingdom. See *Science Teaching . . . and Nature Study: Report of the Conference and Exhibition Held at the Hartley College, Southampton on Friday June 13th and Saturday June 14th, 1902* (Southampton, UK: H. M. Gilbert and Son, 1902). Nature study was also taken seriously in the most prestigious schools, including Eton, Charterhouse, and Harrow. Such institutions were not renowned for their cultivation of sentimentality, and indeed in these contexts nature study was pursued as a means to develop military ability. See Charles Kingsley, "The Study of Natural History for Soldiers," (1872), in *The Works of Charles Kingsley*, vol. 19, *Scientific Lectures and Essays* (London: Macmillan, 1880), 181–98. This justification formed no significant part of the North American discourse. By 1902 nature study was recognized by name in Britain, and was acknowledged to have originated and developed most in the United States. J. Arthur Thomson and Miall both supported the movement from the professional biological end. On nature study in Britain, see also Kohlstedt, "Nature Study in North America and Australasia," 446 and relevant footnotes.

51. Source of data: *Nature-Study Review* 11 (1915): 297–99.

52. Warning printed in *Nature-Study Review* 3 (1907): 190.

53. M. A. Bigelow, "Best Books for Nature-Study," *Nature-Study Review* 2 (1906): 168–69.

54. Macmillan papers.

55. See, for example, Theodesia Hadley (chairman of membership committee), letter to ANSS members, 3 June 1933, ANSS papers.

56. Quoted in ANSS Newsletter, Winter 1954, 3.

57. See, for example, the editorial to the *Nature-Study Review* for 9 (1913): 226–27.

58. Carroll Lane Fenton, "Memorandum on Popularization of Science for Children, Juniors and Adults," 15 September 1938, ANSS papers.

59. David Allen describes a little of the theory and practice of nature study for children in *idem, The Naturalist in Britain: A Social History* (London: Allen Lane, 1976), 202–4. The suitability of nature for study by children was expressed widely; similarly the trope of naturalist as perpetual child, or the tracing of the naturalist's vocation to an incident in childhood, was quite common in this period.

60. Quoted in ANSS Newsletter, Spring 1945, 8.

61. John Burroughs, *Birds and Bees, Sharp Eyes and Other Papers* (Boston: Houghton Mifflin, 1887), 3.

62. Harold W. Fairbanks, in "Nature Study and Its Relation to Science," by H. W. Fairbanks, C. F. Hodge, T. H. MacBride, F. L. Stevens, and M. A. Bigelow, *Nature-Study Review* 1 (1905): 2–22; 6.

63. Letters from David Fairchild to W. M. Wheeler; quotations from letters dated 7 July 1927 and 26 May 1927, Wheeler papers.

64. W. M. Wheeler, *Demons of the Dust: A Study in Insect Behaviour* (London: Kegan Paul, Trench, Trubner, 1931), ix.

65. Notes found in Wheeler's desk after death, Wheeler papers.

66. Maurice Bigelow, letter, *Nature-Study Review* 3 (1907): 236–39.

67. Clayton F. Palmer, letter, *Nature-Study Review* 4 (1908): 28–29.

68. Willard N. Clute, letter, *Nature-Study Review* 4 (1908): 30.

69. The Central Association of Science and Mathematics Teachers was formed in 1903. Sally Gregory Kohlstedt notes that a few women attained positions of leadership in the field of nature study. *Idem*, "Nature Study in North America and Australasia," 446 and footnote.

70. An anonymous article in *Nature-Study Review* 4 (1908): 4, expresses pleasure at interest shown in the subject by some men of science, though others have been inclined to regard it as a fad.

71. Letter to Mary E. Ware of Miss Pierce's School, 5 March 1913, Wheeler papers. Wheeler's records contain a number of requests to go and speak to children, to none of which he apparently acceded. He did give a few popular lectures to adults and children during his first year or two at the American Museum of Natural History, but even by then he was uninterested in the educational ideas of his friend and museum colleague H. C. Bumpus (Evans and Evans, *William Morton Wheeler*, 138, 155).

72. Nina E. Lerman, "The Uses of Useful Knowledge: Science, Technology and Social Boundaries in an Industrializing City," *Osiris* 12 (1997): 39–59.

73. M. A. Bigelow, in "Educational Values and Aims of Nature-Study," by S. Coulter, H. W. Fairbanks and M. A. Bigelow, *Nature-Study Review* 1 (1905): 49–57; 54.

74. S. Coulter, in "Educational Values and Aims of Nature-Study," by Coulter, Fairbanks, and Bigelow, 50.

75. *Nature-Study Review* 3 (1907): 52–53.

76. Marion H. Carter, *Nature Study with Common Things: An Elementary Laboratory Manual* (New York: American Book Co., 1904), 3.

77. H. W. Fairbanks, C. F. Hodge, T. H. MacBride, F. L. Stevens, and M. A. Bigelow, "Nature Study and Its Relation to Science," *Nature-Study Review* 1 (1905): 2–22; 16.

78. Quoted in Evans and Evans, *William Morton Wheeler*, 46.

79. Insofar as nature study books emphasized pictures (one of Anna Comstock's particular skills) rather than words, they were compatible with Pestalozzian methods of learning.

80. McMurry, "Advisable Omissions from the Elementary Curriculum, and the Basis for Them," *Educational Review* 27 (1904): 478–93. Reprinted in full in *Nature-Study Review* 1 (1905): 22–26; 23.

81. H. R. Royston, *The Unity of Life: A Book of Nature Study for Parents and Teachers* (London: Harrap, 1925), 14. Royston's purpose for nature study was to teach children the facts of life, thus avoiding darkly hinted-at dangers of physical, moral, and eugenic disease.

82. Stanley Coulter (Purdue University, Lafayette, Indiana), C. F. Hodge (Clark University, Worcester, Massachusetts), and C. R. Mann (University of Chicago) wrote on the relation of nature study and science teaching in *Nature-Study Review* 4 (1908): 10–24. For a different perspective on the scientificity of observation in ornithology, see Helen Macdonald, "'What Makes You a Scientist Is the Way You Look at Things': Ornithology and the Observer 1930–1955," *Studies in History and Philosophy of Biological and Biomedical Sciences* 33 (2002): 53–77.

83. Talk by McMurrich to American Society of Naturalists, reprinted in *Nature-Study Review* 4 (1908): 90. As far as the story of entomology was concerned, the most significant scientific supporter, besides John Comstock was Vernon Kellogg.

84. John M. Coulter, "Principles of Nature Study," *Nature-Study Review* 1 (1905): 57–61; 59.

85. In 1936 the New York chapter of the ANSS conducted a survey on how and why to teach nature study, focusing on "some mooted questions of the ANSS discussed during the past generation;" one question—receiving no clear answer—asked whether nature study could be taught through "fairy tales, such as 'how the robin got his red breast'?" ANSS papers.

86. *Science Teaching*, 43–44.

87. W. M. Wheeler, "A Study of Some Young Ant Larvae with a Consideration of the Origin and Meaning of Social Habits Among Insects," *Proceedings of the American Philosophical Society* 57 (1918): 293–343; 293.

88. John Burroughs, "Real and Sham Natural History," *The Atlantic Monthly* 91 (1903): 298–309; W. M. Wheeler, "The Obligations of the Student of Animal Behavior," *The Auk* 21 (1904): 251–55; idem, "Woodcock Surgery," *Science* 19 (1904): 347–50. This was the so-called "Nature-Fake Furor." For an account of the textual details of this episode, see Lutts, *The Nature Fakers*.

89. See reviews in *Nature-Study Review* 3 (1907): 240–42.

90. C. F. Holder and David Starr Jordan, *Fish Stories* (New York: Holt, 1909). Offending passages included the following: "Then there were many more little salmon with him, some larger and some smaller, but they all had a very merry time, Those who had been born soonest and had grown largest used to chase the others around and bite off their tails, or still better, take them by the heads and swallow them whole; 'for,' said they, 'even young salmon are good eating.'" *Nature-Study Review* 5 (1909): 113. Jordan was a professional correspondent of Wheeler's.

91. Letter from Vernon Kellogg, 2 April 1895, Comstock papers.

92. J. Dean Simkins, *Ants and the Children of the Garden. Relating the Habits of the Black Harvester Ant and Giving Considerable Information about Ants in General* (manuscript, 1920), Wheeler papers. Simkins sent the manuscript to Wheeler, whose work was courteously cited by the well-educated ant-narrator.

93. *Nature-Study Review* 4 (1908).

94. S. Silcox, "Nature Study in Primary Grades," *Nature-Study Review* 5 (1909): 124–27.

95. Allen Walton Gould, *Mother Nature's Children* (Boston: Ginn and Co., 1900), vi.

96. See Henson, "Through Books to Nature."

97. See Marion Hill in the *Nature-Study Review* 7 (1911), extracted from *American Magazine*, July 1910.

Six • Writing Elite Natural History

1. The magazine was founded in 1900; Ternes' collection *Ants, Indians and Little Dinosaurs* (New York: Charles Scribner's Sons, 1975) celebrated its 75th anniversary with articles selected from throughout its lifetime.

2. See Mary A. Evans and Howard E. Evans, *William Morton Wheeler, Biologist* (Cambridge, MA: Harvard University Press, 1970), 307, for an account of Wheeler's populist students and colleagues.

3. W. M. Wheeler, "On Instinct," *Journal of Abnormal Psychology* 15 (1921): 295–318. Wheeler's student J. G. Myers used anthropomorphism knowingly and wittily in his introduction to idem, *Insect Singers: A Natural History of the Cicadas* (London: George Routledge, 1929), "proving" that—*pace* Aesop—cicadas were better moral models than ants.

4. W. M. Wheeler, "The Ant-Colony as an Organism," *Journal of Morphology* 22 (1911): 301–25; 307–8. Wheeler was at the height of his Bergsonian phase around this time. Later in his life he described his focus on living complexity as a Romantic approach to nature. W. M. Wheeler, "A Notable Contribution to Entomology," *Quarterly Review of Biology* 11 (1936): 337–41.

5. Jean-Henri Fabre, *The Wonders of Instinct: Chapters in the Psychology of Insects* (London: T. Fisher Unwin, 1918), 20–21. Fabre's scorn seems to be directed principally against Anton Dohrn's Stazione Zoologica in Naples.

6. Wheeler, "On Specimen Collecting," MS, n.d.: 24 pp.; 19–20. Wheeler papers.

7. Philip Pauly highlights the awkward organization of biology both by biological subject (botany, zoology, and further specializations) and by function (morphology, pathology, ecology, or genetics). Philip J. Pauly, "Modernist Practice in American Biology," in *Modernist Impulses in the Human Sciences 1870–1930*, ed. Dorothy Ross (Baltimore: Johns Hopkins University Press, 1994), 272–89.

8. W. M. Wheeler, "'Natural History,' 'Œcology,' or 'Ethology'?" *Science* 15 (1902): 971–76.

9. W. M. Wheeler, "Jean-Henri Fabre," *Journal of Animal Behavior* 6 (1916): 74–80; 79.

10. Compare this definition to the professionalizing masculinization of botany, another discipline dogged by amateur status thanks to its many female participants. See Ann B. Shteir, "Gender and 'Modern' Botany in Victorian England," *Osiris*, 2nd ser., 12 (1997): 29–38.

11. Wheeler "On Specimen Collecting," 21–22. Wheeler included in his list the "dried up old professor who would almost sell his soul for a piece of fossil bone," thus feminizing practitioners of non-field sciences along with "mere naturalists."

12. C. V. Legros, *Fabre, Poet of Science* (London: T. Fisher Unwin, 1913); Abbé Augustin Fabre, *The Life of Jean Henri Fabre the Entomologist* (London: Hodder and Stoughton, 1921). For the context of educational secularism see Nicole Hulin, *Les femmes et l'enseignement scientifique* (Paris: PUF, 2002).

13. W. M. Wheeler, "Some Attractions of the Field Study of Ants," *Scientific Monthly* 34 (1932): 397–402.

14. Evans and Evans, *William Morton Wheeler*, 298. See also George H. Parker's autobiography, *The World Expands: Recollections of a Zoologist* (Cambridge, MA: Harvard University Press, 1946), and his biography of Wheeler: *idem*, "Biographical Memoir of William Morton Wheeler," *National Academy of the Sciences of the USA Biographical Memoirs* 19 (1938): 203–41. Wheeler found their examination of the Book of Job especially helpful.

15. Wheeler, "On Specimen Collecting," 10.

16. Wheeler, "On Specimen Collecting," 11–12.

17. W. M. Wheeler, *Demons of the Dust: A Study in Insect Behaviour* (London: Kegan Paul, Trench, Trubner, 1931), viii. On Fabre's "discovery," see Legros, *Fabre, Poet of Science*.

18. Matthew 18, vv. 2–4.

19. When Howard tried to persuade the government that efforts were required to combat insect-borne disease among troops in the Great War, he was told that the United States did not need "men trained to count the spots on a mosquito's wing." Leland O. Howard, *Fighting the Insects: The Story of an Entomologist. Telling the Life and Experiences of the Writer* (New York: Macmillan, 1933), 137–40. See also *idem*, "Entomology and the War," *Scientific Monthly* 8 (1919): 109–17.

20. Fabre, *The Wonders of Instinct*, 36.

21. Deville, "Ernest-Marie-Louis Bedel," footnote to Maurice Bedel, "My Uncles, Louis Bedel and Henri d'Orbigny," trans. W. M. Wheeler, *Quarterly Review of Biology* 8 (1933): 325–30.

22. Simon Ryan, *The Cartographic Eye: How Explorers saw Australia* (Cambridge: Cambridge University Press, 1996).

23. One might persuasively argue that the switch in scale entailed by entomology means it is virtually impossible to resist metaphorizing the project as a Lilliputian journey. Fabre's title was often "Homer of the insects"; D. C. Peattie called C. V. Riley a "Gulliver."

24. William Beebe, *The Book of Naturalists* (New York: Knopf, 1944), 88–89.

25. "Jean-Henri Fabre," in *Société Entomologique de France: Livre du centenaire* (Paris: Au Siège de la Société, 1932). Jean Rostand, "Sur J.-H. Fabre," in *Société Entomologique de France: Livre du centenaire* (Paris: Au Siège de la Société, 1932), 101–5. Jean Rostand, "Jean-Henri Fabre," *Hommes de verité*, ser. 2 (1948): 109–68. Charles Ferton, *La vie des abeilles et des guêpes* (Paris: Chiron, 1923), 345–73.

26. Fabre, *Social Life in the Insect World*, 79–85. See Frederick R. Prete and M. Melissa Wolfe, "Religious Supplicant, Seductive Cannibal, or Reflex Machine? In Search of the Praying Mantis," *Journal of the History of Biology* 25 (1992): 91–136.

27. Fabre, *Social Life in the Insect World*, 106–7. Forel's work on mutual regurgitation is discussed in ibid., chap. 7.

28. Jean-Henri Fabre, *Souvenirs entomologiques: Etudes sur l'instinct et les mœurs des insectes* (10 vols., Paris: Delagrave, 1879–1907), vol. ix, 79–80.

29. Fabre, *Souvenirs entomologiques*, vol. vi, 67.

30. This assessment would cover the Raus. Although they were evolutionists, they extended an experiential psychology to the insects, because not to do so would be an act of arrogance. As they wrote,

Why should we apologize for the use of terms which seem anthropomorphic? . . . We surely cannot see the manifestations of the inner nature of these creatures and think of them as automatic machines any more then we can look at the statue of The Thinker and characterize Rodin with Bethe terminology . . . We realize that the student of animal behavior must be on his guard against making faulty ejects; yet, when confronted with abundant confirmatory evidence, to refuse to ascribe a certain psychic trait to an animal merely because it is not a human being seems to us as great an anthropomorphism as those of which the older comparative psychologists were guilty. (Phil Rau and Nellie L. Rau, *Wasp Studies Afield* [Princeton, NJ: Princeton University Press; London: Oxford University Press, 1918], 367–68.)

31. Rostand, "Jean-Henri Fabre," 157.

32. Ludwig Büchner, *Mind in Animals*, trans. Annie Besant (London: Freethought Publishing Co., 1880), viii.

33. Büchner, *Mind in Animals*, 129.

34. Fabre, *The Wonders of Instinct*.

35. Fabre, *The Wonders of Instinct*, 128–47.

36. Jean-Henri Fabre, *Social Life in the Insect World* (London: T. Fisher Unwin, 1912), 124.

37. John Burroughs, *Birds and Bees, Sharp Eyes and Other Papers* (Boston: Houghton Mifflin, 1887), 54–55.

38. Burroughs, *Birds and Bees*, 3.

39. George W. Peckham and Elizabeth G. Peckham, *Wasps Social and Solitary* (Boston: Houghton Mifflin; London: Constable, 1905), xiii.

40. In order to expose the heresy that reasoning abilities or a psychic life were possessed by insects, Fabre recounted Condillac's famous discussion of a statue which, by gradually acquiring senses, progressively forms an impression of the world. Fabre compared the statue to the grub of the Capricorn beetle which, despite having the impressions and interior life of a "bit of intestine that crawls about," acts purposively for the future, by burrowing to the right part of the tree and constructing a cell to enable successful pupation. Actually, Fabre presents an absurd version of Condillac's argument. He says that the statue "when endowed with the sense of smell, inhales the scent of a rose and out of that single impression creates a whole world of ideas." Nevertheless, this argument shows that Fabre was never anthropomorphic in a psychological sense. The experience of most insects underdetermines the complex behaviors which they reliably manifest. (Fabre, *Wonders of Instinct*, 49–64.)

41. Maurice Maeterlinck, *The Life of the Bee* (London: George Allen, 1901); idem, *The Life of the White Ant* (London: George Allen, 1927); idem, *The Life of the Ant* (London: Cassell, 1930).

42. Josef Čapek and Karel Čapek, *R.U.R. and The Insect Play* (London: Oxford University Press, [1921 and 1923] 1961).

43. The counterposition of these amateurs to Wheeler's "elite natural historians," and their demotion by the latter to the rank of "mere natural historians," invites the question of how, in positive terms, they saw themselves. Certainly, they did not constitute a ready-

made group like professors connected by the web of academia. Many would have identified themselves as allied with some branch of the nature study movement, whether through teaching, writing, reading, or simple ideological affinity. Others were principally identified with their local scientific societies, with correspondingly locally oriented aspirations. Some, like Adele Fielde and Charles Turner, aspired to full academic, professional inclusion but were frustrated in their ambitions.

44. See Arnold Mallis, *American Entomologists* (New Brunswick, NJ: Rutgers University Press, 1971), 348–51, and Marcia Myers Bonta, *American Women Afield: Writings by Pioneering Women Naturalists* (College Station: Texas A&M University Press, 1995), 75–77. George W. Peckham and Elizabeth G. Peckham, *Ant-Like Spiders of the Family Attidae* (Milwaukee: published by the authors, 1892); idem, *On the Instincts and Habits of the Solitary Wasps* (Madison, WI: published by the state, 1898); idem, *Wasps Social and Solitary* (Boston: Houghton, Mifflin; London: Constable, 1905); George W. Peckham, Elizabeth G. Peckham, and W. M. Wheeler, "The Spiders of the Sub-Family *Lyssomanae*," *Transactions of the Wisconsin Academy of Sciences, Arts, and Letters*, 7 (1888): 222–56.

45. Mallis, *American Entomologists*, 490–92; Bonta, *American Women Afield*, 203–5. Phil Rau, "Field Studies in the Behavior of the Non-Social Wasps," *Transactions of the Academy of Science of St. Louis* 25 (1928): 321–489; idem, *The Jungle Bees and Wasps of Barro Colorado Island* (St. Louis, MO: Kirkwood, 1933). Phil Rau and Nellie L. Rau, "The Biology of *Stagmomantis carolina*," *Transactions of the Academy of Science of St. Louis* 22 (1913): 58 pp.; idem, "Longevity in Saturniid Moths and Its Relation to Their Function of Reproduction," *Transactions of the Academy of Science of St. Louis* 23 (1914): 78 pp.; idem, "The Sex Attraction and Rhythmic Periodicity in Giant Saturniid Moths," *Transactions of the Academy of Science of St. Louis*, 26 (1929): 80–221.

46. Helen Norton Stevens, *Memorial Biography of Adele M. Fielde, Humanitarian* (Seattle: Pigott Printing Concern, 1918). Unhampered by the dubious privilege of professionalism on account of her sex, Fielde was able to attack her many "big questions." On Edith Patch, see Bonta, *American Women Afield*, 171–74; Patch's papers may be found in the Special Collection Department, Raymond H. Folger Library, University of Maine, Orono.

47. David Elliston Allen, *The Naturalist in Britain: A Social History* (London: Allen Lane, 1976).

48. W. Conner Sorensen, *Brethren of the Net: American Entomology, 1840–1880* (Tuscaloosa: University of Alabama Press, 1995), 4, 12.

49. Editorial, *Nature-Study Review* 7 (1911): 191–92.

50. Marion H. Carter, letter, *Nature-Study Review* 1 (1905): 266–68. "It may be pertinent, and of interest to those for whom the infant-discovery fetish is still persistent to state that in nearly twenty years as a teacher of all grades . . . I have never known *one* to make any *original* discovery; nor one who even seemed competent *at the time* and with the materials at hand to make a discovery." The exception had been one extremely bright boy who thought he was making new discoveries due to previous encouragement. Upon being given books his interest waned; "there didn't seem to be any use trying to discover new things for everything was already discovered." The Agassiz method was therefore very dangerous; it developed self dependence but at a fearsome price. The "plain cold fact" was that one needed to know the literature before making one's way in science. See also Marion H.

Carter, *Nature Study with Common Things: An Elementary Laboratory Manual* (New York: American Book Co., 1904).

51. T. D. A. Cockerell, letter, *Nature-Study Review* 1 (1905): 163–64.

52. Cockerell also had a different background, being British and an associate of Alfred Russel Wallace.

53. Familial insects are found in Edith M. Patch, *Dame Bug and Her Babies* (Orono, ME: Pine Cone Publishing, 1913), civic ones in Henry C. McCook, *Ant Communities and How They are Governed: A Study in Natural Civics* (New York: Harper, 1909), and Adele M. Fielde, "Artificial Mixed Nests of Ants," *Biological Bulletin* 5 (1903): 320–25.

54. Fabre, *Souvenirs entomologiques*, vol. i, 67–179 and 207–20; vol. ii, 14–37.

55. Auguste Forel, *Fourmis de la Suisse* (Basel: H. Georg, 1874), i.

56. Margaret Floy Washburn, then associate professor of philosophy at Vassar College, covered the "method of anecdote" in her book, *The Animal Mind* (New York: Macmillan, 1909), 4–9, reaching a more charitable conclusion about the status of the Raus' work, at the expense of Romanes and Darwin. On the textual manufacture of scientific authority, see Charlotte Sleigh, "'This Questionable Little Book': Narrative Ambiguity in Nineteenth-Century Literature of Science," in *Unmapped Countries: Biological Visions in Nineteenth-Century Literature and Culture*, ed. Anne-Julia Zwierlein (London: Anthem Press, 2005), 15–30.

57. Rau and Rau, *Wasp Studies Afield*, 3. Phil Rau later commented that the first stage in the solution of the problem of insect behavior was a recognition of the need "to substitute for the dead specimens in cabinets and for theories concocted in easy chairs the patient observation on [sic] the organisms in their native haunts, among their friends, enemies and natural surroundings." This was done especially well by Fabre, Lubbock, and Thorndike, thought Rau. P. Rau, *Jungle Bees and Wasps*, 269.

58. Rau and Rau, *Wasp Studies Afield*, 5.

59. Charles Henry Turner, "Behavior of the Common Roach, *Periplaneta orientales* on an open maze," *Biological Bulletin* 25 (1913): 348–65; idem, "Do Ants Form Practical Judgements?" *Biological Bulletin* 12 (1907): 333–43; idem, "The Homing of Ants," *Journal of Comparative Neurology and Psychology* 17 (1907): 367–434; idem, "The Homing of Burrowing Bees," *Biological Bulletin* 15 (1908): 247–58; idem, "The Homing of the Hymenoptera," *Transactions of the Academy of Science of St. Louis* 24 (1923): 27–45; idem., "The Locomotion of Surface-Feeding Caterpillars are not Tropisms," *Biological Bulletin* 34 (1918): 137–48; idem, "The Mating of *Lasius niger*," *Journal of Animal Behavior* 5 (1915): 337–40. "The Psychology of 'Playing Possum,'" *Transactions of the Academy of Science of St. Louis* 24 (1923): 46–54; idem, "Tropisms in Insect Behavior," *Transactions of the Academy of Science of St. Louis* 24 (1923): 19–26.

60. Phil Rau, "The Scientific Work of Dr. Charles Henry Turner," *Transactions of the Academy of Science of St. Louis* 24 (1923): 10–16; 14–15. The obituary was read at Sumner High School, 25 May 1923. Rau had come under Turner's influence at the school.

61. Rau, "The Scientific Work of Dr. Charles Henry Turner," 16.

62. A. G. Pohlman, "Charles Henry Turner: An Appreciation," *Transactions of the Academy of Science of St. Louis* 24 (1923): 7–9; 8.

63. W. M. Wheeler, "The Organization of Research," *Science* 53 (1921): 53–67; 62–63.

64. Edward C. Ash, *Ants, Bees and Wasps: Their Lives, Comedies and Tragedies* (London: Robert Holden, 1925).

65. Letter from Vernon Kellogg to John Comstock, 12 April 1896, Comstock papers.

66. E. Laurence Palmer, "Fifty Years of Nature Study and the American Nature Study Society," *Nature Magazine*, November 1957, pp. 473–80; 475.

67. Gerard Piel, "The comparative psychology of T. C. Schneirla," in *Development and Evolution of Behavior: Essays in memory of T. C. Schneirla*, ed. Lester R. Aronson, Ethel Tobach, Daniel S. Lehrman, and Jay S. Rosenblatt (San Francisco: W. H. Freeman, 1970), 1–13.

68. Wheeler's kindred spirit, Mencken, also accused Babbitt and More of elitism. See Joan Shelley Rubin, *The Making of Middlebrow Culture* (Chapel Hill: University of North Carolina Press, 1992).

69. This period of reconsideration coincided with a change in national park policy from managed estates to "wild" spaces. See Karen R. Jones, *Wolf Mountains: A History of Wolves Along the Great Divide* (Calgary: University of Calgary Press, 2002), passim.

70. The problematic place in of childhood in Victorian analogical hierarchies has been highlighted by Sally Shuttleworth, "The Psychology of Childhood in Victorian Literature and Medicine," in *Literature, Science, Psychoanalysis, 1830–1970: Essays in Honour of Gillian Beer*, ed. Helen Small and Trudi Tate (Oxford: Oxford University Press, 2003), 86–101.

Seven • Ants in the Library

1. Auguste Forel, *The Senses of Insects* (London: Methuen, [1878–1906] 1908), 88–89. The experiment did not work equally well with all species.

2. Auguste Forel, *Out of My Life and Work* (London: George Allen & Unwin, 1937), 193. Forel began studying Volapük before switching to Esperanto.

3. Forel papers. A key to the code exists in the papers, although it is a matter of simple symbol substitution that would in any case be easy to crack. The content of the messages is not, however, always obvious, since the code extended to private phraseology. "There has been some butchery today," wrote Blanche darkly in one letter.

4. Ernest Thompson Seton nurtured children's fascination with codes by encouraging his scouts to understand natural signifiers (for example, in tracking animals). Scouting of all kinds has also emphasized the importance of military signaling, secret codes in which children also took pleasure.

5. Edouard Bugnion, *The Origin of Instinct: A Study of the War between the Ants and the Termites* (London: Kegan Paul, Trench and Trubner, 1927). Bugnion wrote to Donisthorpe on 3 August 1926 asking him, as a mutual friend, to talk to Ogden and find out how his brother-in-law Forel's translation was coming along, and whether or not Ogden was interested in the article Bugnion proposed: "Les cellules sexuelles et la théorie de l'hérédité."

6. The review originally appeared in Clifton Fadiman, *Reading I've Liked*, and is reprinted in P. Sargant Florence and J. R. L. Anderson, *C. K. Ogden: A Collective Memoir* (London: Elek Pemberton, 1977), 192–212. The review also mentions ants as exemplars of modern industrial efficiency.

7. Letter from W. M. Wheeler to C. K. Ogden, 26 July 192[6], Ogden papers. The letter is dated 1924, but this appears to be a mistake; its contents place it in 1926.

8. Ogden collected curious coincidences, which he called "pernambucos" after an incident in which he and I. A. Richards were walking down the street conversing; in the process of saying the word "Pernambuco," Ogden happened to notice a scrap of paper in the wind bearing this word and nothing else. Florence and Anderson, *C. K. Ogden*.

9. Psyche, the goddess for whom the journals were named, is most obviously associated with the butterfly. But in the myth concerning her affair with Cupid, she is also the recipient of help from the Formicidae. Thomas Bulfinch, *The Age of Fable; or, Stories of Gods and Heroes* (New York: Heritage Press, 1942), 81–92.

10. Norbert Wiener commented in his autobiography, "Ogden . . . had succeeded in prolonging an undergraduate career over an unheard-of period of years." Norbert Wiener, *Ex-Prodigy: My Childhood and Youth* (New York: Simon and Schuster, 1953), 188.

11. Florence and Anderson, *C. K. Ogden*, 55.

12. See John Paul Russo, *I. A. Richards: History, Life and Work* (London: Routledge, 1989), for an account of Cambridge in this period.

13. Letter from Joseph Needham to C. K. Ogden, 5 November 1926, Ogden papers. An H. G. Wells article advised that one could earn between £50 and £400 per article for the popular press. H. G. Wells, "Journalism for the Scientific Worker," *Scientific Worker*, June 1928, pp. 61–62. *Psyche*'s advertised rate of pay in the 1921 *Writers' and Artists' Year Book* was up to £1 per 1,000 words, with a maximum length of 3,000–5,000 words. The Ogden papers contain several letters of retraction from authors on discovering the poor remuneration; sometimes payment did not even reach the advertised rates, or it was nonexistent.

14. Letter from Ogden to Raymond Pearl, 19 February 1936, Ogden papers.

15. See Michael H. Whitworth, *Einstein's Wake: Relativity, Metaphor, and Modernist Literature* (Oxford: Oxford University Press, 2001), chap. 1, for a useful and plausible account of journals as "virtual cultures," which I extend here to Ogden's Library.

16. Florence and Anderson, *C. K. Ogden*, 122.

17. Russo, *I. A. Richards*, 92–93. See also Max Eastman, *The Literary Mind: Its Place in an Age of Science* (New York: Scribner, 1931).

18. C. K. Ogden and I. A. Richards, *The Meaning of Meaning: A Study of the Influence of Language upon Thought and of the Science of Symbolism* (London: Kegan Paul, Trench, Trubner, 1923). See Russo, *I. A. Richards*, 64–65, for the myth of the book's genesis.

19. Chris Baldick, *The Social Mission of English Criticism 1848–1932* (Oxford: Clarendon Press, 1987), 226. Richards' vision for this literary natural history also seems to map rather well onto entomology; the mind/community analogy of Plato's Republic, constitutive of modernist myrmecology, was for Richards the "founding metaphor . . . of Western philosophy." (Baldick, *The Social Mission of English Criticism*, 218.) Richards mentioned the analogy in the Macy cybernetics meeting that he attended (chapter 8). See Lewis for a history of analogies between mind, body, and community, especially in science fiction. Arthur O. Lewis, "Introduction," in *Clockwork Worlds: Mechanized Environments in SF*, ed. D. Erlich and T. P. Dunn (Westport, CT: Greenwood Press, 1983), 3–18.

20. Russo, *I. A. Richards*, 100–101.

21. Herbert Spencer, *Principles of Psychology* (2 vols., New York: Appleton, 1883), vol. ii, 627–30; Grant Allen, *Physiological Aesthetics* (London: H. S. King, 1877); sources quoted in Russo, *I. A. Richards*, 100–101.

22. In 1908, McDougall was obliged to forestall moral criticism of his instinct-based introduction to human psychology by anticipating and countering the opinion that instincts were the "troublesome vestiges of [man's] pre-human state." William McDougall, *An Introduction to Social Psychology* (3rd ed., London: Methuen, [1908] 1910), 23.

23. D. H. Lawrence, *Fantasia of the Unconscious* and *Psychoanalysis and the Unconscious* (Harmondsworth: Penguin, [1923] 1971), 212. On these books see also N. Katherine Hayles, *The Cosmic Web: Scientific Field Models and Literary Strategies in the Twentieth Century* (Ithaca: Cornell University Press, 1984), 85–110. Hayles neatly identifies the paradox that, for Lawrence, experience and description are mutually exclusive activities, but analyses this state of affairs from a distinctly literary rather than historical perspective, allying Lawrence's theory of consciousness with Maxwell's electromagnetic fields.

24. Lawrence, *Fantasia of the Unconscious* and *Psychoanalysis and the Unconscious*, 104.

25. Lawrence, *Fantasia of the Unconscious* and *Psychoanalysis and the Unconscious*, 92.

26. Lawrence, *Fantasia of the Unconscious* and *Psychoanalysis and the Unconscious*, 87, original emphasis.

27. Clark A. Elliott and Margaret W. Rossiter, eds., *Science at Harvard University: Historical Perspectives* (London: Associated University Presses, 1992).

28. On the extent to which "civilized morality" had an impact on doctors, scientists, and academics see Nathan J. Hale Jr., *Freud and the Americans: The Beginnings of Psychoanalysis in the United States, 1876–1917* (2 vols., New York: Oxford University Press, 1971), vol. 1. Hale claims that the instinct was no longer regarded as a credible guide after the First World War, but this is clearly untrue unless one interprets "instinct" in the narrow Brahmin, Unitarian sense.

29. "The Victory of Art over Humanity," in David Bradshaw, ed., *The Hidden Huxley: Contempt and Compassion for the Masses 1920–36* (London: Faber and Faber, 1994), 81.

30. For example, "The psycho-analysts, who trace all interest in art back to an infantile love of excrement, would doubtless offer some simple fecal explanation for the varieties in our aesthetic passions. One man loves masses, another lines: the explanation in terms of coprophily is so obvious that I may be excused from giving it here." Aldous Huxley, *Along the Road: Notes and Essays of a Tourist* (London: Chatto & Windus, [1928] 1930), 164–65.

31. Aldous Huxley, *Music at Night* (London: Harper Collins, [1931] 1994), 24.

32. I. A. Richards, *Science and Poetry* (London: Kegan Paul, Trench and Trubner, 1926), 27 The same point is made in idem, *Principles of Literary Criticism* (London: Routledge, [1924] 1995), 137. Compare with S. Alexander, *Art and Instinct* (Oxford: Clarendon Press, 1927), 18: "When we hear, we are thrown back into the aesthetic or constructive passion from which the artist wrote, and at second hand reproduce the conditions of the poem's origin . . . that creation in turn serves us to recover the passion which is our means to apprehending the poem." Despite an apparent misunderstanding of Richards' position (he claims Richards thinks the aesthetic attitude is "nothing but ordinary feelings . . . towards the subjects of art, in a certain condition of refinement"), Alexander ends up concurring with his ambivalent attitude: To be a poet one *must* have a special aesthetic emotion (an

outgrowth of instinct), but that is not what the poet writes about, or what the reader thinks about.

33. Richards, *Principles of Literary Criticism*, 248.

34. See Richards, *Science and Poetry*, 26–40. Compare this with the remarkably similar attitude of the Leavises toward D. H. Lawrence; they praised him for his gift for experiencing life, judging that his pro-instinctual writing was the outward sign of his inner quality. This holds despite the fact that F. R. Leavis differed violently from Richards on the value of science. On this matter, see Wolf Lepenies, *Between Literature and Science: The Rise of Sociology* (Cambridge: Cambridge University Press, 1988), 180–81, and Russo, *I. A. Richards*, 534–40. Using language that recalled the biological notions of plasticity and fixity, Richards hoped for a future mind that could adjust and adapt to a variety of experiences. This would be of evolutionary significance, leading a favorable biological transmutation of the species. (See Baldick, *The Social Mission of English Criticism*, 220.)

35. Richards, *Science and Poetry*, 40. E. M. Forster's *Howards End* (London: E. Arnold, 1910) contains the bathetic figure of Leonard Bast, a lowly clerk whose attempts to penetrate the world of high culture end in tragedy. The problem is that he has mechanically learned about true art, without having the innate quality necessary to internalize it.

36. In George Gissing's novel *Born in Exile* (3 vols., London: A&C Black, 1892), a Cockney boy gives an excruciating poetry recital, performing it "like a machine." See John Carey, *The Intellectuals and the Masses: Pride and Prejudice among the Literary Intelligentsia, 1880–1939* (London: Faber and Faber, 1992), 96–97.

37. See Lepenies, *Between Science and Literature*, 155–95; Stefan Collini, "On Highest Authority: The Literary Critic and Other Aviators in Early Twentieth-Century Britain," in *Modernist Impulses in the Human Sciences 1870–1930*, ed. Dorothy Ross (Baltimore: Johns Hopkins University Press, 1994), 152–70; and Baldick, *The Social Mission of English Criticism*, 134–61.

38. D. L. LeMahieu, *A Culture for Democracy: Mass Communication and the Cultivated Mind in Britain Between the Wars* (Oxford: Clarendon Press, 1988), 103–37.

39. Ogden and Richards, *The Meaning of Meaning*, 195–96.

40. Joan Shelley Rubin, *The Making of Middlebrow Culture* (Chapel Hill: University of North Carolina Press, 1992), 44–60.

41. Compare with Wheeler's attempt to define himself as unaffectedly natural but not base (chapter 4).

42. Richards, *Science and Poetry*, 41

43. See Richards' appendix, "The Poetry of T. S. Eliot," added to the second edition of *Principles of Literary Criticism* (1926), 231–35.

44. Richards identified sex as the "problem of [his] generation as religion was the problem of the last." Idem, *Principles of Literary Criticism*, 233.

45. Matthew Arnold, *Culture and Anarchy: An Essay in Political and Social Criticism* (London: Smith, Elder & Co., 1869), chap. 6.

46. Aldous Huxley, "To the Puritan All Things are Impure," in *idem, Music at Night*, 120.

47. For Pareto, this conviction was a "sentiment." Ogden and Richards discuss the emotive power of words in *The Meaning of Meaning*.

48. These people were responding to the emotive quality of the words but making the mistake of thinking that they had an objective referent, rather than communicating the (inferior) experience of the writer or speaker. Richards, "Poetry and Beliefs," in *Principles of Literary Criticism*.

49. Richards, *Principles of Literary Criticism*, 81. Haldane was a friend of Wiener's, and a friend also of Caryl Haskins. The significance of this in the context of cybernetic linguistics and myrmecology comes to the fore in chapters 8 and 9.

50. W. M. Wheeler, *Holy Bluff* (manuscript, n.d.), Wheeler papers.

51. William Morton Wheeler was appointed one of its referees.

52. Auguste Forel, *The Social World of the Ants Compared with that of Man* (2 vols., London: G.P. Putnam's Sons, [1921–2] 1928), vol. 1, v–vi.

53. Ogden drew attention to ant communication in Forel, *Social World of the Ants*, vol. 1, 239–40, 447–50.

54. C. K. Ogden, *Basic English: A General Introduction with Rules and Grammar* (London: Kegan Paul, 1930). A dictionary followed in 1940: idem, *General Basic Dictionary* (London: Evans Bros., 1940).

55. See, for example, C. K. Ogden, "Can Basic English be a World Language?" *Picture Post* 21 (23 October 1943): 23–25. Fink asserts that Basic "began to be used as a means of communication between the polyglot Allies" during the war, and that "it spread widely, both in military and cultural use." Howard Fink, "Newspeak: The Epitome of Parody Techniques in *Nineteen Eighty-Four*," *Critical Survey* 5 (1971): 155–63; 155.

56. Richards also accused Ogden of creating his International Library merely as a means to flank *The Meaning of Meaning* with books designed to establish an appearance of its intellectual worth. W. Terrence Gordon, *C. K. Ogden: A Bio-Bibliographic Study* (London: Scarecrow, 1990), 29.

57. See Raymond B. Fosdick, *The Story of the Rockefeller Foundation* (New York: Harper, 1952), 249–51.

58. See, for example, *New York Times* 18 September 1935; Ivy Low, "Ivy Litvinov," *Life*, 1942. (The article was written by Litvinov under her maiden name.)

59. Letter from H. L. Mencken to C. K. Ogden, 23 November [c. 1932–3], Ogden papers

60. Letter from Ivy Litvinov to C. K. Ogden, 21 June 1935, Ogden papers.

61. Letters from Ivy Litvinov to C. K. Ogden, 28 March 1934 and 28 April 1934, and from W. G. Keble to Ivy Litvinov, 6 April 1934, Ogden papers.

62. Letter from C. K. Ogden to Ivy Litvinov, n.d. January 1937; C. K. Ogden to Mrs. Maisky, 15 January 1937, Ogden papers. A piece of evidence to be weighed against this claim is the fact that the New Testament was the best-selling publication in Basic English, moving more than 1,000 copies per day when it was first issued in 1941.

63. For a partial history see Fosdick, *The Story of the Rockefeller Foundation*.

64. Letter from A. J. Svyadosheh to C. K. Ogden, 14 December 1932, Ogden papers.

65. Letter from C. K. Ogden to Raymond Pearl, 19 February 1936; letter from Raymond Pearl to C. K. Ogden, 16 January 1936, Ogden papers.

66. Letter from C. K. Ogden to Raymond Pearl, 19 February 1936, Ogden papers.

67. C. K. Ogden, *Basic for Science* (London: Kegan Paul, 1942). An earlier and less comprehensive book by Ogden to cover the same topic was *Basic English Applied—Science* (Lon-

don: Kegan Paul, 1931). A later volume to cover similar ground was E. C. Graham, *The Science Dictionary in Basic English* (London: Evans Bros., 1965).

68. The Rockefeller Foundation pioneered monitoring for propaganda of broadcasts originating in Europe and received in the United States during the Second World War. The task was taken over by federal agencies after the United States entered the war. See Fosdick, *The Story of the Rockefeller Foundation*, 247.

69. See Aldous Huxley, *Brave New World Revisited* (London: Heron Books, [1959] 1968), chapter titled "Propaganda in a Democratic Society"; F. C. Bartlett, *Political Propaganda* (Cambridge: Cambridge University Press, 1940), 65. The goal of maximizing efficiency is alive and well: a recent news report announced that British Gas was teaching all its employees to send abbreviated text messages, finding that "a text in time saves 8333 minutes a day." (British Gas Web site, story posted 8 March 2004.) This last mode of propaganda—as framework to constrain thought—was, historically speaking, established primarily in the realm of advertising, and only secondarily came to inform political technique, despite Orwell's early discussion of such possibilities in *Nineteen Eighty-Four*. John Watson's transfer to advertising was a key early example of the translation of psychological science into the business of consumption; the Saatchis' journey into United Kingdom politics in the 1980s took advertising into the heart of politics, where it has remained ever since.

70. See Carey, *The Intellectuals and the Masses*, 3–90. Lawrence's *Fantasia of the Unconscious* supplements his widely known eulogy of the instinctual side of human nature with an account of it as a *mass* proclivity. Ibid., 87. See also Baldick, *The Social Mission of English Criticism*, 162–95.

71. Russo, *I. A. Richards*, 458. On the Rockefeller Foundation and its involvement in U.S. media, see Fosdick, *The Story of the Rockefeller Foundation*, 245–49.

72. *Esperanto The Aggressor Language*, Department of the Army Field Manual FM 30-101-1 (Washington, DC: Headquarters, Department of the Army, 1962). Quotations from 215 and 216.

73. Diary entry for 20 October 1939. From T. H. White, *The Book of Merlyn: The Unpublished Conclusion to* The Once and Future King (Austin: University of Texas Press, 1977), xiv. Back in 1934, White had mused over the possibility of staging *Henry V* as communist propaganda. Henry was to be festooned with British flags, and all his speeches were to be delivered into a microphone. Sylvia Townsend Warner, *T. H. White: A Biography* (London: Jonathan Cape with Chatto & Windus, 1967), 75–76.

74. White, *The Book of Merlyn*, xvi.

75. Warner, *T. H. White*, 193–95. White owned Ogden's translation of *Social World of the Ants*, and Julian Huxley's *Ants* (Richards archive), and corresponded with the latter in order to improve his scientific endeavors.

76. White, *The Book of Merlyn*, 57.

77. The *Sword in the Stone* was edited to accommodate the wishes of the American Book Club. See Warner, *T. H. White*, 104–105; the relevant ant passages are White, *The Book of Merlyn*, 40–67; idem, *The Once and Future King* (London: Fontana, [1958] 1962), 119–28.

78. See Fink, "Newspeak," 155–63; W. F. Bolton, *The Language of* Nineteen Eighty-Four:

Orwell's English and Ours (Knoxville: University of Tennessee Press, 1984); and Anthony Burgess, *Language Made Plain* (New York: Crowell, 1965).

79. George Orwell, *Nineteen Eighty-Four* (London: Penguin, [1949] 1989), 312.

Part III • Communicational Ants

1. Steve J. Heims, *John von Neumann and Norbert Wiener: From Mathematics to the Technologies of Life and Death* (Cambridge, MA: MIT Press, 1980), 9–16. Wheeler had arrived at the Bussey Institute one year earlier, just as it was undergoing reorganization as a graduate school.

2. Letter from McCulloch to Lewin, 15 November 1946, McCulloch papers.

3. Wiener himself was a fan of the homeostatic models that suffused Wheeler's work. Among his most cited books were two by Wheeler's close colleagues: L. J. Henderson, *The Fitness of the Environment: An Inquiry into the Biological Significance of the Properties of Matter* (New York: Macmillan, 1913), and Walter B. Cannon, *The Wisdom of the Body* (London: Kegan Paul, 1932).

4. Ellenberger describes the transition from neuroanatomy to anatomoclinical neurology to a dynamic concept of neuroses as a path followed by Freud, Forel, and Forel's student Adolf Meyer. In fact, Ellenberger credits Forel as the pioneer who led psychiatry back from a purely organicist model to a version that incorporated dynamic factors. See Henri F. Ellenberger, *The Discovery of the Unconscious: The History and Evolution of Dynamic Psychiatry* (London: Allen Lane, 1970), 480, 287–89, and for a description of some of the chief varieties of dynamic psychiatry, 289–91.

5. Ronald R. Kline, "What is Information Theory a Theory of? Boundary Work among Information Theorists and Information Scientists in the United States and Britain, 1948–1974," prepared for the Second Conference on the History and Heritage of Scientific and Technical Information Systems, November 16–17, 2002, Chemical Heritage Foundation, Philadelphia.

6. An insider's account has been written: Ullica Segerstråle, *Defenders of the Truth: The Battle for Science in the Sociobiology Debate and Beyond* (Oxford: Oxford University Press, 2000).

Eight • The Macy Meanings of Meaning

1. Karl von Frisch, *A Biologist Remembers* (Oxford: Pergamon Press, [1957] 1967). On von Frisch, see also Eileen Crist, "Can an Insect Speak? The Case of the Honeybee Dance Language," *Social Studies of Science* 34 (2004): 7–43, and Tania Munz, "Birds and Bees Behaving on Film: Karl von Frisch, Konrad Lorenz and the Burden of Proof on the Motion Picture," paper given at British Society for the History of Science Annual Conference, University of Leeds, July 2005.

2. Von Frisch, *A Biologist Remembers*, 140.

3. *Aus dem Leben der Bienen* was published in German in 1950. The text of von Frisch's American lectures was published in English in 1950 as *Bees—Their Vision, Chemical Senses, and Language* (Ithaca: Cornell University Press). In 1954, *Aus dem Leben der Bienen*

was translated from the fifth German edition as *The Dancing Bees: An Account of the Life and Senses of the Honey Bee* (London: Methuen) and published successfully for a general audience.

4. Von Frisch, *A Biologist Remembers*, 111–16; 160–5; letter from von Frisch to Schneirla, Schneirla papers. Von Frisch's report of the interests of Weaver and co. echoes almost word for word former Rockefeller President Fosdick in his claim that "If cultural interests are to be given a wider currency, and if the imperative need of mutual understanding between races is to be met, something must be done to break down the insularity created by ignorance of other languages." Raymond B. Fosdick, *The Story of the Rockefeller Foundation* (New York: Harper, 1952), 249.

5. Von Frisch, *A Biologist Remembers*, 176.

6. Martin Lindauer's major summaries of research were Martin Lindauer, *Communication among Social Bees* (Cambridge, MA: Harvard University Press, 1961), and *idem*, "Recent Advances in Bee Communication and Orientation," *Annual Review of Entomology* 12 (1967): 439–70.

7. Donald R. Griffin, "Expanding Horizons in Animal Communication Behavior," in *How Animals Communicate*, ed. Thomas A. Sebeok (Bloomington: Indiana University Press, 1977), 26–32; 27. See also Griffin's introduction to von Frisch, *Bees*; W. H. Thorpe, "Orientation and Methods of Communication of the Honey Bee and Its Sensitivity to Polarization of Light," *Nature* 164 (1949): 11–14.

8. "During the past theory years we have been repeatedly surprised by discoveries about animal behavior, especially in the area of orientation and communication." Griffin, "Expanding Horizons," 26. Another reason for the ready acceptance of Von Frisch's communicative bees was that his presentation of scientific methodology appealed to his audience. His retelling of his celebrated discovery follows the Peircian, abductive reasoning of the detective story: an inspired, imaginative guess that explains an unexpected and noteworthy phenomenon. An identical style of reasoning or plot was expounded by von Frisch regarding his discovery that polarized light was the cue used by bees to orientate the direction of their dance/foraging flights. Such an account went down well with the Peircian semioticians and pragmatic cyberneticians. See Nancy Harrowitz, "The Body of the Detective Model: Charles S. Peirce and Edgar Allan Poe," in *The Sign of Three: Dupin, Holmes, Peirce*, ed. Umberto Eco and Thomas Sebeok (Bloomington: Indiana University Press, 1983), 179–97.

9. E. O. Wilson, "Pheromones," *Scientific American* 208 (1963): 100–14; 110. John Lubbock, *On the Senses, Instincts and Intelligence of Animals with Special Reference to Insects* (3rd ed., London: Kegan Paul, Trench, [1888] 1891), 192.

10. Skinner, for example, had spent the war attempting to train pigeons to guide missiles. Skinner conversation with Wilson, 29 November 1987, 11 pp.; transcript in Wilson papers. See also Nikolas Rose, *Governing the Soul: The Shaping of the Private Self* (New York: Routledge, 1989), 16–7.

11. Coolidge was curator of mammals at the Museum of Comparative Zoology at the time of his sociology course; during the war he reached the rank of major.

12. Letter from Arthur D. Hasler to T. C. Schneirla, 18 December 1952, Schneirla papers.

13. In the decade following the war, well over 90 percent of funding for research in the physical sciences came from military agencies. Jessica Wang, *American Science in the Age of Anxiety: Scientists, Anticommunism, and the Cold War* (Chapel Hill: University of North Carolina Press, 1999), 38. See also Daniel S. Greenberg, *The Politics of Pure Science* (new ed., Chicago: University of Chicago Press, 1999).

14. J. P. Scott, "Animal Behavior and Social Organization," Remarks on the 20th Anniversary of the Founding of the Animal Behavior Society (manuscript, 1976), Wilson papers.

15. Thomas A. Sebeok, ed., *Animal Communication: Techniques of Study and Results of Research* (Bloomington: Indiana University Press, 1968), 459.

16. Moles in Sebeok, *Animal Communication*, 627.

17. See Ethel Tobach and Lester R. Aronson, "T. C. Schneirla: A Biographical Note," in *Development and Evolution of Behavior: Essays in memory of T. C. Schneirla*, ed. Lester R. Aronson, Ethel Tobach, Daniel S. Lehrman, and Jay S. Rosenblatt (San Francisco: W. H. Freeman, 1970), xi–xviii.

18. On the history of entomology at the American Museum of Natural History, see Joseph Wallace, *A Gathering of Wonders: Behind the Scenes at the American Museum of Natural History* (New York: St Martin's Press, 2000), 131–43. Margaret Mead's autobiography *Blackberry Winter: My Early Years* (New York: Morrow, 1972) gives a good sense of the museum's atmosphere and politics in Schneirla's day.

19. Interview with Ethel Tobach.

20. Interview with Ethel Tobach.

21. T. C. Schneirla, "Second Semiannual Report for 1953 to Biology Branch, Office of Naval Research," 1 February 1954, Schneirla papers.

22. Letter from Schneirla, 17 March 1948, Schneirla papers.

23. Subsequent mentions occurred in 1951, 1953, and 1957. See Gregg Mitman, *The State of Nature: Ecology, Community, and American Social Thought, 1900–1950* (Chicago: University of Chicago Press, 1992), 166 and footnote to unpublished essay.

24. Interview with Ethel Tobach.

25. Wang, *American Science in the Age of Anxiety*, 79–81.

26. Letter from Lester R. Aronson to Wayne Faunce, 19 February 1948, Schneirla papers. Schneirla had great difficulty in getting a Geiger counter for use in Germany, suggesting both the low secrecy of the majority of his work and the impossibility of it ever being otherwise. (Wilson, by contrast, had no trouble in getting hold of a Geiger counter for a similar purpose; Wilson papers.)

27. Schneirla's discovery of the "suicidal ant mill" has been retold in popular histories of the AMNH. See Wallace, *A Gathering of Wonders*, 132–33. The original paper is T. C. Schneirla, and Gerard Piel, "The Army Ants," *Scientific American* (1948), reprinted in *Selected Writings of T. C. Schneirla*, ed. Lester R. Aronson, Ethel Tobach, Jay S. Rosenblatt, and Daniel S. Lehrman (San Francisco: W. H. Freeman, 1972), 750–67. Quotations in the following two paragraphs are drawn from this paper. Needless to say, like numerous "accidents" in the history of science, the incident would not have been noticed, nor would there have been a vocabulary to describe it, were it not for the context of Schneirla's program of research.

28. The "herd instinct," by contrast, was not an integral part of normal mammalian behavior and was therefore a poor analogy for the phenomenon, according to Schneirla.

29. T. C. Schneirla, "Bees," *Ecology* 32 (1951): 562–65.

30. Letter from William Creighton to Bob Gregg, 15 December 1956, Creighton papers. Another critic of von Frisch's work was Adrian Wenner; see *idem*, "Honey Bees: Do They Use the Distance Information Contained in their Dance Maneuver?" *Science* 155 (1967): 847–49.

31. See D. de Solla Price, *Little Science, Big Science* (New York: Columbia University Press, 1963). A recent and highly readable introduction to the topic is given by Jeff Hughes, *The Manhattan Project: Big Science and the Atom Bomb* (Cambridge: Icon, 2002). See also Jon Agar, *The Government Machine: A Revolutionary History of the Computer* (Cambridge, MA: MIT Press, 2003).

32. Thomas P. Hughes and Agatha C. Hughes, eds., *Systems, Experts and Computers: The Systems Approach to Management and Engineering, World War II and After* (Cambridge, MA: MIT Press, 2000). See especially chapters by Thomas P. Hughes and Agatha C. Hughes, Erik P. Rau and Gabrielle Hecht. See also Wang, *American Science in the Age of Anxiety*; Stuart Leslie, *The Cold War and American Science: The Military-Industrial-Academic Complex at MIT and Stanford* (New York: Columbia University Press, 1993); Lily E. Kay, *The Molecular Vision of Life: Caltech, The Rockefeller Foundation, and the Rise of the New Biology* (New York: Oxford University Press, 1993); and Mark Solovey, ed., *Social Studies of Science* 31 (2001), Special Issue on Science in the Cold War.

33. Steven J. Heims, *Constructing a Social Science for Postwar America: The Cybernetics Group 1946–1953* (Cambridge, MA: MIT University Press, 1991), 2.

34. Letter from McCulloch to Lewin, 15 November 1946, McCulloch papers.

35. See Donna Haraway, "The High Cost of Information in Post-World War II Evolutionary Biology: Ergonomics, Semiotics, and the Sociobiology of Communication Systems," *Philosophical Forum* 13 (1981): 244–78; *idem*, "Signs of Dominance: From Physiology to a Cybernetics of a Primate Society," *Studies in the History of Biology* 6 (1983): 129–219. The former of these papers in particular gives a convincing textual reading of the cybernetic elements of Wilson and Altmann's discourse; it demonstrates a fit between cybernetics and postwar zoology, but does not go as far as one might hope in terms of explaining how this came about. This book adds historical depth to Haraway's argument, noting the economic aspects of the discourse between the historic periods of "homeostatic" and "cybernetic" representations of nature. It also problematizes the smoothness of the transition to cybernetic discourse implied by Haraway, since Schneirla, a key player in postwar myrmecology, was not amenable to this approach. Haraway's invocation of cybernetics as a "command-control" system is here complemented by a discussion of cybernetics as autopoiea (again, a Schneirlarian take on the issue).

36. Weaver, from a 1930s Rockefeller report, quoted in Fosdick, *The Story of the Rockefeller Foundation*, 166.

37. Weaver's development of the Rockefeller Foundation's program in natural sciences is described in Fosdick, *The Story of the Rockefeller Foundation*, 156–66, and Kay, *The Molecular Vision of Life*, 41–50.

38. Weaver, quoted in Fosdick, *The Story of the Rockefeller Foundation*, 158.

39. T. C. Schneirla, "Ant Learning as a Problem in Comparative Psychology," in *Twentieth Century Psychology*, ed. P. L. Harriman (New York: Philosophical Library, 1946), 276–305. The paper was given at the second Macy conference on feedback. Reprinted in Aronson et al., eds., *Selected Writings of T. C. Schneirla*, 556–79.

40. Letter from Filmer Northrop to Norbert Wiener, 5 May 1947, McCulloch papers.

41. Ibid.

42. Norbert Wiener, "Time, Communication and the Nervous System" (manuscript, n.d.), McCulloch papers. The comment "task!" refers to the purpose which the conferences were supposed to fulfill.

43. Morris and his work on signs was another recipient of Rockefeller largesse. Fosdick, *The Story of the Rockefeller Foundation*, 264.

44. Heims, *Constructing a Social Science for Postwar America*, 95.

45. Letter from G. E. Hutchinson to McCulloch, McCulloch papers.

46. Letter from McCulloch to Hutchinson, 28 November 1949, McCulloch papers.

47. Letter from Frank Fremont-Smith to McCulloch, 15 March 1949, McCulloch papers.

48. See also the letter from Schneirla to von Bonin, 24 January 1951, McCulloch papers.

49. Letter from Gerhardt to von Bonin, 22 January 1951, McCulloch papers.

50. Claude Shannon, "Presentation of a Maze-Solving Machine," in Heinz von Förster, *Cybernetics: Circular Causal and Feedback Mechanisms in Biological and Social Systems. Transactions of the Eighth Conference*, ed. Margaret Mead and Hans Lukas Teuber (New York: Josiah Macy, Jr. Foundation, c. 1952), 173–80.

51. Charlotte Sleigh, "'The Ninth Mortal Sin': The Lamarckism of W. M. Wheeler," in *Darwinian Heresies*, ed. Abigail Lustig, Robert Richards, and Michael Ruse (Cambridge: Cambridge University Press, 2004), 151–72.

52. Norbert Wiener, *The Human Use of Human Beings: Cybernetics and Society* (rev. ed., Boston: Houghton Mifflin, [1950] 1954), 96.

53. W. M. Wheeler, "The Ant-Colony as an Organism," *Journal of Morphology* 22 (1911): 301–25; 308.

54. Norbert Wiener, *Cybernetics: Or Control and Communication in the Animal and the Machine* (New York: Wiley; Paris: Hermann and Co., 1948), 56.

55. Wiener, *Cybernetics*, 56.

56. Such charges are described in Mary A. Evans and Howard E. Evans, *William Morton Wheeler, Biologist* (Cambridge, MA: Harvard University Press, 1970), 224.

57. I choose "purposivist" rather than "teleological" since the latter, like "vitalist," has negative connotations, notably embodied in Richard Dawkins' critique of Paley in *The Blind Watchmaker* (Harlow: Longman, 1986). Zoologists and psychologists discussed their self-conscious use of "teleology" at the first Macy conference on group processes. See Bertram Schaffner, ed., *Group Processes: Transactions of the First Conference* (New York: Josiah Macy, Jr. Foundation, 1955), 82–87.

58. See Charles S. Peirce, "Pragmatism in Retrospect: A Reformulation," c. 1906, in *The Collected Papers of Charles Sanders Peirce*, ed. C. Hartshorne and P. Weiss, (6 vols., Cambridge MA: Harvard University Press, 1931), vol. v. Peirce's co-author Lady Welby had also

corresponded with Ogden on the matter of "significs" at the turn of the twentieth century (Ogden papers).

59. Quoted in Sebeok, *Animal Communication*, 7. Charles Morris, *Foundations of the Theory of Signs* (Chicago: University of Chicago Press, 1938).

60. Jacob von Uexküll, *Theoretical Biology* (London: Kegan Paul, Trench, Trubner, 1926), 147. On Uexküll, see Anne Harrington, *Reenchanted Science: Holism in German Culture from Wilhelm II to Hitler* (Princeton, NJ: Princeton University Press, 1996), 34–71.

61. C. K. Ogden and I. A. Richards, *The Meaning of Meaning: A Study of the Influence of Language upon Thought and of the Science of Symbolism* (London: Kegan Paul, Trench, Trubner, 1923), 390. Ogden and Richards suggested that Bergson's theory was not so bad, if one incorporated into it mnemic theory, thus removing its mystical elements. Ibid., 264.

62. Account from Thomas A. Sebeok, "Exordium, 'The Estonian Connection,'" at www.ut.ee/SOSE/sebeok.htm, n.d. (c. 1997–2003), accessed 20 August 2003. See also Thomas A. Sebeok, *The Sign & Its Masters* (Austin: University of Texas Press, 1979), chap. 10.

63. Thomas A. Sebeok, *Semiotics in the United States* (Bloomington: Indiana University Press, 1991), 104.

64. Sebeok, *Animal Communication*. A post-*Sociobiology* rethink of this work was published nearly ten years later: Thomas A. Sebeok, ed., *How Animals Communicate* (Bloomington: Indiana University Press, 1977). See also Thomas A. Sebeok, ed., *Approaches to Animal Communication* (The Hague: Mouton, 1969).

65. Sebeok, *Animal Communication*, 4.

66. Sebeok, *Animal Communication*, 5.

67. Sebeok, *Animal Communication*, 12.

68. Sebeok, *Animal Communication*, 6. Meanwhile a similar thought had struck Noam Chomsky in the apparently unlike context of a military-funded electronics lab. He had started out by being given a fellowship to work at MIT in 1955, although, as he confessed, he "hardly knew the difference between a tape recorder and a telephone." So freely was the money forthcoming from the Pentagon, and with so few strings attached, that he was able to develop a theory of universal grammar, treating circuits of the mind instead of electric circuits. Chomsky today sees this as having been revolutionary:

> [Universal grammar] was obvious to me. And it was very counter to the prevailing doctrines at the time, in philosophy and psychology, but they were simply and demonstrably wrong. That language is a biologically-based capacity is so obvious there is hardly any point arguing it; that it is a specific human capacity is also self-evident.

Chomsky's recollections from Tim Adams, "Profile," *Observer Magazine* 30 November 2003, pp. 54–59; 59. Chomsky's claim to priority in regard to discerning the biological basis of language should, by now, hardly require negation. Compare, for instance, to Ogden and Richards, *The Meaning of Meaning*, 390.

69. Auguste Forel, *Ants and Some Other Insects* (London: Kegan Paul, Trench, Trubner, 1904), 35; Auguste Forel, *Hypnotism, or Suggestion and Psychotherapy: A Study of the Psychological, Psycho-Physiological and Therapeutic Aspects of Hypnotism* (London: Rebman, [1889/1905] 1906), 43.

70. Auguste Forel, *Out of My Life and Work* (London: George Allen & Unwin, 1937), 167.

71. At the same time, Forel became impressed by the associative powers of insects; their ability to create a topochemical map, analogous to our own more visual memories. There were two places where such "memories" might be stored: in a centralized brain (of which ants had a plainly inadequate one to permit such feats) or dispersed around the ganglia or even in the muscles themselves as "muscular memory." Subscribers to the ganglia theory included Carl Jung; see *idem*, "Synchronicity, an Acausal Connecting Principle," 1952, in Carl Jung, *Collected Works*, ed. Herbert Read, Michael Fordham, and Gerhard Adler (20 vols., London: Routledge and Kegan Paul, 1953–79), vol. 8, §§ 955–7. See also Charlotte Sleigh, "Inside Out: The Unsettling Nature of Insects," in *Insect Poetics*, ed. Eric Brown, (Minneapolis: University of Minnesota Press, 2006). Subscribers to the theory of muscular memory included Piéron. Forel refused to commit himself to either physical possibility, preferring to keep the process of memory formation (whether ontogenetic or phylogenetic) a quasi-functional one.

72. Alasdair MacIntyre, *The Unconscious: A Conceptual Analysis* (London: Routledge and Kegan Paul, 1958). See also Henri F. Ellenberger, *The Discovery of the Unconscious: The History and Evolution of Dynamic Psychiatry* (New York: Basic Books, 1970), 480. Erik H. Erikson discussed his translation of Freud's energetic speculations at the first Macy conference on group processes. Schaffner, ed., *Group Processes*, 205. Freud's potential energy bears comparison with the psychology of James Rowland Angell. Angell transformed Chicago University into a new center of American psychology, formalizing its character in 1906 in a pragmatic, comparative-evolutionary manner. He conceived of mind "as primarily engaged in mediating between the environment and the needs of the organism." Bypassing the structuralist preoccupation with the nature of consciousness, Angell affirmed the experimental outlook whereby "both psychologists and biologists . . . treat[ed] consciousness as substantially synonymous with adaptive relations to novel situations." John M. O'Donnell, *The Origins of Behaviorism: American Psychology, 1870–1920* (New York: New York University Press, 1985), 11–2.

73. Sigmund Freud, "Dreams and Occultism," *SE* 22: 55. See Pamela Thurschwell, "Ferenczi's Dangerous Proximities: Telepathy, Psychosis and the Real Event," *differences* 11 (1999): 150–78. For Jung, extrasensory perception and other instances of supernatural psychic phenomena were also relics of the intuitive communications that existed among social insects. He read with interest von Frisch's research establishing that bees danced to show their nestmates the exact location of good food sources. Rejecting the notion that the bees' was a conscious communication, Jung concluded that they used instead some kind of parallel, dispersed nervous system to achieve the same effect as humans can using the cortex of the brain. He wondered whether similar communications operated in humans via the sympathetic nervous system, which, unlike consciousness and the cortex, did not sleep. Jung, "Synchronicity."

74. Compare with Ellenberger's discussion of Nietzsche as forerunner of dynamic psychology. Ellenberger, *The Discovery of the Unconscious*, 272–78. Wheeler himself was a self-proclaimed fan of Schopenhauer.

75. Steve J. Heims, *John von Neumann and Norbert Wiener: From Mathematics to the Technologies of Life and Death* (Cambridge, MA: MIT Press, 1980), 155.

76. Heims, *John von Neumann and Norbert Wiener*, 304–5.

77. An early paper making this point was Julian Bigelow, Arturo Rosenblueth and Norbert Wiener, "Behavior, Purpose and Teleology," *Philosophy of Science* 10 (1943): 18–24.

78. Letter from Northrop to Wiener, 5 May 1947, McCulloch papers. Ogden—perhaps surprisingly, given his scientific literacy—was unconcerned with the importance of energy, and did not include it in his list of Basic English words. The absence troubled J. B. S. Haldane as he attempted to translate two of his books into Basic. He complained, "The main difficulties arose from the absence of 'energy' in basic. Force, power, and work, all mean something else. Can it not be used? Energy is as good a word as 'microscope'." Letter from J. B. S. Haldane to Miss [Leonora] Lockhart, n.d. [1935], Ogden papers. Despite having met Ogden during his Cambridge days, Wiener omitted him from his personal history of philology; only Richards' later contributions to his thoughts on linguistics seem to have registered in his memory. See history of philology given in Wiener, *The Human Use of Human Beings*, 93–94.

79. J. B. S. Haldane, "Animal Communication and the Origin of Human Language," *Science Progress* 43 (1955): 385–401; 385.

80. Heinz von Förster, Margaret Mead and Hans Lukas Teuber, eds., *Cybernetics: Circular Causal and Feedback Mechanisms in Biological and Social Systems. Transactions of the Eighth Conference* (New York: Josiah Macy, Jr. Foundation, c. 1952), 134–5.

81. Schneirla and Piel, "The Army Ants," 766–7.

82. Gregory Bateson, *Steps to An Ecology of Mind* (New York: Ballantine, 1972). "If it is wet we are furnished with an organ commonly called an umbrella." Samuel Butler, *Erewhon* (Harmondsworth: Penguin, [1872] 1935), 184.

83. "[Wilson] evidently got the raspberry when he proposed [altruism] to the Royal Entomological Society in 1966." Letter from Creighton to Gregg, 4 January 1969, Creighton papers.

84. Julian Huxley, "A Discussion of the Ritualization of Behaviour in Animals and Men," *Philosophical Transactions of the Royal Society of London* 251 (1966): 249–71; 258.

85. Thomas A. Sebeok, "Zoosemiotic Components of Human Communication," in Sebeok, ed., *How Animals Communicate*, 1055–77.

86. Von Frisch, *A Biologist Remembers*, 105.

Nine • From Pheromones to Sociobiology

1. For biographical information, see E. O. Wilson, *Naturalist* (Harmondsworth: Penguin, 1995), and *idem*, "In the Queendom of the Ants," in *Leaders in the Study of Animal Behavior: Autobiographical Perspectives*, ed. D. A. Dewsbury (Lewisburg, PA: Bucknell University Press, 1985), 464–84.

2. His father, Ed Wilson, Sr., committed suicide, a way out from chronic ill health and alcoholism. His uncle Herbert was, in Ed Jr.'s words, a "dope" and "feeble-minded." For comments on his Uncle Herbert, see Wilson's annotations of Barbour [Babs] Wilson Minhinnette, *White Sands of Time* (manuscript, 1985), 5, Wilson papers.

3. Older colleagues with whom Wilson feuded accused him of dodging the draft, which perhaps explains Wilson's latter-day inability to let the topic lie.

4. Ernst Mayr remarked of Wilson in 2000, "there are two kinds of scientists: media scientists and scientists' scientists," placing Wilson in the former category. Michael Shermer and Frank J. Sulloway "The Grand Old Man of Evolution: An Interview with Evolutionary Biologist Ernst Mayr," *The Skeptic* 8 (2000): 76–82; 79. Wilson achieved press coverage very early in his career—a slew of newspaper articles attended his remote "discovery" of the ants' "missing link" in 1955–6.

5. Letter from William Creighton to Robert Gregg, 15 November 1953, Creighton papers.

6. Wilson's project on the genus ultimately resulted in the publication of a book summarizing his work over the years: E. O. Wilson, Pheidole *in the New World: A Dominant, Hyperdiverse Ant Genus* (Cambridge, MA: Harvard University Press, 2003). A good comparator for the significance of *Pheidole* is the nineteenth-century dispute between Roderick Murchison and Adam Sedgwick over the naming of the "Silurian" or "Cambrian" rock that covered large parts of Europe. See James A. Secord, "King of Siluria: Roderick Murchison and the Imperial Theme in Nineteenth Century British Geology," *Victorian Studies* 25 (1981–2): 413–42.

7. For an excellent discussion of this conflict and the culture of taxonomy, see Joshua Blu Buhs, "Building on Bedrock: William Steel Creighton and the Reformation of Ant Systematics, 1925–1970," *Journal of the History of Biology* 33 (2000): 27–70.

8. Letter from William Creighton to Robert Gregg, 14 December 1948, Creighton papers.

9. E. O. Wilson and W. L. Brown, "The Subspecies Concept and Its Taxonomic Application," *Systematic Zoology* 2 (1953): 97–111.

10. Wilson, *Naturalist*, 203–4.

11. Wilson, *Naturalist*, 204.

12. Letter from William Creighton to Arthur Cole, 11 December 1952, Creighton papers.

13. Letter from William Creighton to Robert Gregg, 20 October 1953, Creighton papers. Gregg was inclined to agree with Creighton's ascription of personal motives to Wilson and Brown: "I am convinced that Brown's and Wilson's ideas are motivated by pure envy for your achievements, and hardly merit . . . attention." Letter from Gregg to Creighton, 2 October 1954, Creighton papers. Brown, meanwhile, accused Creighton of dirty dealings in taxonomy. Wheeler's collection at the MCZ had not only been depleted through the AMNH taking its "rightful share," he claimed. "Raids have since been made by Creighton, Mann . . . and many others." Letter from William Brown to Arthur Cole, 2 February 1950, Wilson papers.

14. Letter from William Creighton to Robert Gregg, 24 September 1954, Creighton papers.

15. Letter from William Creighton to Robert Gregg, 13 April 1956, Creighton papers.

16. Letter from Robert Gregg to Frank Carpenter, 21 November 1954, Wilson papers.

17. Letter from William Creighton to E. O. Wilson, 16 October 1954; letter from E. O. Wilson to William Creighton, 19 October 1954, Creighton papers.

18. Wilson, *Naturalist*, 204.

19. William Creighton to Arthur Cole, 22 September 1956, Creighton papers. Creighton

commented that there were some at Harvard who would also be happy to see the "Happy Harvard Team" fold. No wonder Darlington was "suffering from a severe nervous upset."

20. Letter from William Creighton to Robert Gregg, 6 December 1954, Creighton papers. Gregg commented sourly, "I envy him his opportunity to go to New Guinea and Australia. Is Harvard subsidizing his trip?? It must be nice to lay out one's colleagues in correspondence, and then escape the Army by being kicked half way around the world on a trip designed to let one do exactly what he wants . . . Bless me, I don't see how he manages it." Letter from Robert Gregg to William Creighton, 20 December 1954, Creighton papers.

21. Letters from William Creighton to Robert Gregg, 2 January 1955 and 30 April 1955, Creighton papers.

22. Letter from William Creighton to Robert Gregg, 28 April 1955, Creighton papers.

23. "I wrote you some months back that Carpenter said he was through with Brown." Letter from William Creighton to Robert Gregg, 25 April 1956, Creighton papers.

24. Letter from William Creighton to Arthur Cole, 19 October 1956, Creighton papers.

25. Letters from Robert Gregg to William Creighton, 1 September 1958 and 3 August 1960, Creighton papers.

26. Letter from William Creighton to Robert Gregg, 2 November 1956, Creighton papers: "it would appear that Wilson has finally reached the conclusion that Brown's intolerable manners are detrimental to him (Wilson)."

27. Letter from Creighton to Arthur Cole, 21 January 1956, Creighton papers.

28. Wilson maintained a quiet connection with Brown, for whom he continued to have warm professional—and seemingly personal—regard. He spoke at Brown's memorial service at Cornell in 1997, and their joint work was chronicled in Erich Hoyt, *The Earth Dwellers: Adventures in the Land of Ants* (New York: Simon and Schuster, 1996).

29. Letter from William Creighton to Robert Gregg, 1 January 1958, Creighton papers. Creighton claimed that Wilson had accounted for his change of direction to him by citing Carpenter's judgment that taxonomy was "too narrow" and animal behavior "broad" by comparison. Creighton doubted that Carpenter had made this judgment, which seems reasonable given that Carpenter himself was a taxonomist.

30. *American Scientist* 36 (1948), 564. Hockett originally proposed "sociobiology" as a linguistic science. Its first use as a term designating comparative sociology appears to have been in 1946, at a conference on genetics and social behavior. See Gregory Radick, *Professor Garner's Phonograph: The Origin of Language as a Scientific Problem After Darwin* (Chicago: University of Chicago Press, forthcoming 2007), chap. 8, "Simian Semantics."

31. J. P. Scott, "Methodology and Techniques for the Study of Animal Societies," *Annals of the New York Academy of Sciences* 51 (1950): 1001–122; 1004–5.

32. See J. P. Scott, "Animal Behavior and Social Organization," remarks on the 20th anniversary of the founding of the Animal Behavior Society (manuscript, 1976), Wilson papers. Sebeok was also a founding member of the society.

33. Letter from Schneirla to Grassé, n.d., Schneirla papers.

34. An exception was made for Gustav Kramer of the Max-Planck Institute, despite Schneirla's strong objections that his interests in parapsychology amounted to pseudoscience.

35. T. C. Schneirla, "Final report, Project N-onr552, ONR contract RR165-297" (n.d.; 1953-4), Schneirla papers.

36. The Air Force Research Division was to support an interdisciplinary symposium in communication theory, more amenable to those of Rosenblith's opinion, at the University of Oklahoma's Department of Speech in 1961.

37. American Institute of Biological Sciences ad hoc advisory committee on status and future trends or research in the general field of biological orientation. Minutes of meeting held 29 December 1956, Schneirla papers.

38. Letter from Galambos to Schneirla, 9 February 1953, and reply, 25 March 1953, Schneirla papers.

39. P.-P. Grassé, "La reconstruction du nid et les coordinations interindividuelles. La théorie de la stigmergie," *Insectes Sociaux* 6 (1959): 41–84.

40. Robert A. Hinde, *Ethology: Its Nature and Relations with Other Sciences* (London: Fontana, 1982), 182–87.

41. W. M. Wheeler, "Ethological Observations on an American Ant (*Leptothorax emersoni Wheeler*)," *Journal für Psychologie und Neurologie* 2 (1903); O. Heinroth, "Beiträge zur Biologie, namentlich Ethologie und Psychologie der Anatiden," *Verhandllungen des V Internationalen Ornithologen-Kongresses in Berlin*, 1910: 589.

42. Bertram Schaffner, ed., *Group Processes: Transactions of the First Conference* (New York: Josiah Macy, Jr. Foundation, 1955), 212. The cult of dianetics focused on experiences imprinted early in life—even prenatally—and the scientific research discussed by participants appeared to give this some credibility.

43. Frank Fremont-Smith, "The Josiah Macy, Jr. Foundation Conference Program," in *Group Processes: Transactions of the First Conference*, ed. Bertram Schaffner (New York: Josiah Macy, Jr. Foundation, 1955), 7–8.

44. Transcript in Wilson papers.

45. Franz Alexander et al., editorial, *Behavioral Science* 1 (1956): 1–5.

46. Letter from Robert Morison to T. C. Schneirla, 14 February 1957, Schneirla papers. Schneirla finally achieved NIMH funding in 1966.

47. Letter from Lester Aronson to T. C. Schneirla, 10 March 1948, Schneirla papers.

48. Cybernetic solutions to auditory and visual deficits in humans dated back at least to the Second World War, and was a field in which Wiener had been heavily involved.

49. Typed notes, 6 June 1967, Schneirla papers. Schneirla's stubbornness was legendary. At BCI, he was once observed accidentally to put Kool-Aid into his coffee instead of milk powder. On being apprised of his mistake, Schneirla immediately retorted that he had intended to do so—indeed that he always took Kool-Aid in his coffee—and drank the whole cupful. (Anecdote from Howard Topoff.)

50. William Creighton to Arthur Cole, 19 October 1956, Creighton papers. Creighton had agreed to participate on condition that William Brown would not be present.

51. Letter from William Creighton to Bob Gregg, 15 December 1956, Creighton papers. There is evidence from the Creighton-Gregg correspondence that Schneirla sided with Gregg in the Brown-Wilson acrimony and saw him as a match for Brown. Schneirla was, however, a little more circumspect and politic than Creighton in letting his feelings be known.

52. "Morgan's Canon" required that the most metaphysically parsimonious explanation be given for any behavior observed in animals. Thus, if an action could be explained as a reflex rather than as a reasoned act, it should be.

53. See W. M. Wheeler, *The Social Insects: Their Origin and Evolution* (London: Kegan Paul, Trench and Trubner, 1928), 234.

54. T. C. Schneirla, "Theoretical Considerations of Cyclic Processes in Doryline Ants," *Proceedings of the American Philosophical Society* 101 (1957): 106–33.

55. Letter from T. C. Schneirla to Neal Weber, 26 September 1956, Weber papers.

56. Wilson was not the only one to take the new direction. Having struggled to get his earlier and well-regarded work on ant larvae published, George C. Wheeler (no relation to William Morton Wheeler) applied for a research grant to the U.S. Department of Health, Education and Welfare. He had plans for a new study linking trophallaxis in an extended sense (which he attributed to both Wheeler and Schneirla) and the "trophorhinium." This organ was thought to be, perhaps, for stridulation. The project boiled down to the question of whether larvae squeak to attract their nurses, which then feed them. Letter from William Creighton to Arthur Cole, 17 September 1964, Creighton papers.

57. Wilson's essay on Lysenko is in the Wilson papers.

58. Letter from Caryl Haskins to Neal Weber, 25 May 1937, Weber papers.

59. C. P. Haskins, *Of Ants and Men* (London: Allen and Unwin, 1945), 229.

60. Haskins' whole description of this is remarkably like Dawkins' account of memes, right down to the selectivity of the cultural environment.

61. Letter from Caryl Haskins to T. C. Schneirla, 17 February 1960, Schneirla papers. In 1965 Creighton sneered to Gregg about those (not, for once, Wilson) who "resort to computers" to "cover up basic gaps in their knowledge." Letter from Creighton to Gregg, 31 March 1969, Wilson papers.

62. Letter from Caryl Haskins to Schneirla, n.d., Schneirla papers.

63. Letter from Charles Michener to Wilson, 13 April 1954, Schneirla papers. Schneirla had fallen seriously ill and was replaced by Michener as American editor of the journal. Schneirla had evidently judged papers on his own prior to his illness.

64. Interview with Ethel Tobach.

65. Letter from Schneirla to Grassé, 5 March 1957, Schneirla papers.

66. See, for example, Haskins to Schneirla, 11 April 1955, Schneirla papers.

67. E. O. Wilson, *The Insect Societies* (Cambridge, MA: Belknap Press of Harvard University Press, 1971), 272.

68. Early accounts of allometry are given in D'Arcy Thompson, *On Growth and Form* (1917) and Julian Huxley, *Problems of Relative Growth* (1932). See Wilson, *Naturalist*, 312–14. The topic of allometry was discussed by Emerson, Schneirla, and notable French entomologists at a colloquium in 1950 (published as G. LeMasne, "Discussion sur la fecondité des ouvrières de fourmis," in *Le polymorphisme sociale et son déterminisme chez les fourmis: Colloque internationale CNRS, structure et physiologie des sociétés animales*, 34 (1952), ed. F. Bernard, 123–41; 138–40. Schneirla had been managing projects on caste allometry since at least 1954; a survey of his work is given in R. R. Gianutsos, B. S. Pasternak, and T. C. Schneirla, "Comparative Allometry in the Larval Broods of Three Army-Ant Genera, and Differential Growth as Related to Colony Behavior," *American Naturalist* 102 (1968): 533–54.

69. Eventually they published together: W. Bossert and E. O. Wilson, *A Primer of Population Biology* (Sunderland, MA: Sinauer, 1971).

70. Mayr claimed priority in the interview for *Skeptic* magazine conducted by Shermer and Sulloway. Haskins proposes an experiment in island in colonization in *Of Ants and Men*, 164–65.

71. William Creighton to Arthur Cole, 4 May 1955, Creighton papers.

72. Wilson, *Naturalist*, 115–17. On the fire ants, see Joshua Blu Buhs, "The Fire Ant Wars: Nature and Science in the Pesticide Controversies of the Late Twentieth Century," *Isis* 93 (2002): 376–400; idem, *The Fire Ant Wars: Nature, Science, and Public Policy in Twentieth-Century America* (Chicago: University of Chicago Press, 2004).

73. Buhs, "The Fire Ant Wars," 386.

74. Murray S. Blum complained to William Creighton in a letter of 21 September 1971 that people who knew little about fire ants had jumped on the USDA's band-wagon. Creighton papers. Buhs reviews arguments that the threat posed by the fire ant was talked up by various interested parties (notably the USDA) in order to bolster their authority in the context of the cold war in *The Fire Ant Wars*, 40.

75. E. O. Wilson, "The Fire Ant," *Scientific American* 198 (1958): 36–41; 38.

76. Letter from Murray S. Blum to William Creighton, 26 January 1969, Creighton papers. Wilson's claims about the ants' behavior had also come under attack: asked in 1968 to explain to a committee why no one else had observed damage on the scale indicated by his 1949 report, he replied—to nobody's satisfaction—that their behavior must have changed since then. (Letter from William Creighton to Murray Blum, 28 April 1968, Creighton papers.) To complete Wilson's beleaguered position, Creighton was highly skeptical about most of his identifications and distribution work, leveling the now-familiar charge that Wilson simply did not have the field experience to make his claims credible. (Letter from William Creighton to Murray Blum, 14 May 1968, Creighton papers.) On critiques of Wilson's classification, see William F. Buren, "The Importance of Fire Ant Taxonomy," *Proceedings of the Tall Timbers Conference on Ecological Animal Control by Habitat Management* 7 (1978): 61–6; idem, "Revisionary Studies on the Taxonomy of the Imported Fire Ants," *Journal of the Georgia Entomological Society* 7 (1972): 1–26; Buhs, "Building on Bedrock."

77. With his usual media savvy, Wilson got his stories into the papers, both back home and in Australia. See, for example, "'Missing link' in Ant Family Is Located," *Standard Times [Mass.]*, 20 March 1955; "Missing Ant Link Being Forged," *Perth Daily News*, 19 February 1955.

78. "Find Fire Ant Trail Depends on Odor" [sic], *Worcester [Mass.] Gazette*, 16 March 1959.

79. E. O. Wilson, "Pheromones," *Scientific American* 208 (1963): 100–14; 110.

80. Wilson, *Naturalist*, 308–12.

81. Wilson, *The Insect Societies*, 253. This work was developed further by Haldane and Spurway in 1954.

82. Letter from Schneirla to Wilson, 1 August 1962, Schneirla papers.

83. Wilson, *The Insect Societies*, 234–35. Wilson goes on to mention that this is now a whole new discipline, which was named zoosemiotics by Sebeok in 1965.

84. Steven J. Heims, *Constructing a Social Science for Postwar America: The Cybernetics Group 1946–1953* (Cambridge, MA: MIT Press, 1991), 8.

85. Heims, *Constructing a Social Science*, 9.

86. E. O. Wilson, "Behavior and Organization in Insect Societies," application to NSF for a research grant in biology, Wilson papers. A very similar claim is made in E. O. Wilson, "The Superorganism Concept and Beyond," in *L'Effet de Groupe chez les Animaux*, Colloques Internationaux Centre National de la Recherche Scientifique (Paris) 173 (1967): 27–39; 27.

87. T. C. Schneirla, corrections to Macy "Group Processes" transcript (manuscript, n.d. [1954]), Schneirla papers.

88. William D. Hamilton, "The Genetical Evolution of Social Behavior," I, II, *Journal of Theoretical Biology* 7 (1964): 1–52.

89. Wilson, *The Insect Societies*, 262.

90. Wilson, "The Superorganism Concept and Beyond." See also Wilson, *The Insect Societies*, 281–92. Here trophallaxis is revealed as a startling demonstration of the worker's lack of selfishness and a means for the efficient distribution of liquid food through the colony. See also E. O. Wilson, *Sociobiology: The New Synthesis* (Cambridge, MA: Belknap Press of Harvard University Press, 1975), 29–30.

91. Letter from William Creighton to Robert Gregg, 4 January 1969, Creighton papers.

92. See, for example, Steven Johnson, *Emergence: The Connected Lives of Ants, Brains, Cities and Software* (London: Allen Lane, 2001).

93. Letter from Robert Gregg to William Creighton, 14 January 1973, Creighton papers.

94. Manuscript in Wilson papers.

95. Letter from William Creighton to the editors of *Encyclopaedia Britannica*, 29 May 1967, Creighton papers. At this time, a new printing of the *Encyclopaedia* was produced annually with minor revisions from the previous year.

96. Background history from Carroll M. Williams, *The Jubilee of the Harvard Biological Laboratories: Retrospections* (Pamphlet, Harvard, 1982), 23 pp., Wilson papers.

97. Wilson's account of this period is covered in the chapter "The Molecular Wars" of his autobiography, *Naturalist*.

98. Harvard University News Office press release, Wilson papers.

99. Letter from E. O. Wilson to Franklin Ford, 22 July 1968, Wilson papers.

100. Letter from R. P. Levine, Lawrence Bogorad, J. W. Hastings, George Wald, Winslow R. Briggs, A. M. Pappenheimer, Jr., Matthew Meselson, and John R. Raper to Franklin Ford, 1 August 1968, Wilson papers.

101. Letter from E. O. Wilson to unknown recipient(s), 14 August 1968, Wilson papers.

102. Letter from Ernst Mayr to E. O. Wilson, 23 September 1968, Wilson papers.

103. Letter from E. O. Wilson to Franklin Ford, 26 September 1968, Wilson papers.

104. Report by John Torrey, 22 January 1969, Wilson papers.

105. Letter from Carpenter to unnamed recipients, 21 February 1969, Wilson papers.

106. Weaver quoted in Raymond B. Fosdick, *The Story of the Rockefeller Foundation* (New York: Harper, 1952), 157.

107. Letters between Mayr and Griffin, 1962–3, Wilson papers.

108. Letter from E. S. Barghoorn, F. M. Carpenter, G. L. Clarke, A. W. Crompton,

R. A. Howard, E. Mayr, R. C. Rollins, J. G. Torrey, and E. O. Wilson to Franklin Ford, 26 September 1969, Wilson papers.

109. Letter from E. S. Barghoorn, F. M. Carpenter, G. L. Clarke, A. W. Crompton, R. A. Howard, E. Mayr, R. C. Rollins, J. G. Torrey, and E. O. Wilson to Dean Franklin Ford, 26 September 1969, Wilson papers.

110. Letter from E. O. Wilson to Paul Levine, n.d., marked *"not sent!"* Wilson papers.

111. *MCZ Newsletter* 2, no. 3 (Spring 1973).

112. *Harvard Crimson,* 8 January 1970.

113. Center for Environmental and Behavioral Biology: statement signed by Ernst Mayr, E. O. Wilson, and Herbert Levi (n.d.), "Jurisdiction for the New Wing of the MCZ," Wilson papers.

114. *MCZ Newsletter* 2, no. 3 (Spring 1973).

115. Letter from E. O. Wilson to John T. Dunlop (dean, Faculty of Arts and Sciences), 17 January 1973, Wilson papers.

116. Letter from Creighton to G. C. Wheeler, 25 November 1972, Creighton papers.

117. Letter from Creighton to Robert Gregg, 15 December 1972, Creighton papers.

118. Letter from Creighton to G. C. Wheeler, 25 November 1972, Creighton papers.

119. Letter from Creighton to Robert Gregg, 15 December 1972, Creighton papers.

120. It has also been perceived by Wilson's enemies that he has made use of misinformation. See letters from Creighton to Robert Gregg, 10 February 1973 and 21 March 1973; letter from Robert Gregg to Creighton, 24 May 1973, Creighton papers.

121. Notes for Tanner Lecture, University of Michigan, 1979, Wilson papers.

122. *Time,* 15 September 1975. The books are by Michael Korda (Random House) and Robert J. Ringer (Funk & Wagnalls). George Homans, a lasting sociological connection from Wheeler's days, also had his *Social Behavior* published in revised edition at this time. George Homans, *Behavior: Its Elementary Forms* (New York: Harcourt Brace Jovanovich, 1974).

123. Letter from Sebeok to Wilson, 10 June 1975, Wilson papers. Myrmecologist Joan M. Herbers reflects on the power of words in her field today in "The Loaded Language of Science," *U.S. Chronicle of Higher Education,* 24 March 2006, B5. She argues that the language of ant "slavery" may deter African American students from pursuing myrmecology.

Conclusion

1. George C. Wheeler, "Don't Go to the Ant," *Bios* 28 (1957): 94–103.

2. Quoted in E. O. Wilson, "The Coherence of Knowledge," Phi Beta Kappa oration given at Harvard University, 2 June 1998 (manuscript), Wilson papers.

3. Ezra Pound, *Pisan Cantos* (1948) no. 81.

4. Ronald Bush, "Modernism, Fascism, and the Composition of Ezra Pound's *Pisan Cantos,*" *Modernism/Modernity* 2 (1995): 69–87.

5. W. M. Wheeler, *The Social Insects: Their Origin and Evolution* (London: Kegan Paul, Trench, Trubner, 1928), 234.

6. Auguste Forel, *The Social World of the Ants Compared with that of Man* (2 vols., London: G. P. Putnam's Sons, [1921–2] 1928), vol. i, 462–3.

7. Forel, *The Social World of the Ants*, vol. i, 443

8. Giard too had based his evolution of social behavior on the mother-child parasitism of breast-feeding. See Marion Thomas, "Rethinking the History of Ethology: French Animal Behaviour Studies in the Third Republic (1870–1940)," PhD thesis, University of Manchester, 2003, 113, and Alfred Giard, "Sur les parasites bopyriens et la castration parasitaire," *Comptes Rendus de la Société de Biologie* 39 (1887): 371–72. Charlotte Haldane used Wheeler's focus on parasitism as a summary of his entire philosophy. She has a character from her novel *Man's World* (London: Chatto & Windus, 1926) quote him thus: "Man furnishes the most striking illustration of the ease with which both the parasitic and host roles may be assumed by a social animal." (Ibid., 182).

9. Thomas Belt, *The Naturalist in Nicaragua* (London: John Murray, 1874), is full of reflections on the intelligence of the ants (including the railway-burrowers; ibid., 83–84, 151, 329–30). Lubbock gives reported anecdotes on the subject in John Lubbock, *Ants, Bees and Wasps: A Record of Observations on the Habits of the Social Hymenoptera* (London: Kegan Paul, Trench and Trubner, [1882] 1929), 179–81, before detailing his own more skeptical experiments. Hingston's traveling-naturalist books also remarked critically on such tales. See also Julian Huxley, *Ants* (London: Dennis Dobson, [1930] 1949), 39–46, and Forel, *Social World of the Ants*, vol. ii, 202–3.

10. Thomas A. Sebeok, "Zoosemiotic Components of Human Communication," in *How Animals Communicate*, ed. Thomas A. Sebeok (Bloomington: Indiana University Press, 1977), 1055–77; 1068. On "Clever Hans," see Robert A. Boakes, *From Darwin to Behaviourism: Psychology and the Minds of Animals* (Cambridge: Cambridge University Press, 1984), 78–81.

11. A copy of the pamphlet is found in Orwell's collection donated to British Museum.

12. Joe Haldeman, *The Forever War* (London: Millennium, [1974] 1999), 72.

13. Haldeman, *The Forever War*, 249–50.

14. A rather beautiful and empathetic account of the relativity of language, exemplified by ants, is given in Ursula LeGuin's 1974 story, "The Author of the Acacia Seeds and Other Extracts from the Journal of Therolinguistics," in *idem, Buffalo Gals and Other Animal Presences* (London: Gollancz, 1990), 167–78.

15. W. M. Wheeler, *Holy Bluff* (manuscript, n.d.), Wheeler papers.

16. Mead in discussion following Lawrence S. Kubie, "Communication between Sane and Insane: Hypnosis," in *Cybernetics: Circular Causal and Feedback Mechanisms in Biological and Social Systems. Transactions of the Eighth Conference*, ed. Heinz von Förster, Margaret Mead and Hans Lukas Teuber (New York: Josiah Macy, Jr. Foundation, c. 1952), 92–132; 107.

17. Discussion following Kubie, "Communication between Sane and Insane," 110–11.

18. Steve J. Heims, *John von Neumann and Norbert Wiener: From Mathematics to the Technologies of Life and Death* (Cambridge, MA: MIT Press, 1980). On Augustinianism, see John Forrester, *Truth Games: Lies, Money, and Psychoanalysis* (Cambridge, MA: Harvard University Press, 1997), 12–14, 49–53.

19. "The integrity of the channels of internal communication is essential to the welfare of society"; this included clear and direct communication from government to populace, and feedback in the democratic form of the vote. Norbert Wiener, *The Human Use of*

Human Beings: Cybernetics and Society (rev. ed., Boston: Houghton Mifflin, [1950] 1954), 130.

20. J. B. S. Haldane, "Animal Communication and the Origin of Human Language," *Science Progress* 43 (1955): 385–401; 386, 393. See also J. B. S. Haldane and Helen Spurway, "A Statistical Analysis of Communication in *Apis mellifera* and a Comparison with Communication in Other Animals," *Insectes Sociaux* 1 (1954): 247–83.

21. E. O. Wilson, *On Human Nature* (Cambridge, MA: Harvard University Press, 1978), 159. Wilson refers the reader to Sissela Bok, *Lying: Moral Choice in Public and Private Life* (Pantheon: New York, 1978).

22. E. O. Wilson, "The Coherence of Knowledge," Phi Beta Kappa oration given at Harvard University, 2 June 1998 (manuscript), Wilson papers.

23. Letter from William Creighton to Bob Gregg, 13 April 1969, Creighton papers; also recounted in letter from Creighton to Howard Evans, 25 March 1968, Creighton papers.

24. W. M. Wheeler, "On Instincts," *Journal of Abnormal Psychology* 15 (1921): 295–318; 311.

25. Letter from Wheeler to E. O. Essig, 24 July 1930, Wheeler papers.

26. W. M. Wheeler, "The Termitodoxa, or Biology and Society," *Scientific Monthly* 10 (1920): 113–24; 119.

27. Letter from Wheeler to Irving Fisher, 19 January 1924, Wheeler papers.

28. The entomologists Vernon Kellogg and Anne H. Morgan each took a place on the committee, as did at least two other acquaintances of Wheeler's: Robert Yerkes and William McDougall. Interestingly, Charles Davenport, a prominent eugenist and member of the Advisory Council, was registered as a member of the Entomological Society in 1910. (Various literature of Eugenics Society of the United States of America, Wheeler papers.) In 1933 Wheeler spoke at a "Biology and Society" symposium organized by the American Society of Naturalists. While Wheeler's paper was an abstract discussion of the issues that all societies needed to solve, the other papers at this conference were overt discussions of applied eugenics. They considered which races were inferior, and which components of American society were of similarly low hereditary caliber. In context, Wheeler did precisely what he claimed to be chary about: he contributed to debate about human society. The three papers were printed in *Scientific Monthly* 39 (1934): 289–322. Wheeler also contributed a chapter titled "Societal Evolution" to the eugenic *Human Biology and Racial Welfare*, ed. E. V. Cowdry (New York: Hoeber, 1930), 139–155.

29. W. M. Wheeler, *Foibles of Insects and Men* (New York: Alfred A. Knopf, 1928), xxiv.

30. In the cold war sci-novel *Starship Troopers*, galactic infantrymen have to model themselves on ants in order to take on their antlike foes. Robert A. Heinlein, *Starship Troopers* (New York: Penguin Putnam, [1959] 1987).

31. Bert Hölldobler and E. O. Wilson, *Journey to the Ants: A Story of Scientific Exploration* (Cambridge, MA: Belknap Press of Harvard University Press, 1994), 9.

ESSAY ON SOURCES

A discipline so undisciplined as myrmecology presents a considerable challenge when it comes to sources. Its participants are scattered around the world in a variety of institutions, or none, and its context varies from popular science to academic zoology, via psychology, linguistics, and psychiatry (to name but three related fields). My starting point for the whole project was Julian Huxley's book *Ants* (London: Dennis Dobson, [1930] 1949). In reading this I stumbled on a fascinating topic and a bibliography of authors utterly unknown to me. Thus I began tracing out a network, first through published sources, and then, having established the most influential figures, through archives. In the case of each major figure, a different context suggested itself, entailing in turn its own specific contextual research, published and unpublished. Two useful resources for beginning any entomological hunt are Pamela Gilbert, *A Compendium of Biographical Literature on Deceased Entomologists* (London: British Museum [Natural History], 1977); and Pamela Gilbert and Chris J. Hamilton, *Entomology: A Guide to Information Sources* (London: Mansell Publishing, 1983).

A list of pre-twentieth-century primary sources relating to insects might begin with Bernard Mandeville's 1714 satire on society, *Fable of the Bees; or, Private Vices, Publick Benefits* (2 vols., Oxford: Clarendon Press, 1924). Shortly after this, the polymath René Antoine Ferchault de Réaumur conducted somewhat more consistent research on insectan life cycles, anatomy, and behavior, publishing these investigations as *Mémoires pour servir à l'histoire des insectes* (6 vols., Paris: Imprimerie Royale, 1734–42). These volumes, particularly the writings on bees, were bowdlerized and published under various titles by numerous authors over the next fifty to one hundred years. William Morton Wheeler found and translated additional unpublished Réaumur manuscripts on ants, which were brought out as *Tome septième* to the series *Histoire des fourmis* (Paris: Paul Lechevalier, 1928). British entomology of Réaumur's period included Moses Harris, *The Aurelian or Natural History of English Insects namely Moths and Butterflies* (London, 1766) and the Rev. William Gould's *An Account of English Ants* (London: A. Millar, 1747).

In the nineteenth century, some remarkable research on ants was carried out by the blind Swiss natural historian François Huber, assisted by his son Pierre. Huber the younger published his conclusions in *Recherches sur les mœurs des fourmis indigènes* (Paris: Chez J. J. Paschoud, 1810)—the book that Forel's grandmother was given by its author, her unsuccessful suitor. Meanwhile, in Great Britain, the standard insect text was for many years William Kirby and William Spence, *An Introduction to Entomology, or, Elements of the Natural History of Insects* (4 vols., London: Longman, Hurst, Rees, Orme & Brown, 1815–26).

A nice sense of nineteenth-century natural theology is given by the anonymous *Lessons Derived From the Animal World* (London: Society for the Promotion of Christian Knowledge, 1851). The second of its two volumes is devoted entirely to insects and their moral virtues. Many traveling naturalists of the nineteenth century devoted a good part of their writings to insects. Thomas Belt's *The Naturalist in Nicaragua* (London: John Murray, 1874) is particularly strong on ants. John Lubbock's, *Ants, Bees and Wasps: A Record of Observations on the Habits of the Social Hymenoptera* (London: Kegan, Paul, Trench, 1882) overlaps with the time span covered by this book, but in a sense is worth reading as background for its distinctively high Victorian take on the subject.

There is also a growing collection of secondary sources on pre-twentieth-century insects. An early foray into the field was Charles L. Hogue's article "Cultural Entomology," *Annual Review of Entomology* 32 (1987): 181–99. David Freedberg's *The Eye of the Lynx: Galileo, His Friends, and the Beginnings of Modern Natural History* (Chicago: University of Chicago Press, 2002) gives, among other things, an excellent account of the entomological visions of Linceian natural philosophers (including a wonderful exploration of the significance of bees). One of the best essays I have ever read on insects was Jonathan Sheehan's "The Mind and Metaphysics of Early Modern Ants" (1999). Despite my nagging on several occasions, he has never published the piece, but I would like to acknowledge his generosity in making it available for me to draw on. Bees, not ants, were really the *insectes du jour* of the sixteenth to eighteenth centuries, and this is reflected in the secondary literature. See Frederick R. Prete, "Can Females Rule the Hive? The Controversy over Honey Bee Gender Roles in British Bee-Keeping Texts of the Sixteenth-Eighteenth Centuries," *Journal for the History of Biology* 24 (1991): 113–44, and Jean-Marc Drouin, "L'Image des sociétés d'insectes en France à l'epoque de la Révolution," *Revue de Synthèse* 4 (1992): 333–45.

The history of early economic entomology in North America is well served by both primary and secondary sources. The first generation of professional entomologists, most especially L. O. Howard, were prolific in promoting and latterly celebrating their importance. On early legislative change, see Howard, "Legislation Against Injurious Insects; a Compilation of the Laws and Regulations in the United States and British Columbia," *Bureau of Entomology Bulletin* 33 (1895). For nationalist appeals regarding the importance of entomology, see A. S. Packard, introduction to "First Annual Report on the Injurious and Beneficial Insects of Massachusetts," in Howard, *A History of Applied Entomology (Somewhat Anecdotal)*, Smithsonian Miscellaneous Collections. 84 (Washington, D.C. Smithsonian Insitution, 1930), 207; E. O. Essig, *History of Entomology* (New York: Hafner Publishing, [1931] 1965), 48–53; W. P. Flint and C. L. Metcalf, *Insects: Man's Chief Competitors* (Baltimore: Williams and Wilkins Co. in association with The Century of Progress Exposition, 1932), 106–11. On local persuasion regarding the importance of entomology, see Howard, *A History of Applied Entomology*, 184; and for a detailed account of machinations at the local level, see Essig, *History of Entomology*, 54–81. On publication and dissemination, see *ibid.*, 59–65; and Herbert H. Ross, *A Textbook of Entomology* (New York: John Wiley, 1948), 16–17 and 21–22.

Early histories and defenses of entomology include Essig, *History of Entomology*; L. O. Howard, "A Brief Account of the Rise and Present Condition of Official Economic Entomology," *Insect Life* 7 (1894): 55–108; *idem, A History of Applied Entomology*; *idem, A Fifty-Year Sketch of Medical Entomology*, Smithsonian Report (Washington, DC: American Public Health Association, 1921); *idem, The Insect Menace* (London: D. Appleton, 1931); *idem, Fighting the Insects: The Story of an Entomologist. Telling the Life and Experiences of the Writer* (New York: Macmillan, 1933); Herbert Osborn, *Fragments of Entomological History: Including Some Personal Recollections of Men and Events* (Columbus, OH: published by the author, 1937); *idem, Fragments of Entomological History: Part II* (Columbus, OH: published by the author, 1946); Harry B. Weiss and Grace M. Ziegler, *Thomas Say, Early American Naturalist* (Springfield, IL: Charles E. Thomas, 1931); and Flint and Metcalf, *Insects: Man's Chief Competitors*.

When it comes to secondary sources, Patricia Tyson Stroud's *Thomas Say: New World Naturalist* (Philadelphia: University of Pennsylvania Press, 1992) is strong in making clear the nationalist dimensions of U.S. entomology, connecting Say with Jeffersonian ideals. W. Conner Sorensen's *Brethren of the Net: American Entomology, 1840–1880* (Tuscaloosa: University of Alabama Press, 1995) nicely blends the traditions of natural history and economic entomology in its history. Several books and articles specifically chart the rise of professional, applied entomology. The best of these is Paulo Palladino's exemplary *Entomology, Ecology and Agriculture: The Making of Scientific Careers in North America, 1885–1985* (Amsterdam: Harwood Academic Publishers, 1996). Shorter pieces on the subject include J. F. M. Clark, "Beetle Mania: The Colorado Beetle Scare of 1877," *History Today* 42 (1992): 5–7, and W. Conner Sorensen, "The Rise of Government Sponsored Applied Entomology, 1848–1870," *Agricultural History* 62 (1988): 98–115. Meanwhile, the story north of the border is told by P. W. Reingart in *From Arsenic to DDT: A History of Entomology in Western Canada* (Toronto: University of Toronto Press, 1980).

Earlier and less satisfactory histories of economic entomology include Arnold Mallis, *American Entomologists* (New Brunswick, NJ: Rutgers University Press, 1971); George Ordish, *The Constant Pest: A Short History of Pests and their Control* (London: Peter Davies, 1976); David Pimentel, ed., *Insects, Science, and Society* (New York: Academic Press, 1975); and Thomas R. Dunlap, "Farmers, Scientists, and Insects," *Agricultural History* 54 (1980): 93–107.

Sources on European applied entomology are much thinner on the ground, and are more often woven into histories of colonial medicine, reflecting the perspective of Europeans themselves around 1900. J. F. M. Clark has made valuable inroads into the British situation; see his "Eleanor Ormerod (1828–1901) as an Economic Entomologist: 'Pioneer of Purity Even More than Paris Green,'" *British Journal for the History of Science* 25 (1992): 431–52. The situation in Germany has been well served by Sarah Jansen. She has focused on the construction of the insect as "pest" and connected this with the discourse and treatment of human "pests" and "parasites" identified by early twentieth-century Germany. See *idem*, "Chemical-Warfare Techniques for Insect Control: Insect 'Pests' in Germany before and after World War I," *Endeavour* 24 (2000): 28–33; "An Imperial Insect in Imperial Germany: Visibility and Control in Making the Phylloxera in Germany, 1870–1914," *Science in Context* 13 (2000): 31–70; and *"Shädlinge" Geschichte eines wissenschaftlichen und politischen Konstrukts, 1840–1920* (Frankfurt: Campus, 2001).

Moving now beyond background and on to the focus of this book, Auguste Forel is well-known in the history of psychiatry, and as such his life and work are relatively well documented. The entomological side of Forel's life is reflected in the collection held near his Vaudois home of latter years, at the Fonds du Département des Manuscrits de la Bibliothèque Cantonal-Universitaire, Lausanne. Because Forel's life was so closely woven into the locale and its culture, there is a strong legacy of his personality and work at the Museum of Natural History, Palais de Rumine, Lausanne. I would especially like to record my thanks to Daniel Cherix at the Palais for his introduction to the Forel collection and his recollections of oral history, which have come to him from Forel's protégé Heinrich Kütter.

A large amount of Forel's correspondence has been carefully edited into one large volume, which forms a very useful source in itself: Hans H. Walser, ed., *August Forel: Briefe, Correspondance: 1864–1927* (Berne: Hans Huber, 1968). To get the flavor of Forel's obsessions, myrmecological and otherwise, one should read both his autobiography, *Out of My Life and Work* (London: George Allen & Unwin, [1935 in German] 1937), and his two-volume *Social World of the Ants Compared with that of Man* (London: G. P. Putnam's Sons, [1921–22 in French] 1928. There is no recent biography of Forel, but an early account is given by Alex von Muralt in *Auguste Forel* (Bern: Editions Hans Huber, 1931).

The theme of instinct and intelligence, central to Forel's understanding of ants, is pretty much inescapable as soon as one dips into myrmecological—or even entomological—literature of the era. It would be pointless to recycle all the texts cited in the discussion of this theme, but a good summary may be found in R.W. G. Hingston's *Problems of Instinct and Intelligence* (London: Edward Arnold, 1928).

The best comparison to Forel's theory of instinct as concretized intelligence is found in Eugène Bouvier, *The Psychic Life of Insects* (London: T. Fisher Unwin, [1918] 1922). Forel borrowed his vocabulary and more from Ricard Semon's *The Mneme* (London: Ruskin; New York: Macmillan, [1904] 1921). Laura Otis gives an excellent account of Semon and his contemporaries in *Organic Memory: History and the Body in the Late Nineteenth and Early Twentieth Centuries* (Lincoln: University of Nebraska Press, 1994). More general accounts of theories of animal mind and behavior may be found in Robert J. Richards, *Darwin and the Emergence of Evolutionary Theories of Mind and Behavior* (Chicago: University of Chicago Press, 1987); and Robert A. Boakes, *From Darwin to Behaviourism: Psychology and the Minds of Animals* (Cambridge: Cambridge University Press, 1984).

The troublesome disciplinary background to Forel's "myrmecology" can be pieced together from a number of sources. On the French context, where insect studies formed part of psychology, see Laurent Loty, "Sens de la discipline . . . et de l'indiscipline: Réflexions pour une pratique paradoxale de l'indisciplinarité," *Bulletin de la Société Française pour l'Histoire des Sciences de l'Homme* 20 (2000): 3–16; and Richard W. Burkhardt, "Le comportement animal et la biologie française, 1920–1950," in *Les sciences biologiques et médicales en France, 1920–1950*, ed. Claude Debru, Jean Gayon, and Jean-François Picard (Paris: CNRS Editions, 1994), 99–111. Marion Thomas has recently completed an excellent study, "Rethinking the History of Ethology: French Animal Behaviour Studies in the Third Republic (1870–1940)" (PhD thesis, University of Manchester, 2003) which contains a good deal of material on insects and their human analogies. I would like to thank Marion for sharing her expertise and criticism during the writing of this book.

On the American situation, where Wheeler and other entomologists reinterpreted Forel in a zoological context, see Gregg Mitman, *The State of Nature: Ecology, Community, and American Social Thought, 1900–1950* (Chicago: University of Chicago Press, 1992). For a slightly later period, see Richard W. Burkhardt and Gregg Mitman, "Struggling for Identity: The Study of Animal Behavior in America, 1930–1945," in *The Expansion of American Biology*, ed. Ronald Rainger, Keith Benson, and Jane Maienschein (New Brunswick, NJ: Rutgers University Press, 1991), 164–94.

Accounts of ethology, the continuation of the European style, may be found in Robert A. Hinde, *Ethology: Its Nature and Relations with Other Sciences* (London: Fontana, 1982); Richard W. Burkhardt, "On the Emergence of Ethology as a Scientific Discipline," *Conspectus History* 7 (1981): 62–81; and Philippe Chavot, "A la recherche d'une structure unifiée? Le développement de l'ethologie en France après la seconde guerre mondiale," *Bulletin de la Société d'Histoire et d'Epistomologie des Sciences de la Vie* 2 (1995): 32–40.

Papers relating to William Morton Wheeler's career are held at the Pusey Library, Harvard University. Unfortunately, there is very little of a personal nature in this archive, apart from what can be read between the lines of professional correspondence. Some of his professional correspondents, such as David Fairchild, were also good friends, and their letters do give some dimensionality to Wheeler's life and work. Wheeler's more personal materials—which would have given a fascinating insight into the activities of this forceful character—were held at his home and destroyed after his and his wife's death by their daughter, Adaline Wheeler. As far as I can tell, Dora Emerson Wheeler's material went the same way—a great waste of a historical record concerning her political and nature study interests.

Mary A. Evans and Howard E. Evans wrote their laudatory biography *William Morton Wheeler, Biologist* (Cambridge, MA: Harvard University Press, 1970) in order to bolster the role of whole animal biology (and myrmecology) when it was going through a rough patch in Harvard's departmental structure. Its polemical purposes are not too intrusive, however, and the book gives a good overview of Wheeler's life and work. Of Wheeler's works, *Ants: Their Structure, Development and Behavior* (New York: Columbia University Press, 1910) was of most lasting importance to myrmecology, though it is not exactly a light read. His paper "The Ant-Colony as an Organism," *Journal of Morphology* 22 (1911): 301–325, has had most impact in the wider world and is definitely worth reading. Finally, his *Foibles of Insects and Men* (New York: Alfred A. Knopf, 1928) collects a number of his less technical papers (including the bizarre "Termitodoxa"), illuminating both Wheeler's sense of humor and the nature of analogies that he was inclined to draw between the two- and the six-legged.

For the generic context of Wheeler's myrmecology-natural history and nature studies, I visited the Department of Manuscripts and University Archives, Cornell University Libraries. Here there are a number of relevant archives, including those of prominent promoters of the nature study movement, John Henry Comstock and Anna Botsford Comstock, and Liberty Hyde Bailey. There are also papers and records of the American Nature Study Society, including all kinds of pamphlets, letters, newsletters, and so forth. Nature study publications are also represented in commercial form by the Cornell University Press and Comstock Publishing Co. records, 1880–1935, and the Macmillan papers.

There are many enjoyable published sources relating to nature study, some more ephemeral than others. No one should deny themselves the pleasure of dipping into the ten volumes of Jean-Henri Fabre's *Souvenirs entomologiques: Etudes sur l'instinct et les mœurs des insectes* (Paris: Delagrave, 1879–1907), either in the original or in one of its many translated, re-edited forms. Other notable authors in the field include Edith M. Patch, John Burroughs, Ernest Thompson Seton, Liberty Hyde Bailey, and Donald Culross Peattie. Three popularizations of social insect science were published around the world by Maurice Maeterlinck, eliciting interest from the public and professional discomfort from the entomologists. They appeared in English as *The Life of the Bee* (London: George Allen, 1901), *The Life of the White Ant* (London: George Allen, 1927), and *The Life of the Ant* (London: Cassell, 1930).

Secondary sources on American nature writing include Joseph Wood Krutch, ed., *Great American Nature Writing* (New York: Sloane, 1950), and Peter A. Fritzell, *Nature Writing and America: Essays upon a Cultural Type* (Ames: Iowa State University Press, 1990). Little has been published on nature study and its problematic disciplinary status (though see the Henson pieces cited in chapter 5), but worth seeing on the "nature-fake furor" is Ralph H. Lutts' *The Nature Fakers: Wildlife, Science and Sentiment* (Charlottesville: University Press of Virginia, 1990).

C. K. Ogden and I. A. Richards provide the most productive route into the theme of ants as a model for linguistics and semiology. Their archives are scattered in various locations; those that I consulted were the Richards papers at King's College, Cambridge (there is also some useful material in the Keynes papers at the same location). I. A. Richards had a long association with Magdalene College, Cambridge, and his library collection can be found there, together with a large number of letters. There is a good selection of papers relating to Ogden and his publishing enterprises in the Manuscripts section of the University Library, Cambridge. Their joint publication *The Meaning of Meaning: A Study of the Influence of Language upon Thought and of the Science of Symbolism* (London: Kegan Paul, Trench and Trubner, 1923) is a vital starting point for any discussion on semiology. The publications over which Ogden had editorship form an interesting study in their own right from the point of view of social and intellectual history. See the *International Library of Psychology, Philosophy and Scientific Method*, the *Psyche* journal and *Psyche Miniatures*, the *To-day and To-morrow*, the *History of Civilisation* series, and *Science for You*. In terms of secondary sources on the two men, see P. Sargant Florence and J. R. L. Anderson, *C. K. Ogden: A Collective Memoir* (London: Elek Pemberton, 1977); W. Terrence Gordon, *C. K. Ogden: A Bio-Bibliographic Study* (London: Scarecrow, 1990); and John Paul Russo, *I. A. Richards: History, Life and Work* (London: Routledge, 1989).

It remains for someone to work more on the connections between the Orthological Society, the Basic English Foundation, and their various funding bodies, particularly the latter's sponsorship by the Rockefeller foundation. Richards' work for Disney is also an intriguing topic that, so far as I am aware, has not been touched by historians. Thomas Sebeok is another figure who deserves a good deal of further research. How, precisely, did his career interweave the natural sciences with the humanities? Putting all these questions together sketches out a project to reveal the science, politics, and ideology of intra- and international communication during World War II and the beginnings of the cold war.

Specifically entomological sources for the 1930s and the World War II period are scarce. There is Julian Huxley's *Ants* (London: Dennis Dobson, [1930] 1949), but the most important writing on insect behavior is Karl von Frisch's work on bees: *Aus dem Leben der Bienen* was published in German in 1950. The text of von Frisch's American lectures was published in English in 1950 as *Bees—Their Vision, Chemical Senses, and Language* (Ithaca: Cornell University Press). In 1954, *Aus dem Leben der Bienen* was translated from the fifth German edition as *The Dancing Bees*. See also the autobiography *A Biologist Remembers* (Oxford: Pergamon Press, 1967).

Studying the history of cybernetics is a frustrating business as there is, astonishingly, no central repository or even bibliography of primary sources. The best secondary sources are those by Steven J. Heims, especially *Constructing a Social Science for Postwar America: The Cybernetics Group 1946–1953* (Cambridge, MA: MIT Press, 1991). The transcripts of the Macy Conferences on Circular Causal and Feedback Mechanisms in Biological and Social Systems (1946–53) were not published until after they were retitled as the Cybernetics conferences from 1950. These were all published by the Josiah Macy, Jr. Foundation, as was the first *Group Processes* transcript; the subsequent *Group Processes* transcripts, edited by Bertram Schaffner, were published by the Macy Foundation in Princeton, N.J. Notes on the earlier Circular Causal and Feedback conferences can be found in the Margaret Mead papers at the Library of Congress, Washington, D.C.—though researchers should be warned that deciphering her handwriting is a herculean task. The papers of Walter McCulloch, a prominent organizer of the conferences, were consulted at the American Philosophical Society, Philadelphia.

T. C. Schneirla's mid-career work also dates from this era. The official Schneirla papers are held at the Archives of the History of American Psychology at the University of Akron, Ohio. A large number of his most important papers have been collected and conveniently reprinted in Lester R. Aronson, Ethel Tobach, Jay S. Rosenblatt, and Daniel S. Lehrman, eds., *Selected Writings of T. C. Schneirla* (San Francisco: W. H. Freeman, 1972). Schneirla's famous paper dealing with the suicidal Ecitons ("The Army Ants," co-written with Gerard Piel) is included in the collection. A good sense of Schneirla's alliances, colleagues, and themes can be gleaned from another substantial edited volume: Lester R. Aronson, Ethel Tobach, Daniel S. Lehrman, and Jay S. Rosenblatt, eds., *Development and Evolution of Behavior: Essays in memory of T. C. Schneirla* (San Francisco: W. H. Freeman, 1970).

E. O. Wilson has periodically been packaging up boxes of papers, press cuttings, and other materials from his office and sending them to the Library of Congress, Washington, D.C. Since the collection is incomplete, they are completely uncataloged at present. Nevertheless, it is possible to consult the collection if one has plenty of patience. In among this lot are the Creighton papers, a well-ordered but also uncatalogued collection of professional and catty letters from the mid-twentieth century. One or two later myrmecologists have their materials lodged with the American Philosophical Society. Here, those pertaining to Neal A. Weber were useful.

E. O. Wilson is a superb writer, and it is a pleasure to read his many books. The most important from the point of view of this study are *The Insect Societies* (Cambridge, MA:

Belknap Press of Harvard University Press, 1971); *Sociobiology: The New Synthesis* (Cambridge, MA: Belknap Press of Harvard University Press, 1975); and *On Human Nature* (Cambridge, MA: Harvard University Press, 1978). Wilson's autobiography *Naturalist* (Harmondsworth: Penguin, 1995) is likewise an enjoyable and fairly insightful account of his own life.

INDEX

Agricultural Experiment Stations, 3
Alcock, Alfred, 8, 9
alcohol, 15, 22, 24, 29–30, 32–33, 36, 42, 65, 221
Allee, W. C., 51, 110
allometry, 204, 207
Altmann, Stuart, 207
amateurism, 40, 54, 98, 118–19, 128–29, 134, 136
American Association of Economic Entomologists (formerly Association of Economic Entomologists), 4, 136–37
American Museum of Natural History, 67, 109, 116, 119, 168, 170, 197
American Nature Study Society (ANSS), 108–9, 111, 113, 115–16, 130, 135–37
American Society of Naturalists, 97, 115
analogy, 18–19, 32, 34, 98, 101–3, 173, 225–28
Animal Behavior Society, 182, 195
animal language, 140, 154, 170. *See also* ants; bees
animal psychology, 39–40, 43–49, 61; disciplinary questions surrounding, 39, 43–49; in France, 16, 45–49, 63, 169
ANSS. *See* American Nature Study Society
ant, castes of: queen, 57, 79, 117, 220 (*see also* maternity); soldier, 32, 72; worker, 77–80, 82–83, 134, 159, 220
ant, names of: army, 6, 171; Dolichoderinae, 206; Dorylinae, 78, 201; *Eciton*, 91, 172, 179, 222; fire, 205–8, 211; *Formica pratensis*, 27; *Formica sanguinea*, 27; *Lasius*, 57; Myrmicinae, 78–79, 206; *Nothomyrmecia macrops*, 205; *Pachycondyla montezumae*, 78; *Paedalgus*

termitolestes, 79; *Pheidole*, 191; *Pogonomyrmex*, 200; Ponerinae, 78–79
anthropomorphism, 45, 96, 116–19, 125–28, 131, 135–36, 208, 210, 228, 261n30, 261n40
ant-human comparisons: cybernetic, 169; division of scientific labor, 64; economic, 90; evolution of, 32; Fabre's unwillingness to compare, 126; familial and civic, 131; by A. H. Forel, 27–28; mental, 49–50; Pareto, 89; perfectibility, 34–35, 50; Schneirla's unwillingness to compare, 173, 199; by traveling naturalists, 9–11; trophallactic equivalence, 80, 82, 86; by Victorians, 9–10, 17, 61; in war, 73; Wilson's unwillingness to compare, 225
antisociality, 67, 76, 82, 90–95, 138, 159, 220
ant-like machines, 17, 169, 178, 186, 203, 228
ant mill, 172, 185, 221
ants: communication of, 140, 169; feeding of, 76–79, 82–85, 88, 126, 184, 220 (*see also* trophallaxis); missing link of, 204–5, 207; odor of, 80, 169, 206 (*see also* miscibility of ants' nests)
apes, 49, 53, 154, 177
aphids, 6
Aquinas, Thomas, 16, 49
Aronson, Lester, 172, 199
Association of Economic Biologists, 5
Avebury, Lord. *See* Lubbock, John

Babbitt, Irving, 149
Baha'i movement, 36
Bailey, Liberty Hyde, 99–101, 103, 107–11, 122, 136

Barro Colorado Island, 171–72
Basic English, 18, 140, 143, 154–59, 161, 222
Bates, H. W., 7, 48
Bateson, Gregory, 186, 188, 222, 224
Beebe, William, 97–98, 104, 106, 119, 124, 137, 190
Beer, Gillian, 14, 101
Beerbohm, Max, 219
bees: anthropomorphic, 127; and cybernetics, 164; dance language of, 12, 19, 167–69, 172–73, 177, 200, 209, 222 (*see also* Frisch, Karl von); T. S. Eliot's, 150–51; as experimentally convenient organism, 160; as instinctual paradigm, 49; Smithian, 75, 89
behavior: evolutionist natural historical accounts of, 44; experiments on, 39; mechanistic accounts of, 43–44; names for science of, 43, 197; paradigmatic of entomology, 9; plasticity of, 31, 33, 59, 221; provisioning, 52–53, 58–59, 77, 131, 221; taxonomy, considered in combination with, 11–12, 21–22, 38
behaviorism, 63, 183, 187, 201
Belt, Thomas, 7, 10, 49–50, 54, 56
Bergson, Henri, 51, 55, 76, 104, 147, 180, 197
Bethe, Albrecht, 44, 48, 55
Bigelow, Julian, 186, 224
Bigelow, Maurice A., 67, 108, 112–14
Birch, Herbert, 177–79, 185–86, 222
Blum, Murray S., 205
Bohn, Georges, 45–47, 55, 61–62
Bonnet, Charles, 25
Bossert, William H., 204–5
Bouvier, Eugène, 38, 45–50, 55, 60; *La vie psychique des insectes*, 47
Boveri, Marcella, 167
Boveri, Theodor, 65
Brave New World. *See* Huxley, Aldous
Brown, William L., 191–94, 203, 205–6
Brun, Rudolph, 25, 48, 60
Büchner, Ludwig, 126
Bugnion, Edouard (brother-in-law of A. H. Forel), 24–25, 31, 35, 60, 226
Burghölzli asylum, 24
Burroughs, John, 18, 101–4, 116, 122, 127, 132–33, 135–36
Bussey Institute, Harvard University, 67, 137, 211
Butler, Samuel, 186
butterflies, 131

Carnegie Institute, 138, 201
Carpenter, Frank, 191, 193, 212, 214–15
Carter, Marion H., 130
caste. *See* ant: castes of; polymorphism
Ceylon, 8, 168, 205
Chamberlain, Neville, 155
Chapman, Frank M., 110, 116
children, 98, 106–19, 123, 129–32, 137
China, 155, 157, 221–22
Chomsky, Noam, 275n68
Churchill, Winston, 155
Clark, John, 40–42, 51
Clark University, 65, 133
Cockerell, T. D. A., 131
cold war, 139, 158–59, 164, 223
Cole, Arthur C., 191, 193, 200–201
collection of ants, 1, 14, 22, 40–43, 122
Colonial Entomological Research Committee. *See* Imperial Bureau of Entomology
Columbia University Teachers College, 66–67, 108
communication: cyberneticians and, 169, 176, 179, 181–82, 185–88, 197–98; military interest in, 170, 196, 200, 217; Ogden and, 148; Rockefeller Foundation and, 168–69; Schneirla and, 173, 200–201, 206; of science, 157–58, 198; Sebeok and, 182–83, 217, 223; Wilson and, 139, 202, 204–7
comparative psychology. *See* animal psychology
Comstock, Anna, 69, 107, 110, 113–14, 118, 122, 135
Comstock, John Henry, 67, 69, 107, 114, 117, 122, 135–37
Coolidge, Harold J., 76, 170
cooperation in evolution, 51, 76
Cornell University, 69, 99, 107–9, 167, 176, 197
Cornetz, Victor, 48
Crawley, W. C., 40–41, 51
Creighton, William S.: dispute with Wilson, 191–95, 203, 205, 210, 216; *Encyclopaedia Britannica*, 210–11; and Schneirla, 200–201; taxonomy of, 139
crowd. *See* masses
cyberneticians, 16–17, 19, 140, 162–64, 169, 187–88, 197, 201, 207–8, 224. *See also* Macy conferences

cybernetics, 139, 163–64, 173–79, 180–84, 199, 206. *See also* Macy conferences

Darlington, Philip, 193
Darwin, Charles, 44–45, 76, 101, 105. See also *The Origin of Species*
dealation, 57–58, 77, 91, 184
deception, 223–25
degeneration, 17, 87
Delage, Yves, 45
Department of Agriculture, U.S., 3, 13, 205
Depression, the, 82, 85–86, 111
Dethier, Vincent, 170, 200
DeVore, Irven, 215
Dircks, Henry, 101
Donisthorpe, Horace St. J. K., 41, 141
Driesch, Hans, 197
Durkheim, Emile, 74–75

economic accounts of ant life, 82–90
economic entomology, 2–6, 8–9, 63, 107, 123. *See also under* Wheeler, William Morton
education, 22, 30, 51, 114–15, 130. *See also* children; teachers
Eliot, Charles, 114
Eliot, T. S., 150–53, 159, 228
embryology, 66
Emerson, Alfred, 171, 197, 203
Emerson, Ralph Waldo, 100–102, 105, 110, 122
Emery, Carlo, 24, 34–35, 41, 226; links with W. Wheeler, 67
energy, 56, 58–60, 77–78, 181, 183–85, 277n78
engineering, accounts of ant life rooted in, 8, 56–60
engram, 22, 31–32
Entomological Society of America, 67, 128, 136–37, 195, 203
Entomological Society of London, 5, 8
Entomologist's Monthly Magazine, 9
entomology: careers in, 3–6; as discipline, 2–3; domestic, 129–34; types of, 1–2
Escherich, Karl, 65
Esperanto. *See* language: international
Espinas, Alfred, 47–48, 62, 74–75, 89
ethics, 14, 36
eugenics, 15, 22, 24, 36–37, 221, 227
Evans, Howard, 212

Fabre, Jean-Henri: criticized, 45, 126–28, 136; dislike of ants, 127, 131; and domestic entomology, 130–34; instinct theory of, 52–55, 128, 221; and nature study movement, 110; *Souvenirs Entomologiques*, 120, 122; and W. Wheeler, 120–25, 134–36; *The Wonders of Instinct*, 52
Fairchild, David, 91–92, 113, 119, 134, 138, 227
Fairchild, Graham (son of D. Fairchild), 112, 170
Federal Bureau of Entomology, 3
Ferton, Charles, 48, 54
Fielde, Adele M., 80, 129, 131, 140, 169, 222–23
field science, 70, 93, 120–21, 133, 195. *See also* laboratory as locus of biology
flies, 2, 6–8
Ford, Franklin, 212–13
Forel, Alexis (cousin of A. H. Forel), 42
Forel, Alexis (great-uncle of A. H. Forel), 25
Forel, Auguste Henri: as animal psychologist, 47; beliefs of, 22, 28–29, 36–37; correspondents of, 40–43, 46; definition of ant studies, 60–62; *Les Fourmis de la Suisse*, 11, 21; instinct theory of, 30–33; languages of, 46, 141; life of, 23–24; *Le monde social des fourmis*, 12, 24, 28, 31, 41, 126, 141; and nationalism, 28, 33; relations with collectors, 40–43; relations with French animal psychology, 47; relations with W. Wheeler, 67; and religion, 35–36; *The Senses of Insects*, 39; *The Sexual Question*, 24
Forel, Blanche (sister of A. H. Forel; wife of Edouard Bugnion), 35
Forel, Emma (wife of A. H. Forel, née Steinheil), 24, 35, 41
Förster, Heinz von, 178–79
La Fourmilière (home of A. H. Forel), 18, 23–24
fourmilière, technical term, 27–29, 34
Les Fourmis de la Suisse, 21, 23, 27, 43
Franco-Prussian War, 26, 33
Frankenstein, 1
Fremont-Smith, Frank, 197
Freud, Sigmund, 94, 164, 183–85, 224
Frisch, Karl von: bee experiments of, 167–69; communicational difficulties of, 198; compared with Fabre, 136; Macy conference discussions on, 19, 175–79, 188; Rockefeller Foundation sponsorship, 168, 223; Schneirla on, 173; scientists' enthusiasm for, 12, 168–

Frisch, Karl von (*continued*)
69; and Wilson, 214–15. *See also* bees: dance language of
functionalism, 39, 74, 81, 87, 164, 183, 186, 203–4, 209

Galler, Sidney, 172, 196
Giard, Alfred, 45–48, 55, 60–62
Glanville, Eleanor, 1
Grassé, Pierre-Paul, 196, 209
Green, E. E., 8, 9
Gregg, Robert, 191, 193, 203, 210
Griffin, Donald, 173, 176, 214, 222
Gudden, Bernhard von, 23. *See also* Munich

Hachet-Souplet, Pierre, 45
Haeckel, Ernst, 18, 33–34, 74
Haldane, J. B. S., 153, 158, 185, 225
Haldeman, Joe, 223
Harvard University: von Frisch visits, 167–68, 191, 193–94, 201, 222; McDougall at, 72–73; president's enthusiasm for nature study, 114; Richards comes to, 144; W. Wheeler at, 12, 69, 72–73; Wiener at, 163; Wilson at, 12–13, 222; Wilson-era politics of, 165, 211–16
Haskins, Caryl, 201–5, 207, 217
Hatch Act, 3
Hayles, N. Katherine, 14
Hegh, Emile, 8, 56
Heims, Steven J., 174, 208
Henderson, L. J., 92
Hingston, R. W. G., 8, 9, 50, 119
His, Wilhelm, 32
Hockett, C. F., 195
holism, 14, 70–71, 169, 208–9
Hölldobler, Bert, 210, 215–16
Hoover, Herbert, 82, 84–86; economics of, 82–88
Howard, L. O., 3
Huber, François, 25
Huber, Pierre, 25, 154
Hudson, W. H., 100
Hutchinson, G. E., 176–77, 224
Huxley, Aldous, 64, 87–90, 93–95; on instinct, 146–48, 150–51
Huxley, Julian, 50, 64, 87–89, 104, 183, 188, 222, 228
hypnotism, 22, 24, 32, 39

Illiger, J. C. W., 126
Imperial Bureau of Entomology (formerly Colonial Entomological Research Committee), 5, 7–8
Imperial Department of Agriculture for India, 10
India, 8, 10, 157, 168
Indiana University, 167, 182
individualism, 17, 75, 84, 87–88, 90, 159, 208
insect psychology, 22, 34, 39, 59, 128–29, 134, 180, 216
instinct: alien quality of in insects, 10, 38, 126; of amity, 73; contested nature of, 45, 221, 244n85; as drive, 17, 32, 61; dynamic, 47; evolutionary accounts of, 30–32, 44, 55–60, 62; as heterosexuality, 94; human, 71–73; inherited, 30–32, 48; and intelligence, 16, 21–22, 39, 49, 127–28, 221; literary constructions of, 139, 145–54; mechanist accounts of, 44, 54–55, 128; nonevolutionary accounts of, 40, 51–55, 62; as orientation, 48; paradigmatic of insects, 16, 39, 49–51, 60; phylogeny of, 38; reproductive, 95; social, 71–73; traditional accounts of, 16, 49
Institut de Psychologie Animale, 45
Institut Générale Psychologique, 45–46
Institut Psychologique de Paris, 46
intelligence, 49, 127, 187–88, 222
International Congress of Entomology, 2, 9
internationalism, 22, 223
International Union for the Study of Social Insects (IUSSI), 197, 199, 204

Jakobson, Roman, 176
Janet, Charles, 56–58, 77–78, 184
Janet, Pierre, 45
Japan, 155
Jefferies, Richard, 100
Jennings, Herbert Spencer, 45, 55
Johns Hopkins University, 170, 200
Jordan, David Starr, 107, 110, 117, 134
Journal of Animal Behavior, 46, 123, 182

Kellogg, Vernon, 107, 110, 117, 135
Kinsey, Alfred, 168
Klüver, Heinrich, 177
Krutch, Joseph Wood, 104–5, 135
Kütter, Heinrich, 25, 43

laboratory as locus of biology, 13, 32, 40–46 passim, 61–63, 70, 93, 97, 134. *See also* field science
Lacordaire, J. T., 126
Lamarckism, 16, 56, 60, 78, 164, 179, 184, 186, 220
Land Grant Agricultural Colleges, 3
language, 17–18, 222; cybernetic interest in, 164, 176, 181–82; international, 14, 37, 139–41, 154–59, 161; Ogden and Richards on, 139–40, 145, 187; reliability of, 223; W. Wheeler's opinions on, 153
Lawrence, D. H., 146–48, 150–51, 159
Leavis, F. R. and Q. D., 159
Lefroy, Harold Maxwell, 5
Lehrman, Daniel, 177, 197
Levine, Paul, 215
Lévi-Strauss, Claude, 183
Lindauer, 168, 215
linguistics. *See* language
Litvinov, Ivy, 156–57, 161
Loeb, Jacques, 44–46, 54–55, 66, 120
London, Jack, 103, 105, 127
Long, William J., 116–17, 136
Longfellow, Henry Wadsworth, 103
Lorenz, Konrad, 173, 197, 224
Lubbock, John, 10, 44–45, 49, 61, 109, 169

Macy conferences: "Cybernetics: Circular Causal and Feedback Mechanisms," 143, 153–54, 163, 169, 174–79, 185–88, 198, 203, 217; "Group Processes," 197, 199
Macy Foundation, 174–75, 195, 198, 200
Malinowski, Bronislaw, 89, 143, 156
mantis, praying, 53, 125–26
Marchal, Paul, 9, 56
Martin, William, 105
Marx, Karl, 228
masculinity, 64, 97, 104–6, 136–38, 190. *See also* Wheeler, William Morton: masculinity of
Massachusetts Institute of Technology, 12, 202
masses: human, 75, 149, 152, 159, 223, 228; human and insectan, 10, 17, 90, 140, 184; insectan, 6
maternity, 59, 79, 94–95, 131, 134, 184, 220–21, 251n50
mathematics, 165, 190, 204–5, 207
Mauss, Marcel, 90, 183

Mayr, Ernst, 193, 197, 204–5, 213–15
McCulloch, Walter, 163, 177
McDougall, William, 72–73
Mead, Margaret, 188, 197–98, 224
Mencken, Henry L., 87, 147, 156, 158
mentalism, 187–88
Meselson, Matthew S., 211–13
metaphors, 14–15, 101
Micrographia, 1
military significance of animals, 12, 170, 187–88, 196
Milwaukee, 63, 65; Museum, 114
miscibility of ants' nests, 27–29, 140, 221
mneme, 31, 48, 221
molecular biology, 211
monism, 22, 33–36
More, Paul Elmer, 149
Morgan, T. H., 96
Morgan's Canon, 200, 217, 241n43, 281n52
Morin, Pauline (mother of A. H. Forel), 26
Morrill Acts, 3, 69
Morris, Charles, 176, 181
moth, 47
Munich, 26, 168
Muséum d'Histoire Naturelle, Paris, 47
Museum of Comparative Zoology (MCZ), Harvard University, 76, 192, 212–15
Museum of Natural History, Geneva, 42
myrmecology: coined as term, 11, 13, 64, 121; as discipline, 14, 22, 38–43, 62–63, 98, 111, 119, 124, 134. *See also* natural historian as expert

Naples Zoological Station, 65, 121
nationalism, 28, 33
Native Americans, 103
natural historian as expert, 13, 40, 64, 97–98, 104–6, 113, 118–19, 128–29, 134–37, 227
natural history, 39–40, 55; American character of, 97, 99–101, 104–6, 254n33; children and, 112–13; elite (*see* natural historian as expert); vs. laboratory biology, 32, 45, 60 (*see also* field science); as literature, 98–106; "mere," 13, 96, 98, 119, 131, 135, 226; methodological freedom of, 64; nature of, 96–97, 123, 134; vs. nature study, 109; publication market, 99–100, 106–10; W. Wheeler as ideal practitioner of, 98, 106, 134, 137. *See also* nature study movement

Natural History Museum, London, 5
natural theology, 1, 9, 125, 127–28, 130
nature-fake furor, 116–18, 127, 222
nature study movement, 99, 106–19, 130, 137; compared with science, 109, 111, 113–14, 124
The Nature-Study Review, 108, 110, 113, 114, 117, 130, 135
Navy, U.S., 12
nest, architecture of, 56–57
Neumann, John von, 182, 224–25
neurology, 23, 30, 32, 182
New York University, 171
nid, 27–28
Northrop, Filmer S. C., 176, 185
nostalgia, 110, 121

observation, 113–15, 118, 120–22, 124, 128–34
Office of Naval Research (ONR), 170–72, 195–96, 200, 211
Ogden, C. K., 148, 181–83, 187, 198, 202; and ants, 139, 141; and Basic English, 18, 140, 153–61, 222–23; connections of, 142–44; *Meaning of Meaning*, 141, 144–45, 149, 154, 177, 182, 224; publishing enterprises of, 141–44
organization of research, 92–93
orientation, 48–49, 131, 169–70, 187, 195–96, 200, 222
The Origin of Species, 5, 16, 39, 152
Ormerod, Eleanor, 4
Orthological Institute, 154–56, 222
Orwell, George, 17, 161, 179

Packard, Alphaeus S., 60
Palais de Rumine, Lausanne, 26, 42–43
parasitism, 59, 87, 184, 220–23
Pareto, Vilfredo, 64, 71, 86–91 passim, 141, 144, 147, 153, 228
Parker, George H., 168, 211, 227
Parsons, Talcott, 144, 172
Patch, Edith, 129–30
Pearl, Raymond, 158
Peattie, Donald Culross, 104–6, 122, 137, 190
Peckham, Elizabeth G. and George W., 65, 127, 129, 131–33
Peirce, Charles S., 144, 164, 181, 271n8
perfectibility, 35, 50
Perrier, Edmond, 45–48

pheromones, 12, 165, 169, 204–6, 210, 212
Piéron, Henri, 45–46, 48, 61
polymorphism, 34, 78–79, 204, 220
Poulter, Thomas C., 170
Poulton, Edward, 5
Pound, Ezra, 219, 222
propaganda, 158–59, 162, 164, 169, 224
psychiatry, 22–24, 35
psychoanalysis, 164, 184–85, 208
psychology, 16, 71–73, 93, 148, 171. *See also* animal psychology
purpose, 164, 169, 175, 178–80, 186–88, 199, 222, 228. *See also* teleology

Quakers, criminal and lacking fingers, 109

Rabaud, 48–49, 54, 61–62
race, 22, 29, 31, 130
RAND Corporation, 202
Rau, Nellie L. and Philip, 60, 62, 129, 130, 132–33
Red Man. *See* Native Americans
reductionism, 14, 34, 84, 208–9, 246n28
representation, 140, 144, 164, 180–81, 183, 186, 188, 209, 222
Richards, I. A., 18, 139–40, 143, 181, 187, 202, 228; and Basic English, 155–57, 222–24; career, 144; and cybernetics, 177; and instinct, 16, 147–50, 153; *Meaning of Meaning*, 144–45; and Sebeok, 182–83; and White, 160–62
Riley, Charles V., 3, 106
Rivers, W. H. R., 72
Rockefeller Foundation: Basic English sponsor, 18, 155, 157; biology sponsor, 174–75, 211; von Frisch sponsor, 168, 223; nature activities sponsor, 109; and Schneirla, 196–97, 199, 217
Romanes, George, 39, 44–45, 54, 61
Roosevelt, Theodore, 101, 116, 136
Roubaud, Emile, 8, 54, 56, 58–59, 77–78, 83, 220
Rubin, Joan Shelley, 98

Santschi, Félix, 25, 48
Saussure, Henri de, 25
Schneirla, Theodore C., 12, 169, 210; Biological Orientation Conference, 195–96; career of, 170–73; critical of anthropomorphism, 19, 136, 179, 185, 187–88, 217, 222; and cyber-

netics, 164, 175–79, 185–88, 228; on von Frisch, 173; and Grassé, 196–97, 209; and Haskins, 202–4; insect communication symposium, 199–201, 206; outsider status of, 139, 171; un-American activities of, 171–72; on Wilson, 207–8. *See also* ant mill

Scott, J. P., 170, 195

Sebeok, Thomas A., 164, 180–82, 188, 204, 218, 222

semiocracy, 217

semiotics, 144, 182–83

Semon, Richard, 33, 39

senses of insects, 10, 39, 80, 169

sentiment, 96, 115–19, 125, 127–28, 135–36, 198, 226

Seton, Ernest Thompson, 103, 117

Shannon, Claude, 162, 178–79, 181, 183

Shapley, Harlow, 171

Shipley, Arthur, 5

Skinner, B. F., 198

socialism, 22, 30, 34, 36, 140, 160, 221

sociality: evolution of, 9, 79, 84, 95, 132, 184, 221; gradations of, 58–60, 74, 77–78

sociobiology, 181, 195, 217

sociology, 12, 73–76, 87, 89–90, 175, 180; based on trophallaxis, 79–81

Sorensen, W. Conner, 130

Sorokin, Pitirim, 76, 91

Soviet Union, 94, 156–58, 222

Spencer, Herbert, 18, 30, 111, 146

Spencer, John W., 107, 108, 114, 118, 135, 254n33

Spurway, Helen, 225

Stanford University, 107, 117, 135

Steinheil, Edouard (father-in-law of A. H. Forel), 56

stigmergy, 196

structuralism, 164

superorganism, 70, 76, 83, 88, 120, 209–10

Switzerland: as *fourmilière*, 29; history of, 25–26; myrmecologists of, 25

symphiles, 80, 83

taxonomy, 12, 14, 21–22, 38, 40, 60, 70, 139, 165, 190–95, 204

teachers, 108, 110–11, 113, 115–17, 130, 171

teleology, 17, 52, 74, 125, 128, 185. *See also* purpose

termites: *Brave New World*, basis for, 87, 93; destructive effects of, 6–7, 131; economic accounts of, 90; evolution of, 31; instincts of, 72; nests of, 8; as Soviet, 32, 72, 90, 94, 196

Thoreau, Henry David, 99, 122

Thorpe, William, 60

Tinbergen, Nikolaas, 197

Torrey, John, 213–14

Transactions of the Entomological Society of London, 9

traveling naturalists, 6–11

Trivers, Robert, 215

trophallaxis, 79–89, 91, 160, 178, 200–203, 209–10, 220, 248n53; compared with soma, 88–89

tropisms, 44, 47, 54

Trotter, Wilfred, 29, 72

Turner, Charles H., 48, 60, 129, 132–33

two-kingdom model, 49–51, 187

Uexküll, Jacob von, 144, 180–82, 188, 197, 205

"Uncle John." *See* Spencer, John W.

University of Chicago, 65, 113, 171, 182, 198

University of Lausanne, 26, 37

University of Tennessee, 191

University of Texas, 66

Vaud: canton of, 24; character of, 29; history of, 25–26; life in, 36–37

Vietnam War, 223

vitalism, 61, 164, 169, 180, 186–87

Wallace, Alfred Russel, 7

Ward, Henry A., 65, 119

Wasmann, Erich, 52–55, 154, 221

wasp, names of: *Ammophila* (digger), 132; *Eumenes* (potter), 77; *Pelopoeus* (mason), 52, 54; *Philanthus*, 52, 58, 128, 132; *Sphex* (digger), 53; *Synagris cornuta*, 59

wasps: amateur study of, 62, 129–31; ants, evolutionary forerunners of, 132; nests of, 56–59, 184; social evolution of, 57–59, 77–79; wisdom of, 125, 128

Watson, James D., 13, 213

Watson, John B., 46

Weaver, Warren, 168, 174, 197, 214

Weber, Neal, 203

Wenner, Adrian M., 170
Wheeler, Adaline (daughter of W. Wheeler), 90, 142
Wheeler, Dora Emerson (née Dora Bay Emerson, wife of W. Wheeler), 41, 66, 82, 84, 90, 108
Wheeler, George C. (no relation to W. Wheeler), 219
Wheeler, Ralph (son of W. Wheeler), 90
Wheeler, William Morton: "The Ant Colony as an Organism," 70, 75; antisociality of, 90–91; *Ants*, 68; career themes, 62–63; and children, 112–13; compared with Wilson, 208, 216; correspondents of, 66–67, 70–71; *Demons of the Dust*, 123; and economic entomology, 67–71; *Foibles of Insects and Men*, 19; A. H. Forel connection, 41–42; life of, 65–67, 69–70, 114, 122; masculinity of, 94–95, 122; as natural historian, 96–98; Ogden connection, 141, 156, 227; populist colleagues of, 119; and religion, 92, 122; sociological approach to myrmecology, 73–76; students of, 71; "The Termitodoxa," 87, 98, 217, 226–27; University of Paris lectures, 79–80, 86
White, T. H., 139, 160–62, 188, 228
Whitman, C. O., 38, 44, 46, 63, 65
Whitman, Walt, 103, 197
Wiener, Norbert, 142, 163, 173, 176, 179–85 passim, 224–25
Wilson, Edward O.: confidence of, 208–9; departmental politics, 211–18; fire ants, 205; *On Human Nature*, 225; *The Insect Societies*, 207, 216, 225; life of, 190–91; pheromones, 206–8, 210; professional connections of, 201–4, 207; research context, 139, 195–201; Schneirla's critique of, 207–8; *Sociobiology*, 11, 165, 209, 216–17; taxonomy dispute involving, 191–95; "theory of everything," 188, 207; and trophallaxis, 209–10
Woods Hole Marine Biological Laboratory, 65, 96, 120
Wordsworth, William, 100, 105
World War I, 12, 33, 72, 82, 92, 140, 142, 145
World War II, 17, 138, 154–55, 158, 162, 167, 170, 223

Yale University, 201
Yerkes, Robert, 45–46

Ziegler, Heinrich, 34
zoosemiotics, 180–81, 218
Zurich, 23–24, 26